Michael Cole

APPLIED MATHEMATICS SERIES

Edited by

I. S. SOKOLNIKOFF

TENSOR ANALYSIS

THEORY AND APPLICATIONS TO GEOMETRY AND MECHANICS OF CONTINUA

APPLIED MATHEMATICS SERIES

The Applied Mathematics Series is devoted to books dealing with mathematical theories underlying physical and biological sciences, and with advanced mathematical techniques needed for solving problems of these sciences.

TENSOR ANALYSIS

THEORY AND APPLICATIONS TO GEOMETRY AND MECHANICS OF CONTINUA

Second Edition

I. S. SOKOLNIKOFF

PROFESSOR OF MATHEMATICS
UNIVERSITY OF CALIFORNIA
LOS ANGELES

JOHN WILEY & SONS, INC.
NEW YORK · LONDON · SYDNEY

SECOND PRINTING, DECEMBER, 1965

Library of Congress Catalog Card Number: 64-13223

PRINTED IN THE UNITED STATES OF AMERICA

PREFACE TO THE SECOND EDITION

In preparing the Second Edition of this book I have been guided by suggestions kindly made to me by users of the First Edition. There appeared to be no compelling reasons for making major changes in the introductory chapter concerned with linear transformations and matrices, or in the second chapter, devoted to algebra and calculus of tensors.

In Chapter 3 some sections concerned with the uses of calculus of variations in geometry have been expanded, some new illustrative material introduced, and two new sections, on parallel surfaces and the Gauss-Bonnet theorem, have been added. Chapters 2 and 3 in the present edition contain adequate material for an introductory course on metric differential geometry at the beginning graduate level or, for that matter, at the upper-division undergraduate level.

Chapter 4, dealing with analytical mechanics, has been expanded. It contains a distillation of the essentials of classical analytical mechanics and potential theory, which, together with Chapter 5 on relativistic mechanics, should be, but often is not, a part of the equipment of every student of mathematics. A number of illustrative examples that further illuminate the theory have been introduced, and the discussion of nonholonomic dynamical systems, of Hamilton's canonical equations, and of potential theory has been made more detailed.

The concluding chapter, devoted to mechanics of continua, was entirely rewritten. It presents from a unified point of view and, it is hoped, with sufficient clarity, the essentials of the nonlinear theory of mechanics of deformable media. This chapter provides a common basis for a careful development of the mathematical theories of elasticity, plasticity, hydrodynamics, and gas dynamics.

I. S. Sokolnikoff

Pacific Palisades, California
January 1964

PREFACE TO THE FIRST EDITION

This book is an outgrowth of a course of lectures I gave over a period of years at the University of Wisconsin, Brown University, and the University of California. My audience consisted, for the most part, of graduate students interested in applications of mathematics, and this fact shaped both the content and the character of exposition.

Because of the importance of linear transformations in motivating the development of tensor theory, the first chapter in this book is given to a discussion of linear transformations and matrices, in which stress is placed on the geometry and physics of the situation. Although a large part of the subject matter treated in this chapter is normally covered in courses on matrix algebra, only a few of my listeners have had the sort of appreciation of matrix transformations that an applied mathematician should have.

The second chapter is concerned with algebra and calculus of tensors. The treatment in it is self-contained and is not made to depend on some special field of mathematics as a vehicle for the development of tensor analysis. This is a departure from the customary practice of making geometry or relativity a medium for the unfolding of tensor analysis. Although this latter practice has a great deal to commend it because it provides a simple means for motivating the study of tensors, it often leaves an erroneous impression that the formulation of tensor analysis depends somehow on geometry or relativity.

The remaining four chapters in this volume deal with the applications of tensor calculus to geometry, analytical mechanics, relativistic mechanics, and mechanics of deformable media. Thus, Chapter 3 contains a selection of those geometrical topics that are important in the study of analytical dynamics and in such portions of elasticity and plasticity as deal with the deformation of plates and shells. This chapter provides a substantial introduction to the subject of metric differential geometry. In Chapter 4, the essential concepts of analytical mechanics are presented adequately and concisely. An introduction to relativistic mechanics is contained in Chapter 5. The treatment there was intentionally made very brief because some excellent books on relativity have appeared recently and there seems little point in duplicating their contents.

The final chapter of the book is concerned with a formulation of the essential ideas of nonlinear mechanics of continuous media in the most general tensor form. The classical linearized equations of elasticity and fluid mechanics appear as special cases of the general treatment.

Perhaps the best evidence of the remarkable effectiveness of the tensor apparatus in the study of Nature is in the fact that it was possible to include, between the covers of one small volume, a large amount of material that is of interest to mathematicians, physicists, and engineers.

A survey of applied mathematics as broad as that in this book must inevitably reflect contributions of so many scholars that it is futile to attempt to assign proper credit for original ideas or methods of attack. However, in the treatment of geometry, the influence of T. Levi-Civita and A. J. McConnell, whose books (especially McConnell's *Applications of the Absolute Differential Calculus*) I used in my classes for many years as required reading, is clearly discernible. Specific acknowledgments to these and other authors are made in the appropriate places in the text. However, my greatest debt is to my listeners, who have made the job of writing this book seem both enjoyable and worth while.

It is a particular pleasure to single out among my listeners Mr. William R. Seugling, Research Assistant at the University of California at Los Angeles, who gave unstintingly of his time in following this book through press.

I. S. Sokolnikoff

Los Angeles, California
November 1951

CONTENTS

1 LINEAR VECTOR SPACES. MATRICES

2 TENSOR THEORY

3 GEOMETRY

4 ANALYTICAL MECHANICS

1

LINEAR VECTOR SPACES. MATRICES

1. Coordinate Systems

In order to locate a geometrical configuration a reference frame is needed. Among the simplest reference frames used in mathematics are the cartesian coordinate systems. Although the construction of such coordinate systems is familiar to the reader from courses in analytic geometry, we review it here in order to set in relief certain basic notions that underlie the concept of coordinates covering the space of our physical intuition. This review will pave the ground for some far-reaching generalizations of the concept of physical space, which we formulate in Sec. 4.

The cardinal idea responsible for the invention of coordinate systems by Descartes is the identification of the set of points composing a straight line with the totality of real numbers. It consists of the assumption that to each real number there corresponds a unique point on a straight line, and conversely.[1]

We choose a straight line X and a point O on it (Fig. 1). This point O, which we call the origin, divides the line into two half-rays. We

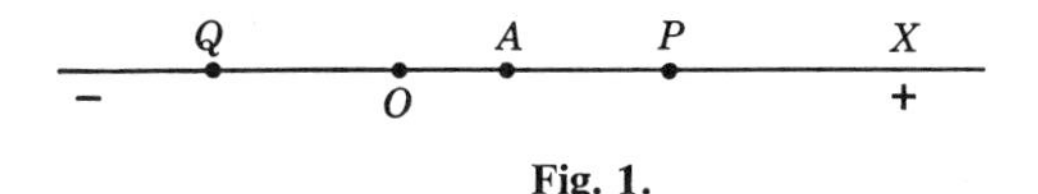

Fig. 1.

[1] Although the idea of one-to-one reciprocal correspondence between the set of points composing a line and the totality of real numbers had it roots in the Eudoxus theory of incommensurables, dating back to the fourth century B.C., the invention of coordinate systems did not come until the first part of the seventeenth century. It should be also noted that a rigorous analysis of the relation between linear sets of points and real numbers was made only during the closing years of the last century, chiefly through the efforts of Dedekind and Cantor. The concept of rigor depends entirely on conventions dictated by prevailing tastes indicative of the degree of mathematical sophistication in a given chronological period. Fruitful intuitive concepts are usually made rigorous by (*a*) making explicit agreements as to which ideas fall into a category of definable concepts and which do not, and (*b*) introducing into mathematical theories new modes of reasoning which (one hopes) are free of contradiction.

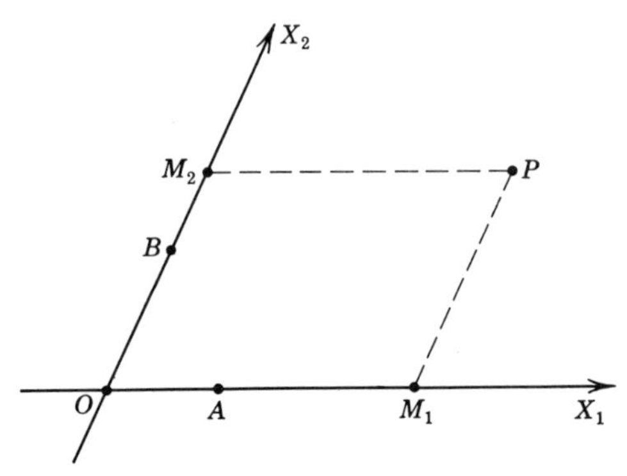

Fig. 2.

designate one of these as the *positive* and the other as the *negative* half-ray. On the positive half-ray we choose a point A and call the length of the line segment OA the *unit length.* We next *coordinate* points on X with a set of real numbers in the following way: If P is any point on the positive half-ray, we define a number x associated with P by the formula

$$x = \frac{\overline{OP}}{\overline{OA}},$$

where $\overline{OP}$ and $\overline{OA}$ are lengths of the line segments OP and OA. The number x is the *coordinate* of P. The coordinate x of the point Q on the negative half-ray is defined by the ratio

$$x = -\frac{\overline{OQ}}{\overline{OA}}.$$

We also assume that each real number x corresponds to one and only one point on X. This association of the set of points on X with the set of real numbers constitutes a *coordinate system* of the *one-dimensional space* consisting of points on X.

The coordination of the set of points lying in the plane with sets of real numbers is accomplished by taking two straight lines X_1 and X_2 intersecting at a single point O (Fig. 2). On each line a coordinate system is constructed as above, but the units on each line need not be equal. A pair of such lines with unit points A and B marked on them form the *coordinate axes* X_1, X_2. With each point P in the plane of coordinate axes we associate an *ordered pair* of real numbers (x_1, x_2) determined as follows. The line through P drawn parallel to the X_2-axis intersects the X_1-axis in a point M_1 with coordinate x_1, and the line through P parallel to the X_1-axis cuts X_2 in a point M_2 with coordinate x_2. The ordered pair of numbers (x_1, x_2) are the coordinates of P in the plane, and the one-to-one correspondence of ordered pairs of numbers with the set of points in the plane X_1X_2 is the *coordinate system* of the two-dimensional space consisting of points in the plane.

The extension of this representation to points in a three-dimensional space is obvious. We take three noncoplanar lines X_1, X_2, X_3 intersecting at the common point O. On each of these lines we establish coordinate systems, and we associate with each point P an ordered triplet of numbers (x_1, x_2, x_3) determined by the intersection with the axes of three planes drawn through P parallel to the *coordinate planes* X_1X_2, X_2X_3, and X_1X_3.

The coordinate systems just described are called *oblique cartesian systems*. Their construction makes use of the notions of length and parallelism of ordinary Euclidean geometry, and the essential feature of it is the concept of one-to-one correspondence of points with ordered sets of numbers. In the event the coordinate axes X_1, X_2, X_3 intersect at right angles, the coordinate system is said to be *orthogonal cartesian*, or *rectangular cartesian*. In applications, orthogonal coordinate systems are generally used because the expression for the length d of the line segment $\overline{AB}$ joining a pair of points with coordinates $A(a_1, a_2, a_3)$ and $B(b_1, b_2, b_3)$ has the simple form

$$d = \sqrt{(b_1 - a_1)^2 + (b_2 - a_2)^2 + (b_3 - a_3)^2}. \tag{1.1}$$

This is the familiar formula of Pythagoras. If the coordinate system is oblique, the formula for the distance d is somewhat more complicated. We will learn in Sec. 9 that one can pass from an orthogonal system of coordinates to an oblique system by making a linear transformation of coordinates. From this fact and from the structure of formula 1.1, it would follow that the length of the line segment joining the points with oblique coordinates (x_1, x_2, x_3) and (y_1, y_2, y_3) is

$$d = \sqrt{\sum_{i,j=1}^{3} g_{ij}(y_i - x_i)(y_j - x_j)}, \tag{1.2}$$

where the g_{ij}'s are constants that depend on the coefficients in the above-mentioned linear transformation of coordinates. We will be concerned in the sequel with a detailed study of quadratic forms appearing under the radical in formula 1.2 and with their bearing on metric properties of space.

2. The Geometric Concept of a Vector

In the preceding section we recalled the construction of coordinate systems in the familiar three-dimensional space where the formula of Pythagoras is used to measure distances between pairs of points. Spaces

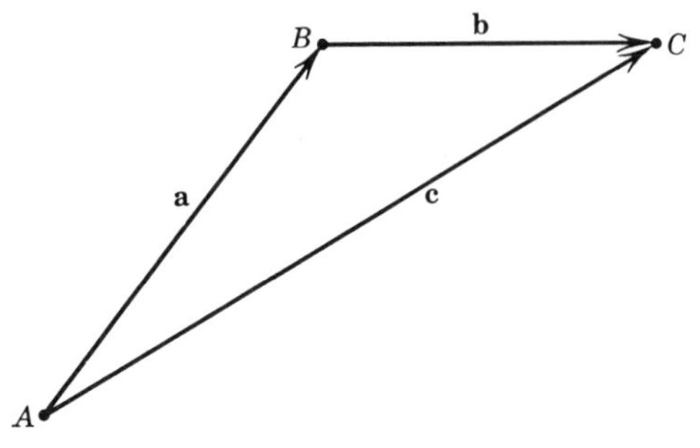

Fig. 3.

where it is possible to construct a coordinate system such that the length of a line segment is given by the formula of Pythagoras are called *Euclidean spaces*. In these spaces the notion of displacement is fundamental. Thus, if a point A is moved to a new position B, the displacement from A to B can be visualized as *directed line segment* $\overrightarrow{AB}$ (Fig. 3). If B is displaced to a new position C, the resultant displacement can be achieved by moving the point A to the position C. These operations can be denoted symbolically by the equation

$$\overrightarrow{AB} + \overrightarrow{BC} = \overrightarrow{AC}.$$

In the elementary treatment of vector analysis, directed line segments are termed *vectors*, and they are usually denoted by a single letter printed in boldface type. Thus the foregoing formula can be written

$$\mathbf{a} + \mathbf{b} = \mathbf{c}, \tag{2.1}$$

where $\overrightarrow{AB} = \mathbf{a}$, $\overrightarrow{BC} = \mathbf{b}$, $\overrightarrow{AC} = \mathbf{c}$.

The rule for the composition of vectors indicated in Fig. 3 was first formulated by Stevinus in 1586 in connection with the experimental study of laws governing the composition of forces. It is known as the *parallelogram law of addition*. The fact that many entities occurring in physics can be represented by directed line segments, whose law of composition is symbolized by formula 2.1, is responsible for the usefulness of vector analysis in applications. We have here an instance of geometrization of physics which had no less important influence on the evolution of this subject than the arithmetization of geometry had on the growth of mathematical analysis.

From the idea of a vector as displacement determined by a pair of points, we are led to conclude that two vectors are equal if the line segments representing them are equal in length and their directions parallel. We shall denote the length of the vector $\mathbf{a}$ by the symbol $|\mathbf{a}|$. We will assume that the concept of length is independent of the chosen reference frame, so that the length $|\mathbf{a}|$ can be calculated (by Pythagorean formula) from the coordinates of the initial and terminal points of $\mathbf{a}$.

The negative of the vector $\mathbf{a}$ (written $-\mathbf{a}$) is the vector whose length is the same as that of $\mathbf{a}$ but whose direction is opposite. We define the vector zero (written $\mathbf{0}$) corresponding to a zero displacement by the formula

$$\mathbf{a} + (-\mathbf{a}) = \mathbf{0}.$$

From the geometrical properties of directed line segments we deduce that

(I) $$\mathbf{a} + \mathbf{b} = \mathbf{b} + \mathbf{a}.$$

(II) $$(\mathbf{a} + \mathbf{b}) + \mathbf{c} = \mathbf{a} + (\mathbf{b} + \mathbf{c}).$$

(III) If $\mathbf{a}$ and $\mathbf{b}$ are vectors, there exists a unique vector $\mathbf{x}$ such that

$$\mathbf{a} = \mathbf{b} + \mathbf{x}.$$

We next define the operation of multiplication of vectors by real numbers. If α is a real number the symbol $\alpha\mathbf{a} \equiv \mathbf{a}\alpha$ is a vector whose length is $|\alpha|\,|\mathbf{a}|$ and whose direction is the same as that of $\mathbf{a}$ if $\alpha > 0$, opposite to $\mathbf{a}$ if $\alpha < 0$. If $\alpha = 0$, then $\alpha\mathbf{a} = \mathbf{0}$.

From this definition and from properties of real numbers we conclude that

(IV) $$(\alpha_1 + \alpha_2)\mathbf{a} = \alpha_1\mathbf{a} + \alpha_2\mathbf{a}$$

(V) $$\alpha(\mathbf{a} + \mathbf{b}) = \alpha\mathbf{a} + \alpha\mathbf{b}$$

(VI) $$\alpha_1(\alpha_2\mathbf{a}) = (\alpha_1\alpha_2)\mathbf{a}, \qquad 1 \cdot \mathbf{a} = \mathbf{a},$$

for any real numbers α_1 and α_2.

We introduce next the definition of *scalar product* of two vectors, which will provide us with a new notation for the length of a vector.

DEFINITION. *The scalar product of two vectors* $\mathbf{a}$ *and* $\mathbf{b}$, *written* $\mathbf{a} \cdot \mathbf{b}$, *is a real number* $|\mathbf{a}|\,|\mathbf{b}| \cos(\mathbf{a}, \mathbf{b})$, *where* $\cos(\mathbf{a}, \mathbf{b})$ *is the cosine of the angle between* $\mathbf{a}$ *and* $\mathbf{b}$.

Stated in the language of geometry, $\mathbf{a} \cdot \mathbf{b}$ is equal to the product of the projection of $\mathbf{a}$ on $\mathbf{b}$ multiplied by the length of $\mathbf{b}$. Thus the length of the vector $\mathbf{a}$ is given by the positive square root of $\mathbf{a} \cdot \mathbf{a}$. We also note that $\mathbf{a}$ and $\mathbf{b}$ are orthogonal if, and only if, $\mathbf{a} \cdot \mathbf{b} = 0$.

From this definition and the properties of real numbers we can easily deduce the following theorems.

(VII) $$\mathbf{a} \cdot \mathbf{a} = |\mathbf{a}|^2 > 0, \quad \text{unless } \mathbf{a} = \mathbf{0}.$$

(VIII) $$\mathbf{a} \cdot \mathbf{b} = \mathbf{b} \cdot \mathbf{a}.$$

(IX) $$\mathbf{a} \cdot (\mathbf{b} + \mathbf{c}) = \mathbf{a} \cdot \mathbf{b} + \mathbf{a} \cdot \mathbf{c}.$$

(X) $$\alpha(\mathbf{a} \cdot \mathbf{b}) = (\alpha\mathbf{a} \cdot \mathbf{b}), \quad \text{where } \alpha \text{ is a real number.}$$

3. Linear Vector Spaces. Dimensionality of Space

We formulate next the definition of *linear dependence* of a set of vectors $\mathbf{a}_1, \mathbf{a}_2, \ldots, \mathbf{a}_n$, which will have an important connection with the concept of dimensionality of space.

Linear Dependence. A set of n vectors $\mathbf{a}_1, \mathbf{a}_2, \ldots, \mathbf{a}_n$ *is called linearly dependent if there exist numbers* $\alpha_1, \alpha_2, \ldots, \alpha_n$, *not all of which are zero, such that*

$$\alpha_1\mathbf{a}_1 + \alpha_2\mathbf{a}_2 + \cdots + \alpha_n\mathbf{a}_n = \mathbf{0}.$$

If no such numbers exist, the vectors are said to be linearly independent.

Consider two vectors $\mathbf{a}$ and $\mathbf{b}$ which are like, or oppositely, directed (Fig. 4). Then there exists a number $k \neq 0$ such that

$$\mathbf{b} = k\mathbf{a}. \tag{3.1}$$

If we set $k = -\alpha/\beta$, we can write this equation as

$$\alpha\mathbf{a} + \beta\mathbf{b} = \mathbf{0},$$

and hence two collinear (or parallel) vectors are linearly dependent since neither α nor β is zero. We will say that the totality of vectors $k\mathbf{a}$ for an arbitrary real k and $\mathbf{a} \neq \mathbf{0}$ forms a one-dimensional real *linear vector space*. The reason for this terminology is clear since every point on the line can be represented by some *position vector* $k\mathbf{a}$.

If $\mathbf{a}$ and $\mathbf{b}$ are two noncollinear vectors, represented by directed line segments with common origin O (Fig. 5), any vector $\mathbf{c}$ lying in the plane of $\mathbf{a}$ and $\mathbf{b}$ can be represented in the form

$$\mathbf{c} = m\mathbf{a} + n\mathbf{b}. \tag{3.2}$$

Formula 3.2 follows at once from the rule for addition of vectors and from the definition of multiplication of vectors by scalars. Equation 3.2 can be rewritten in symmetric form to read

$$\alpha\mathbf{a} + \beta\mathbf{b} + \gamma\mathbf{c} = \mathbf{0},$$

which is the condition for linear dependence of the set of three vectors, since not all constants in this formula vanish. The formula $m\mathbf{a} + n\mathbf{b}$, where $\mathbf{a}$ and $\mathbf{b}$ are two linearly independent vectors and m and n are arbitrary real numbers, defines a *two-dimensional real linear vector space*. We see that in a two-dimensional linear vector space a set of three vectors is always linearly dependent.

Fig. 4.

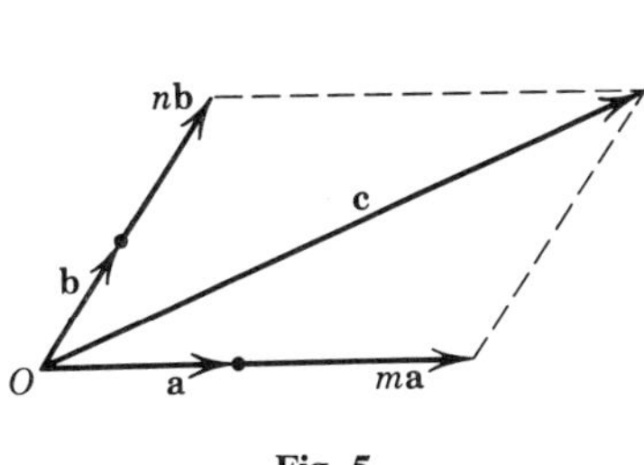

Fig. 5.

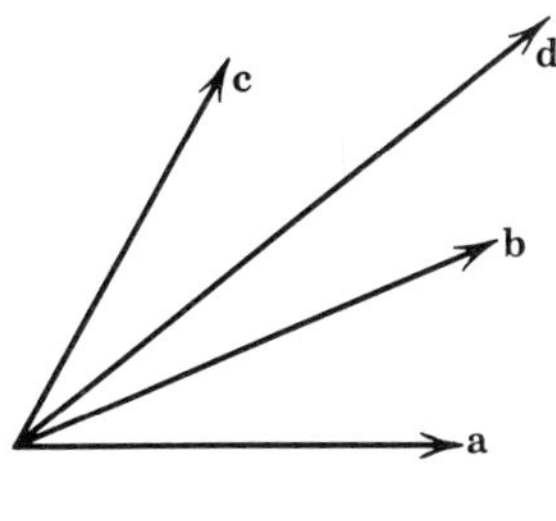

Fig. 6.

If we start with three noncoplanar vectors **a, b, c** issuing from the common origin O (Fig. 6), we can clearly represent every vector **d** in the form

$$\mathbf{d} = m\mathbf{a} + n\mathbf{b} + p\mathbf{c}, \tag{3.3}$$

from which it follows that among four vectors **a, b, c, d** there always exists a nontrivial relation of the form

$$\alpha\mathbf{a} + \beta\mathbf{b} + \gamma\mathbf{c} + \delta\mathbf{d} = \mathbf{0}.$$

Formula 3.3, for an arbitrary choice of real numbers m, n, p, defines a *three-dimensional real linear vector space*. The terminal points of position vectors **d** sweep out a three-dimensional space of points if m, n, and p are allowed to range over the entire set of real numbers. In a three-dimensional linear vector space every set of four vectors is linearly dependent. We will make use of the connection of the number of linearly independent vectors with the dimensionality of space to formulate the concept of dimensionality of a linear vector space of n dimensions.

The vectors **a**, **b**, and **c** in (3.3) are called *base* or *coordinate vectors*, and the numbers m, n, and p are the *measure numbers* or *components* of the vector **d**. Once a set of base vectors is specified, every vector is determined uniquely by a triplet of measure numbers.

A set of three mutually orthogonal vectors in a three-dimensional space is obviously linearly independent, and if we choose as our coordinate vectors three mutually orthogonal vectors $\mathbf{a}_1$, $\mathbf{a}_2$, $\mathbf{a}_3$, each of length 1, the resulting set of base vectors is said to be *orthonormal*.

We can visualize a set of orthonormal vectors directed along the axes of a suitable rectangular cartesian reference frame; in this case every vector **x** has the representation

$$\mathbf{x} = x_1\mathbf{a}_1 + x_2\mathbf{a}_2 + x_3\mathbf{a}_3,$$

where (x_1, x_2, x_3) are called the *physical components* of $\mathbf{x}$, and the terminal points of the base vectors $\mathbf{a}_i$, $(i = 1, 2, 3)$, have the coordinates

$$\begin{aligned} \mathbf{a}_1&: \quad (1, 0, 0), \\ \mathbf{a}_2&: \quad (0, 1, 0), \\ \mathbf{a}_3&: \quad (0, 0, 1). \end{aligned}$$

We conclude this section by noting the rules for the addition and multiplication of vectors when the latter are referred to an orthonormal system of base vectors $\mathbf{a}_i$, $(i = 1, 2, 3)$. If we have two vectors $\mathbf{x}$ and $\mathbf{y}$ whose components are (x_1, x_2, x_3) and (y_1, y_2, y_3), respectively, then the vector $\mathbf{x} + \mathbf{y}$ has the components $(x_1 + y_1, x_2 + y_2, x_3 + y_3)$. If α is a real number, the components of the vector $\alpha\mathbf{x}$ are $(\alpha x_1, \alpha x_2, \alpha x_3)$. From the distributive law of scalar multiplication of vectors it follows at once that the product of

$$\mathbf{x} = x_1\mathbf{a}_1 + x_2\mathbf{a}_2 + x_3\mathbf{a}_3$$

and

$$\mathbf{y} = y_1\mathbf{a}_1 + y_2\mathbf{a}_2 + y_3\mathbf{a}_3$$

is

$$\mathbf{x} \cdot \mathbf{y} = x_1y_1 + x_2y_2 + x_3y_3,$$

since $\mathbf{a}_i \cdot \mathbf{a}_j = \delta_{ij}$, where $\delta_{ij} = 1$ if $i = j$, and $\delta_{ij} = 0$ when $i \neq j$. This follows from the assumed orthonormal nature of the base vectors $\mathbf{a}_i$. The foregoing formula leads at once to the familiar expression for the length $|\mathbf{x}|$ of the vector $\mathbf{x}$ referred to an orthogonal cartesian reference system. Thus

$$\begin{aligned} \mathbf{x} \cdot \mathbf{x} &= x_1^2 + x_2^2 + x_3^2 \\ &= |\mathbf{x}|^2, \end{aligned}$$

so that

$$|\mathbf{x}| = \sqrt{x_1^2 + x_2^2 + x_3^2}.$$

Clearly $|\mathbf{x}| > 0$, unless $x_1 = x_2 = x_3 = 0$.

4. *N*-Dimensional Spaces

In a variety of circumstances one encounters a correspondence of sets of objects with ordered sets of numbers where the number of independent entities exceeds three. For instance, in dealing with the states of gas determined by the pressure (p), the volume (v), the temperature (T), and the time (t), one may wish to coordinate these entities with ordered sets of four real numbers (x_1, x_2, x_3, x_4). Here a diagrammatic representation of the states of gas by points in the three-dimensional physical space is clearly impossible. However, the essential idea in the concept of coordinate systems is not a pictorial representation but the one-to-one reciprocal correspondence of objects with sets of numbers. The notion of distance between pairs of arbitrary points is, likewise, irrelevant. Indeed, the idea of distance becomes devoid of geometric sense even in the familiar representation of the states of gas [the pressure (p) and the volume (v)] by points in the cartesian pv-plane. It is manifestly absurd to speak of the distance between two states characterized by ordered pairs of numbers (p, v).

The utility of analytic treatment of physical problems is so great that we are naturally led to form the concept of spaces of higher dimensions by utilizing the idea of one-to-one correspondence between the sets of numbers and objects. The "objects" here might be of quite diverse sorts. In certain situations they might be pressures, volumes, and temperatures; in others they might be the amounts of electrical charge and the complex potentials produced by the motion of such charge, and so on.

We define[2] *a space (or manifold) of N dimensions as any set of objects that can be placed in a one-to-one correspondence with the totality of ordered sets of N (real or complex) numbers* $x_1, x_2, \ldots, x_N$ *such that*

$$|x_i - A_i| < k_i, \qquad (i = 1, 2, \ldots, N),$$

where $A_1, \ldots, A_N$ *are constants and the* $k_1, k_2, \ldots, k_N$ *are real numbers.*

The inequalities in this definition specify the *range of variation* of the numbers x_i. If the numbers x_i are real, the N-dimensional space is *real*, and we can write the inequalities in the form

$$a_1 \leq x_1 \leq a_2, \qquad b_1 \leq x_2 \leq b_2, \ldots, \qquad t_1 \leq x_N \leq t_2.$$

Some of the equality signs may be omitted, and we may have for the range of variables x_k, for example, $0 \leq x_k < \infty$.

We denote the space of N dimensions by the symbol V_N, and we use the term "points" to mean "objects."

Any particular one-to-one association of the points with the ordered sets of numbers $(x_1, x_2, \ldots, x_N)$ *is called a coordinate system, and the numbers* $x_1, x_2, \ldots, x_N$ *are termed the coordinates of points in the coordinate system.*

There is no implication in these definitions that the concept of distance between pairs of points has any meaning. If a *rule* is specified for the measurement of the distance between points, the space V_N is called *metric.* For the time being we will not assume that our spaces are metrized.

[2] Compare O. Veblen, *Invariants of Quadratic Differential Forms*, p. 13. In speaking of one-to-one correspondence we always have in mind one-to-one *reciprocal* correspondence.

A set of equations of the form

$$x_i = x_i(y_1, y_2, \ldots, y_N), \qquad (i = 1, 2, \ldots, N), \tag{4.1}$$

in which the functions x_i are single-valued and are such that, in the region under consideration, they yield N single-valued solutions

$$y_i = y_i(x_1, x_2, \ldots, x_N),$$

defines a transformation of coordinates.

We will defer a discussion of the general functional transformation (4.1) to Chapter 2. In the remainder of this chapter we are concerned with a detailed study of an important case of linear (or affine) transformations of coordinates

$$y_i = \sum_{j=1}^{N} a_{ij} x_j, \qquad (i = 1, \ldots, N),$$

and with the bearing of such transformations on linear vector spaces.

5. Linear Vector Spaces of *n* Dimensions

A sketch of the rudiments of vector analysis in Sec. 2, based on the concept of directed displacement, contained a set of ten theorems embodied in the formulas identified by the Roman numerals. These theorems can be taken as a point of departure in the generalization of the concept of a vector in the n-dimensional space since the idea of directed displacement and length become devoid of familiar sense whenever n exceeds three. Accordingly, we will postulate that there exist points in the n-dimensional space and that

A. *Every two points in the real n-dimensional space determine an entity which we call a vector.* We denote this entity by the symbol **a**.

B. *Every two vectors* **a** *and* **b** *have a sum* $\mathbf{a} + \mathbf{b}$ *which obeys laws I, II, and III stated in Sec.* 2.

It follows from the third of these laws that the operation of subtraction of vectors is unique and that there exists a vector **0** such that $\mathbf{a} + \mathbf{0} = \mathbf{a}$ for every vector **a**.

C. *For every real number* α *and vector* **a** *there exists a vector* $\alpha\mathbf{a} = \mathbf{a}\alpha$ *obeying laws IV, V, and VI of Sec.* 2.

We retain the definition of linear dependence of the set of n vectors, with respect to the field of real numbers $\alpha_1, \ldots, \alpha_n$, and take as our *axiom of dimensionality* the assumption that

D. *There exist n linearly independent vectors in the n-dimensional space, but every set of* $n + 1$ *vectors is linearly dependent.*

This axiom implies that every vector $\mathbf{x}$ can be represented in the form

$$\mathbf{x} = \alpha_1 \mathbf{a}_1 + \alpha_2 \mathbf{a}_2 + \cdots + \alpha_n \mathbf{a}_n, \tag{5.1}$$

where $\mathbf{a}_1, \mathbf{a}_2, \ldots, \mathbf{a}_n$ is any set of n linearly independent vectors. We say that the totality of vectors determined by formula 5.1, where the α_i are arbitrary real numbers, constitutes a real *linear vector space of n dimensions.*

To give a meaning to the concepts of length and orthogonality of vectors we need a postulate

E. *With every two vectors* $\mathbf{a}$ *and* $\mathbf{b}$ *we can associate a number* $\mathbf{a} \cdot \mathbf{b}$, *called their scalar product, which obeys laws VII, VIII, IX, and X of Sec.* 2.

At this stage we are not concerned with the nature of the formula used to calculate the number $\mathbf{a} \cdot \mathbf{b}$. Suffice to say that the properties embodied in the laws of scalar multiplication lead to a definite rule for computing $\mathbf{a} \cdot \mathbf{b}$ once a coordinate system is introduced for the specification of coordinates of points determining the vectors.

A vector satisfying postulates A through E is said to be defined in the n-dimensional Euclidean space E_n.

We shall use the language of Euclidean geometry and will mean by the *length* of the vector $\mathbf{a}$ the positive square root of the scalar product of the vector $\mathbf{a}$ by itself. Thus the length $|\mathbf{a}| = \sqrt{\mathbf{a} \cdot \mathbf{a}}$. If $|\mathbf{a}| = 1$ the vector $\mathbf{a}$ is called a *unit vector.* Two vectors $\mathbf{a}$ and $\mathbf{b}$ are said to be *orthogonal* whenever $\mathbf{a} \cdot \mathbf{b} = 0$.

We proceed to demonstrate that every set of m linearly independent vectors in E_n $(m \leq n)$ can be orthogonalized. This means that from a given set of m linearly independent vectors $\mathbf{x}_1, \mathbf{x}_2, \ldots, \mathbf{x}_m$ we can construct a set of vectors $\mathbf{a}_1, \ldots, \mathbf{a}_m$ such that $\mathbf{a}_i \cdot \mathbf{a}_j = 0$ whenever $i \neq j$. Moreover, it is possible to choose the vectors $\mathbf{a}_i$ so that they are unit vectors.

Proof. We assume that the set of vectors $\{\mathbf{x}_i\}$, $(i = 1, \ldots, m)$ is linearly independent. Thus the equation

$$c_1 \mathbf{x}_1 + c_2 \mathbf{x}_2 + \cdots + c_m \mathbf{x}_m = \mathbf{0} \tag{5.2}$$

can be satisfied only by choosing $c_1 = c_2 = \cdots = c_m = 0$. It follows that $\mathbf{x}_1 \neq \mathbf{0}$, for, if it were zero, the numbers

$$c_1 = 1,\ c_2 = c_3 = \cdots = c_m = 0$$

would satisfy (5.2) and hence the vectors would be linearly dependent, which is contrary to our hypothesis. Denote by $\mathbf{a}_1$ the product of $\mathbf{x}_1$ by the reciprocal of its length so that

$$\mathbf{a}_1 = \frac{\mathbf{x}_1}{|\mathbf{x}_1|}.$$

Clearly $\mathbf{a}_1 \cdot \mathbf{a}_1 = 1$, so that $\mathbf{a}_1$ is a unit vector.

The set of vectors

$$\mathbf{a}_1, \mathbf{x}_2, \ldots, \mathbf{x}_m$$

is obviously a linearly independent set. Consider next the vector

$$\mathbf{a}_2' = \mathbf{x}_2 - (\mathbf{x}_2 \cdot \mathbf{a}_1)\mathbf{a}_1.$$

The product of this vector by $\mathbf{a}_1$ vanishes since

$$\mathbf{x}_2 \cdot \mathbf{a}_1 - (\mathbf{x}_2 \cdot \mathbf{a}_1)\mathbf{a}_1 \cdot \mathbf{a}_1 = 0.$$

Thus $\mathbf{a}_2'$ is orthogonal to $\mathbf{a}_1$ and $\mathbf{a}_2'/|\mathbf{a}_2'| \equiv \mathbf{a}_2$ is a unit vector.

The set of vectors

$$\mathbf{a}_1, \mathbf{a}_2, \mathbf{x}_3, \ldots, \mathbf{x}_m$$

is linearly independent, and we can define the vector $\mathbf{a}_3'$ by the formula

$$\mathbf{a}_3' = \mathbf{x}_3 - (\mathbf{x}_3 \cdot \mathbf{a}_1)\mathbf{a}_1 - (\mathbf{x}_3 \cdot \mathbf{a}_2)\mathbf{a}_2,$$

which is orthogonal to both $\mathbf{a}_1$ and $\mathbf{a}_2$. The vector $\mathbf{a}_3 \equiv \mathbf{a}_3'/|\mathbf{a}_3'|$ is a unit vector, and the set

$$\mathbf{a}_1, \mathbf{a}_2, \mathbf{a}_3, \mathbf{x}_4, \ldots, \mathbf{x}_m$$

is a linearly independent set of vectors.

A repetition of this procedure will yield a set of m linearly independent unit vectors

$$\mathbf{a}_1, \mathbf{a}_2, \ldots, \mathbf{a}_m, \tag{5.3}$$

each of which is expressed in terms of the $\mathbf{x}_i$. The set of orthogonal unit vectors (5.3) is called an *orthonormal set.*

If $m = n$, the set of orthonormal vectors $\mathbf{a}_1, \mathbf{a}_2, \ldots, \mathbf{a}_n$ is called *complete* because *every vector* $\mathbf{x}$ in E_n can be represented in the form

$$\mathbf{x} = \alpha_1\mathbf{a}_1 + \alpha_2\mathbf{a}_2 + \cdots + \alpha_n\mathbf{a}_n. \tag{5.4}$$

By analogy with the three-dimensional case, a complete set of orthonormal vectors can be taken as a set of coordinate vectors oriented along the axes of the n-dimensional *orthogonal cartesian reference frame.* The terminal points of these vectors then have the coordinates

$$\begin{array}{l} 1, 0, \ldots, 0, \\ 0, 1, \ldots, 0, \\ 0, 0, 1, \ldots, 0, \\ \cdots\cdots\cdots\cdots \\ 0, 0, 0, \ldots, 1. \end{array}$$

The constants $\alpha_1, \alpha_2, \ldots, \alpha_n$ in (5.4) are called the *components* of the vector $\mathbf{x}$. Multiplying (5.4) scalarly by $\mathbf{a}_1, \mathbf{a}_2, \ldots, \mathbf{a}_n$ in turn, and remembering that[3] $\mathbf{a}_i \cdot \mathbf{a}_j = \delta_{ij}$, we obtain

$$\mathbf{a}_1 \cdot \mathbf{x} = \alpha_1, \qquad \mathbf{a}_2 \cdot \mathbf{x} = \alpha_2, \qquad \ldots, \qquad \mathbf{a}_n \cdot \mathbf{x} = \alpha_n.$$

Thus the vector $\mathbf{x}$ can be represented in the form

$$\mathbf{x} = (\mathbf{a}_1 \cdot \mathbf{x})\mathbf{a}_1 + (\mathbf{a}_2 \cdot \mathbf{x})\mathbf{a}_2 + \cdots + (\mathbf{a}_n \cdot \mathbf{x})\mathbf{a}_n. \tag{5.5}$$

If we introduce the notation $\mathbf{a}_i \cdot \mathbf{x} = x_i$, equation 5.5 assumes the form

$$\mathbf{x} = x_1\mathbf{a}_1 + x_2\mathbf{a}_2 + \cdots + x_n\mathbf{a}_n.$$

Using the distributive property of scalar multiplication, we get

$$\mathbf{x} \cdot \mathbf{x} = x_1^2 + x_2^2 + \cdots + x_n^2, \tag{5.6}$$

so that

$$|\mathbf{x}| = \sqrt{x_1^2 + x_2^2 + \cdots + x_n^2}.$$

This is the formula of Pythagoras in E_n.

If $\mathbf{y} = y_1\mathbf{a}_1 + y_2\mathbf{a}_2 + \cdots + y_n\mathbf{a}_n$, then

$$\mathbf{x} \cdot \mathbf{y} = x_1y_1 + x_2y_2 + \cdots + x_ny_n.$$

This formula has the same structure as the expression for the scalar product of two vectors in ordinary three-dimensional space of Euclidean geometry.

We note that in an orthogonal cartesian reference frame a vector $\mathbf{x}$ is uniquely determined by an n-tuple of numbers $(x_1, x_2, \ldots, x_n)$. This property is taken by some authors as the definition of a vector in E_n.

For the sum of two vectors $\mathbf{x}$ and $\mathbf{y}$, with components

$$\begin{array}{ll} \mathbf{x}: & (x_1, x_2, \ldots, x_n), \\ \mathbf{y}: & (y_1, y_2, \ldots, y_n), \end{array}$$

[3] The symbol δ_{ij}, the *Kronecker delta*, means

$$\begin{aligned} \delta_{ij} &= 1, \quad \text{if } i = j, \\ &= 0, \quad \text{if } i \neq j. \end{aligned}$$

we have the formula

$$\mathbf{x} + \mathbf{y}: \quad (x_1 + y_1, x_2 + y_2, \ldots, x_n + y_n)$$

and for the product of $\mathbf{x}$ by the scalar α,

$$\alpha\mathbf{x}: \quad (\alpha x_1, \alpha x_2, \ldots, \alpha x_n).$$

The formula

$$\mathbf{x} \cdot \mathbf{y} = x_1 y_1 + x_2 y_2 + \cdots + x_n y_n$$

serves to define metric properties of vectors in E_n.

The passage from an orthonormal set of vectors $\mathbf{a}_i$ to any other set of base vectors is accomplished by subjecting the elements of the n-tuple to a suitable linear transformation. In essence the approach to vectors by way of n-tuples of numbers reduces the study of vectors to the study of algebraic properties of linear transformations. In this book we prefer to stress the geometric concepts that underlie the idea of a vector and not submerge them in a purely algebraic formalism.

6. Complex Linear Vector Spaces

The considerations of Sec. 5 can be easily extended to the field of complex numbers.

In a complex n-dimensional linear vector space the vector $\mathbf{x}$ is determined by the ordered set of n complex numbers $(x_1, x_2, \ldots, x_n)$, the elements x_i of which are the *components* of $\mathbf{x}$.

We define the sum $\mathbf{x} + \mathbf{y}$ of two vectors

$$\mathbf{x}: \quad (x_1, x_2, \ldots, x_n),$$

$$\mathbf{y}: \quad (y_1, y_2, \ldots, y_n)$$

by the rule

$$\mathbf{x} + \mathbf{y}: \quad (x_1 + y_1, x_2 + y_2, \ldots, x_n + y_n),$$

and the product $\alpha\mathbf{x}$ by

$$\alpha\mathbf{x}: \quad (\alpha x_1, \alpha x_2, \ldots, \alpha x_n).$$

It is customary to define the scalar product $\mathbf{x} \cdot \mathbf{y}$ by the formula

$$\mathbf{x} \cdot \mathbf{y} = \sum_{i=1}^{n} \bar{x}_i y_i, \tag{6.1}$$

where a bar over x_i denotes the conjugate of the complex number x_i.

We note that

$$\mathbf{y} \cdot \mathbf{x} = \sum_{i=1}^{n} \bar{y}_i x_i, \tag{6.2}$$

so that

$$\mathbf{x} \cdot \mathbf{y} = \overline{\mathbf{y} \cdot \mathbf{x}}, \tag{6.3}$$

since the conjugate of the sum is the sum of the conjugates and the conjugate of the product is equal to the product of conjugates.

Formula 6.1 is adopted for the calculation of the scalar product in order to ensure that

$$\mathbf{x} \cdot \mathbf{x} = \sum_{i=1}^{n} \bar{x}_i x_i$$

be a real number. It clearly specializes to (5.6) when the numbers x_i are real.

The vectors $\mathbf{x}$, $\mathbf{y}$ are said to be *orthogonal* if $\mathbf{x} \cdot \mathbf{y} = 0$. As regards the notion of linear independence, we retain the definition given in Sec. 3 with the understanding that the coefficients α_i now belong to the field of complex numbers.

Problems

1. If we start with the definition of a vector $\mathbf{x}$ as an n-tuple of n real or complex numbers $(x_1, x_2, \ldots, x_n)$, and use for the definition of sum and product the formulas

$$\mathbf{x} + \mathbf{y}: \quad (x_1 + y_1, \ldots, x_n + y_n),$$
$$k\mathbf{x}: \quad (kx_1, \ldots, kx_n),$$
$$\mathbf{x} \cdot \mathbf{y} = \sum_{i=1}^{n} \bar{x}_i y_i,$$

then

$$(\mathbf{x} + \mathbf{y}) \cdot \mathbf{z} = \mathbf{x} \cdot \mathbf{z} + \mathbf{y} \cdot \mathbf{z},$$
$$\mathbf{x} \cdot (\mathbf{y} + \mathbf{z}) = \mathbf{x} \cdot \mathbf{y} + \mathbf{x} \cdot \mathbf{z},$$
$$(k\mathbf{x}) \cdot \mathbf{y} = \bar{k}(\mathbf{x} \cdot \mathbf{y}),$$
$$\mathbf{x} \cdot (k\mathbf{y}) = k(\mathbf{x} \cdot \mathbf{y}).$$

2. Prove that, if $\mathbf{a}^{(1)}, \mathbf{a}^{(2)}, \ldots, \mathbf{a}^{(n)}$ is a set of n linearly independent vectors in a complex n-dimensional vector space, then the only vector $\mathbf{x}$ orthogonal to each of the vectors $\mathbf{a}^{(i)}$ is the zero vector.

3. Prove that a set of mutually orthogonal nonzero vectors is always linearly independent.

4. Let the set of vectors $\mathbf{a}^{(i)}$ in E_n: $(a_1^{(i)}, a_2^{(i)}, \ldots, a_n^{(i)})$, $i = 1, 2, \ldots, n$, be linearly dependent, and suppose that r of them, $\mathbf{a}^{(1)}, \mathbf{a}^{(2)}, \ldots, \mathbf{a}^{(r)}$, $r < n$, are linearly independent. Show that every vector $\mathbf{x}$ that is orthogonal to this set of r linearly independent vectors forming the subset of E_n is also orthogonal to the remaining $n - r$ vectors in the given set.

7. Summation Convention. Review of Determinants

It is clear from the developments in preceding sections that the linear forms and matrices associated with them enter prominently in the study of vectors in the n-dimensional manifolds. Since such forms will occur frequently throughout the remainder of this chapter, it is desirable to introduce a compact abridged notation and to rewrite with its aid certain familiar results from the theory of determinants.

From now on we shall adhere to the following summation convention. *If in some expression a certain index occurs twice, we shall mean that this expression is summed with respect to that index for all admissible values of the index.* Thus the linear form $\sum_{i=1}^{4} a_i x_i$ has the index i occurring in it twice; we will omit the summation symbol Σ and write $a_i x_i$ to mean $a_1x_1 + a_2x_2 + a_3x_3 + a_4x_4$. Of course, the range of admissible values of the index, 1 to 4 in this case, must be specified. If the symbol i has the range of values 1 to 3 and j ranges from 1 to 4, the expression

$$a_{ij}x_j, \qquad (i = 1, 2, 3), \quad (j = 1, 2, 3, 4), \tag{7.1}$$

represents three linear forms

$$\begin{cases} a_{11}x_1 + a_{12}x_2 + a_{13}x_3 + a_{14}x_4, \\ a_{21}x_1 + a_{22}x_2 + a_{23}x_3 + a_{24}x_4, \\ a_{31}x_1 + a_{32}x_2 + a_{33}x_3 + a_{34}x_4. \end{cases} \tag{7.2}$$

In expression 7.1 the index i is the *identifying index*. It denotes one of the forms in (7.2), depending on the chosen value of i. The index j, however, since it occurs twice, is the *summation* index. The summation (or *dummy*) index can be changed at will. Thus (7.1) can be written in the form $a_{ik}x_k$ if k has the same range of values as j. The summation index is analogous to a variable of integration in a definite integral, which also can be changed at will.

Unless a statement to the contrary is made, we will assume that the summation and the identifying indices have the ranges of values from 1 to n. Thus $a_i x_i$ will represent a linear form

$$a_1x_1 + a_2x_2 + \cdots + a_nx_n.$$

Although in the last term of this expression the letter n occurs twice, it does not represent the sum, since n here has a fixed value. In order to avoid ambiguity, or when we want to suspend the summation convention, we may enclose the index in parentheses. Thus we can write the linear form as

$$a_1x_1 + a_2x_2 + \cdots + a_{(n)}x_{(n)}.$$

The quadratic form $\sum_{i=1}^{n}\sum_{j=1}^{n} a_{ij}x_ix_j$ will be written $a_{ij}x_ix_j$. An expression $a_{ij}x_iy_j$ represents a bilinear form containing n^2 terms, whereas $a_{ij}a_{jk}$ represents n^2 sums of the type

$$a_{i1}a_{1k} + a_{i2}a_{2k} + \cdots + a_{in}a_{nk},$$

since each of the identifying, or *free*, indices i and k can have values from 1 to n. We will not trouble to enclose the indices in parentheses when the context makes it clear (as in the above expression) that such indices have fixed values. If, however, we wish to discuss a particular term in this sum we will write $a_{i(j)}a_{(j)k}$.

Frequently, it is convenient to identify the different symbols by using superscripts rather than subscripts. For instance, we may write the sequence of terms $x^1, x^2, \ldots, x^n$, where the superscripts are not the powers of the variable but the identifying indices. The typical term in this sequence is x^i, $(i = 1, 2, \ldots, n)$. A linear form in the x^i, with the coefficients a_i, will be written as a_ix^i. A bilinear form, with the coefficients a^{ij}, in the variables x_i and y_i will be written as $a^{ij}x_iy_j$.

A determinant

$$\begin{vmatrix} a_{11} & a_{12} & \cdots & a_{1n} \\ a_{21} & a_{22} & \cdots & a_{2n} \\ \cdots & \cdots & \cdots & \cdots \\ a_{n1} & a_{n2} & \cdots & a_{nn} \end{vmatrix}$$

whose elements are a_{ij} will be written, as is customary, $|a_{ij}|$. If the elements of this determinant are denoted by $a_j{}^i$, where the superscript i indicates the row and the subscript j the column in which this element appears, we will write the determinant as $|a_j{}^i|$. Thus

$$|a_j{}^i| = \begin{vmatrix} a_1{}^1 & a_2{}^1 & \cdots & a_n{}^1 \\ a_1{}^2 & a_2{}^2 & \cdots & a_n{}^2 \\ \cdots & \cdots & \cdots & \cdots \\ a_1{}^n & a_2{}^n & \cdots & a_n{}^n \end{vmatrix}.$$

For the multiplication of two determinants $|a_j{}^i|$ and $|b_j{}^i|$ we have the familiar rule:

$$|a_j{}^i| \cdot |b_j{}^i| = |c_j{}^i|,$$

where $c_j{}^i = a_k{}^ib_j{}^k$. If we deal with determinants $|a_{ij}|$ and $|b_{ij}|$, then the element c_{ij} in the ith row and the jth column of the product of $|a_{ij}|$ and $|b_{ij}|$ is $c_{ij} = a_{ik}b_{kj}$.

The cofactor of the element $a_j^{\ i}$ in $|a_j^{\ i}|$ is denoted by $A_i^{\ j}$. If we write the *Kronecker delta* as $\delta_j^{\ i}$, where

$$\begin{aligned}\delta_j^{\ i} &= 1, \quad \text{if } i = j,\\ &= 0, \quad \text{if } i \neq j,\end{aligned}$$

then for the expansion of $|a_j^{\ i}|$ in terms of cofactors we have the following formulas:

$$a_j^{\ i} A_k^{\ j} = a\delta_k^{\ i}, \tag{7.3}$$

$$a_j^{\ i} A_i^{\ k} = a\delta_j^{\ k}, \tag{7.4}$$

where $a = |a_j^{\ i}|$. These formulas include the familiar simple Laplace developments of $|a_j^{\ i}|$. The first of these then represents the expansion in terms of the elements of the ith row; the second, in terms of the elements of the jth column of $|a_j^{\ i}|$.

If the elements of the determinant a are denoted by a_{ij}, we shall write the cofactor of a_{ij} as A_{ij}. Simple Laplace developments corresponding to (7.3) and (7.4) assume the forms

$$a_{(i)j} A_{(i)j} = a \qquad \text{and} \qquad a_{i(k)} A_{i(k)} = a.$$

We can derive Cramer's rule for the solution of the system of n linear equations

$$a_j^{\ i} x^j = b^i, \qquad (i, j = 1, \ldots, n), \tag{7.5}$$

in n unknowns x^i, where $|a_j^{\ i}| \neq 0$, as follows: Multiply both sides of equations in (7.5) by $A_i^{\ k}$, and sum with respect to i. This yields

$$a_j^{\ i} A_i^{\ k} x^j = b^i A_i^{\ k}.$$

By (7.4) this reduces to

$$a\delta_j^{\ k} x^j = b^i A_i^{\ k},$$

or

$$a x^k = b^i A_i^{\ k}.$$

Thus

$$x^k = \frac{b^i A_i^{\ k}}{a}. \tag{7.6}$$

Frequently, the cofactor of the element a_{ij} in $|a_{ij}|$ is denoted by A^{ij}, so that the Laplace developments (7.3) and (7.4) assume the forms

$$a_{(i)j} A^{(i)j} = a,$$

$$a_{j(i)} A^{j(i)} = a.$$

To gain familiarity with this notation, the reader is advised to derive Cramer's rule when the system of linear equations is written in the form $a_{ij}x^j = b_i$. He will also prove that, if $a_j{}^i b_k{}^j = \delta_k{}^i$, then $|a_j{}^i| = 1/|b_j{}^i|$.

We will return to the subject of determinants in Sec. 41, where a different notation permits us to eliminate references to rows and columns of the determinant and enables us to write it in terms of its elements, without reference to cofactors.

Problems

1. Write out in full the following expressions:

(*a*) $\delta_j{}^i a^j$. (*b*) $\delta_{ij}x^i x^j$. (*c*) $a_{ij}b_{jk} = \delta_{ik}$. (*d*) $a_{ijk}x^k$.

(*e*) $\dfrac{\partial f_i}{\partial x_j}\, dx_j$. (*f*) $\delta_i{}^i$. (*g*) $a^i = \dfrac{\partial x^i}{\partial y^j}\, b^j$. (*h*) $a_{ij(k)}x^j y^{(k)}$.

(*i*) $g_{ij} = \dfrac{\partial y^k}{\partial x^i}\dfrac{\partial y^k}{\partial x^j}$. (*j*) $a_{i(j)}x^{(j)}$. (*k*) $\delta_{ij}\delta^{jk}$.

The symbols $\delta_j{}^i$, δ_{ij}, and δ^{ij} all denote the Kronecker deltas.

2. Verify that (7.6) is the solution of (7.5).

8. Linear Transformations and Matrices

A set of n relations of the form

$$x_i' = a_{ij}x_j, \qquad (i, j = 1, \ldots, n), \tag{8.1}$$

where the a_{ij}'s are constants, is called a *linear homogeneous transformation* of the set of variables x_i into a set x_i'. We shall suppose that the transformation 8.1 is nonsingular, so that the set of n linear equations 8.1 can be solved for the x_i in terms of the x_i'. This implies that the determinant $|a_{ij}|$ of the coefficients of x_j's is different from zero.

The solution of (8.1) for the x's yields

$$x_i = \frac{A_{ji}}{a}\, x_j', \tag{8.2}$$

where A_{ij} is the cofactor of the element a_{ij} in $|a_{ij}| \equiv a$.

The set of equations 8.1 can be interpreted in two essentially different ways:

(*a*) The quantities x_i may be regarded as components of a vector $\mathbf{x}$: $(x_1, x_2, \ldots, x_n)$, and the numbers x_i' as components of another vector $\mathbf{x}'$: $(x_1', x_2', \ldots, x_n')$, where both $\mathbf{x}$ and $\mathbf{x}'$ are referred to a reference frame with the system of base factors $\mathbf{a}_i$; in this case we think of equations 8.1 as representing a transformation of the vector $\mathbf{x}$ into another vector $\mathbf{x}'$.

(*b*) The two sets of numbers $(x_1, x_2, \ldots, x_n)$ and $(x_1', x_2', \ldots, x_n')$ can be regarded as components of the same vector $\mathbf{x}$ when $\mathbf{x}$ is referred to two different sets of cartesian reference frames determined by the base vectors $\mathbf{a}_1, \mathbf{a}_2, \ldots, \mathbf{a}_n$ and $\mathbf{a}_1', \mathbf{a}_2', \ldots, \mathbf{a}_n'$; in this event equations 8.1 give a transformation of coordinate axes.

Before proceeding to a specific discussion of these two interpretations of the set of equations 8.1, it is necessary to review the operations with matrices.

An array of mn numbers, arranged in m rows and n columns, is called an $m \times n$ matrix. We denote the matrix formed from the elements a_{ij} (or $a_j{}^i$) by

$$(a_{ij}) \equiv \begin{bmatrix} a_{11} & a_{12} & \cdots & a_{1n} \\ a_{21} & a_{22} & \cdots & a_{2n} \\ \cdots & \cdots & \cdots & \cdots \\ a_{m1} & a_{m2} & \cdots & a_{mn} \end{bmatrix} \quad \text{or} \quad (a_j{}^i) \equiv \begin{bmatrix} a_1{}^1 & a_2{}^1 & \cdots & a_n{}^1 \\ a_1{}^2 & a_2{}^2 & \cdots & a_{n}{}^2 \\ \cdots & \cdots & \cdots & \cdots \\ a_1{}^m & a_2{}^m & \cdots & a_n{}^m \end{bmatrix}.$$

We shall also write the symbol A for the matrix (a_{ij}). We shall say that the matrix $A = (a_{ij})$ is equal to the matrix $B = (b_{ij})$ if, and only if, $a_{ij} = b_{ij}$ for each i and j. That is, if $A = B$ the elements in the corresponding rows and columns of the matrices must be equal.

By the sum $A + B$ of two matrices $A = (a_{ij})$ and $B = (b_{ij})$ of the same type, that is, containing the same number of rows and columns, we mean the matrix

$$A + B = (a_{ij} + b_{ij}).$$

If we have an $m \times n$ matrix A and an $n \times p$ matrix B, we can define the product of matrices A and B, written AB, by the formula

$$AB = (a_{ij}b_{jk}). \tag{8.3}$$

Thus the product AB is an $m \times p$ matrix; we can multiply two matrices only if the number of columns in the first factor is equal to the number of rows in the second.

For the most part we shall deal with square matrices, that is, matrices containing an equal number of rows and columns.

A matrix all of whose elements are zero is called a *zero matrix*. It is denoted by the symbol O.

We note two peculiarities of matrix multiplication. From the definition 8.3 it follows that, if A and B are two $n \times n$ matrices, then AB is not necessarily equal to BA.

For example, if

$$A = \begin{bmatrix} 0 & 1 \\ 1 & 0 \end{bmatrix} \quad \text{and} \quad B = \begin{bmatrix} -1 & 0 \\ 0 & 1 \end{bmatrix},$$

then

$$AB = \begin{bmatrix} 0 & 1 \\ -1 & 0 \end{bmatrix}, \quad \text{whereas} \quad BA = \begin{bmatrix} 0 & -1 \\ 1 & 0 \end{bmatrix}.$$

Thus the product of matrices, in general, is not commutative. However, if we have two matrices of order n, which contain zero elements everywhere except possibly along the diagonal, then they are commutative, and obey the simple law of multiplication.

$$\begin{bmatrix} \lambda_1 & 0 & \cdots & 0 \\ 0 & \lambda_2 & \cdots & 0 \\ \cdot & & \cdot & \cdot \\ \cdot & & \cdot & \cdot \\ \cdot & & \cdot & \cdot \\ 0 & 0 & \cdots & \lambda_n \end{bmatrix} \begin{bmatrix} \mu_1 & 0 & \cdots & 0 \\ 0 & \mu_2 & \cdots & 0 \\ \cdot & & \cdot & \cdot \\ \cdot & & \cdot & \cdot \\ \cdot & & \cdot & \cdot \\ 0 & 0 & \cdots & \mu_n \end{bmatrix} = \begin{bmatrix} \lambda_1\mu_1 & 0 & \cdots & 0 \\ 0 & \lambda_2\mu_2 & \cdots & 0 \\ \cdot & & \cdot & \cdot \\ \cdot & & \cdot & \cdot \\ \cdot & & \cdot & \cdot \\ 0 & 0 & \cdots & \lambda_n\mu_n \end{bmatrix}$$

Such matrices are called *diagonal* matrices. The diagonal matrices will be found to be of considerable importance in what follows.

A particular diagonal matrix

$$I = \begin{bmatrix} 1 & 0 & \cdots & 0 \\ 0 & 1 & \cdots & 0 \\ \cdot & & \cdot & \cdot \\ \cdot & & \cdot & \cdot \\ \cdot & & \cdot & \cdot \\ 0 & 0 & \cdots & 1 \end{bmatrix}$$

is called the *identity* matrix. We note that, if A is any matrix, then

$$AI = IA = A.$$

We also observe that the product of two matrices may vanish when neither of the matrices is a zero matrix.

Thus, if $A = \begin{bmatrix} 1 & 1 & 0 \\ 0 & 0 & 0 \\ 0 & 0 & 0 \end{bmatrix}$ and $B = \begin{bmatrix} 0 & 0 & 0 \\ 0 & 0 & 0 \\ 1 & 0 & 0 \end{bmatrix}$, then $AB = \begin{bmatrix} 0 & 0 & 0 \\ 0 & 0 & 0 \\ 0 & 0 & 0 \end{bmatrix}$.

However, the *determinant* $|AB|$ of the product of two square matrices is precisely equal to the product of the determinants $|A|$ and $|B|$ of the matrices A and B. This follows at once from the observation that the law of formation of the element in the ith row and kth column of the product of two determinants is identical with the corresponding rule for the product of two matrices. We call an $n \times n$ matrix whose determinant is zero a *singular* matrix.

Finally, we define the multiplication of the matrix $A = (a_{ij})$ by the number k, written kA, as the matrix each of whose elements is multiplied by k. Thus $kA = (ka_{ij})$.

As an exercise the reader will verify the following theorems, which follow directly from the definitions given previously.

(I) $$A + B = B + A.$$

(II) $$(A + B) + C = A + (B + C).$$

(III) $$(A + B)C = AC + BC.$$

(IV) $$C(A + B) = CA + CB.$$

The notation just developed permits us to write the system of equations 8.1 in the form of a vector equation

$$\mathbf{x}' = A\mathbf{x}, \tag{8.4}$$

where $A = (a_{ij})$ and where we agree to interpret $\mathbf{x}$ either as a column matrix

$$\begin{bmatrix} x_1 \\ x_2 \\ \cdot \\ \cdot \\ \cdot \\ x_n \end{bmatrix} \quad \text{or a square matrix} \quad \begin{bmatrix} x_1 & 0 & 0 & \cdots & 0 \\ x_2 & 0 & 0 & \cdots & 0 \\ \cdots & \cdots & \cdots & \cdots & \cdots \\ x_n & 0 & 0 & \cdots & 0 \end{bmatrix}.$$

The inverse transformation 8.2 can be written

$$\mathbf{x} = A^{-1}\mathbf{x}', \tag{8.5}$$

where

$$A^{-1} = \begin{bmatrix} \dfrac{A_{11}}{|A|} & \dfrac{A_{21}}{|A|} & \cdots & \dfrac{A_{n1}}{|A|} \\ \dfrac{A_{12}}{|A|} & \dfrac{A_{22}}{|A|} & \cdots & \dfrac{A_{n2}}{|A|} \\ \cdots & \cdots & \cdots & \cdots \\ \dfrac{A_{1n}}{|A|} & \dfrac{A_{2n}}{|A|} & \cdots & \dfrac{A_{nn}}{|A|} \end{bmatrix}, \tag{8.6}$$

and the A_{ij}'s are the cofactors of the elements a_{ij} in the determinant $|A|$.

The matrix A^{-1} is called the *inverse* of the matrix A, and it is defined for any nonsingular matrix A. From definition 8.6 it follows that the matrices A and A^{-1} are related by the formulas

$$AA^{-1} = I, \qquad A^{-1}A = I,$$

where I is the identity matrix. For, $AA^{-1} = (a_{ik}A_{jk}/|A|)$ and $a_{ik}A_{jk} = \delta_{ij}\,|A|$. The identity matrix I corresponds to an identity transformation $x_i' = x_i$; this transformation when written in the matrix form (8.4) appears as $\mathbf{x}' = I\mathbf{x}$, or

$$\mathbf{x}' = \mathbf{x}.$$

We call the matrix

$$A' = \begin{bmatrix} a_{11} & a_{21} & \cdots & a_{n1} \\ a_{12} & a_{22} & \cdots & a_{n2} \\ \cdots & \cdots & \cdots & \cdots \\ a_{1n} & a_{2n} & \cdots & a_{nn} \end{bmatrix},$$

obtained by interchanging the rows and columns in the matrix

$$A = \begin{bmatrix} a_{11} & a_{12} & \cdots & a_{1n} \\ a_{21} & a_{22} & \cdots & a_{2n} \\ \cdots & \cdots & \cdots & \cdots \\ a_{n1} & a_{n2} & \cdots & a_{nn} \end{bmatrix},$$

the *transpose* of A.

Using the definition of transpose and the laws of addition and multiplication of matrices it is easy to show that

(V) $$(A + B)' = A' + B'.$$

(VI) $$(kA)' = kA'.$$

(VII) $$(AB)' = B'A'. \qquad \text{(Note order.)}$$

If A is nonsingular, then the *matric equations*

$$AX = I \quad \text{and} \quad XA = I$$

have unique solutions $X = A^{-1}$, as can be immediately verified by multiplying them by A^{-1} on both sides and noting that

$$A^{-1}A = AA^{-1} = I.$$

If we take $A^{-1}A = AA^{-1}$ and form the transpose, we get

$$A'(A^{-1})' = (A^{-1})'A'.$$

Multiplying by $(A')^{-1}$ on the left, we obtain

$$\begin{aligned}(A')^{-1}A'(A^{-1})' &= (A')^{-1}(A^{-1})'A' \\ (A^{-1})' &= (A')^{-1}(AA^{-1})' \\ &= (A')^{-1}.\end{aligned}$$

Thus

$$(A^{-1})' = (A')^{-1}.$$

We can also readily show that

$$(AB)^{-1} = B^{-1}A^{-1}. \qquad \text{(Note order.)}$$

If we have two successive linear transformations

$$x_i' = a_{ij}x_j \qquad \text{and} \qquad x_i'' = b_{ij}x_j', \qquad (i, j = 1, \ldots, n),$$

the direct transformation from the variables x_i to the variables x_i'' is

$$x_i'' = b_{ij}a_{jk}x_k, \qquad (i, j, k = 1, \ldots, n);$$

this is called the *product transformation.* Writing these transformations in matrix notation yields

$$\mathbf{x}' = A\mathbf{x} \qquad \text{and} \qquad \mathbf{x}'' = B\mathbf{x}',$$

so that

$$\mathbf{x}'' = BA\mathbf{x}.$$

Since the product BA, in general, is not equal to AB, we see that the order in which the transformations are performed is not immaterial.

It should be observed that the matrix A in the equation $x' = Ax$ can be interpreted as an operator which converts a vector $\mathbf{x}$ into another vector $\mathbf{x}'$. Because of the properties

$$A(k\mathbf{x}) = kA\mathbf{x}$$

and

$$A(\mathbf{x} + \mathbf{y}) = A\mathbf{x} + A\mathbf{y},$$

where k is any scalar, A is frequently called a *linear vector operator* or *linear vector function.* It can be viewed as an apparatus for the manufacture of a new vector from a given vector. We shall expound these points in greater detail by considering a number of examples of the uses of matrices in several situations familiar from analytic geometry and elementary vector analysis.

9. Linear Transformations in Euclidean 3-Space

Let us refer our Euclidean 3-space (E_3) to a system of coordinates with base vectors $\mathbf{a}^{(1)}$, $\mathbf{a}^{(2)}$, $\mathbf{a}^{(3)}$, linearly independent but not necessarily orthogonal. Then any vector $\mathbf{x}$ can be represented in the form

$$\mathbf{x} = x_j \mathbf{a}^{(j)}, \qquad (j = 1, 2, 3), \tag{9.1}$$

where the x_j are appropriate real measure numbers. If we introduce a *real* linear transformation

$$x_i' = a_{ij} x_j \qquad \text{with } |a_{ij}| \neq 0, \qquad (i, j = 1, 2, 3), \tag{9.2}$$

or

$$\mathbf{x}' = A\mathbf{x}, \tag{9.3}$$

we can interpret the resulting vector $\mathbf{x}'$ as a deformed vector produced by the *deformation of space* which is characterized by the operator A. In general, the length of the vector $\mathbf{x}'$ will be different from that of $\mathbf{x}$, and its orientation relative to our fixed reference frame will differ from the orientation of the vector $\mathbf{x}$.

Obviously there are infinitely many reference frames that may be imbedded in our space, and in each frame the vector $\mathbf{x}$ is characterized uniquely by a triplet of numbers. Let us inquire: What is the form of the transformation giving the *same* deformation of space as that characterized by the matrix A, when the vector $\mathbf{x}$ is referred to a new frame of reference in which the base vectors $\boldsymbol{\alpha}^{(1)}$, $\boldsymbol{\alpha}^{(2)}$, $\boldsymbol{\alpha}^{(3)}$ are related to the old base vectors $\mathbf{a}^{(1)}$, $\mathbf{a}^{(2)}$, $\mathbf{a}^{(3)}$ by the formulas

$$\boldsymbol{\alpha}^{(i)} = b_{ij} \mathbf{a}^{(j)}? \tag{9.4}$$

We shall suppose that the matrix $(b_{ij}) \equiv B$ is nonsingular and denote the components of $\mathbf{x}$ relative to the new system by (ξ_1, ξ_2, ξ_3), so that

$$\mathbf{x} = \xi_i \boldsymbol{\alpha}^{(i)}. \tag{9.5}$$

If we insert in (9.5) the expressions 9.4 for the base vectors $\boldsymbol{\alpha}^{(i)}$ in terms of $\mathbf{a}^{(i)}$, we obtain

$$\mathbf{x} = \xi_i b_{ij} \mathbf{a}^{(j)}. \tag{9.6}$$

A comparison of this equation with (9.1) yields the connection between the components ξ_i and x_i, namely,

$$x_j = b_{ij} \xi_i. \tag{9.7}$$

We note that the matrix B in the transformation 9.4 of base vectors $\mathbf{a}^{(j)}$ differs from the matrix B' in the transformation 9.7 of components of the vector $\mathbf{x}$ in that the rows and columns in these matrices are interchanged. Thus the matrix B' is the transpose of the matrix B. We write (9.7) in the form

$$\mathbf{x} = B'\boldsymbol{\xi}. \tag{9.8}$$

The solution of (9.8) for $\boldsymbol{\xi}$ is given by

$$\boldsymbol{\xi} = (B')^{-1}\mathbf{x}. \tag{9.9}$$

To simplify writing we denote $(B')^{-1}$ by C, so that (9.9) becomes

$$\boldsymbol{\xi} = C\mathbf{x}, \tag{9.10}$$

where

$$C \equiv (B')^{-1}. \tag{9.11}$$

Formula 9.10 permits us to calculate the components of the vector $\mathbf{x}$ when it is referred to a new system of base vectors $\boldsymbol{\alpha}^{(i)}$, determined by (9.4). Consequently the components ξ_1', ξ_2', ξ_3' of $\mathbf{x}'$, relative to the reference frame with base vectors $\boldsymbol{\alpha}^{(i)}$, are given by

$$\boldsymbol{\xi}' = C\mathbf{x}', \tag{9.12}$$

and the question of the expression (in the new frame) for the deformation of space characterized by (9.3) amounts to finding the relation connecting the components ξ_1, ξ_2, ξ_3 with ξ_1', ξ_2', ξ_3'. The substitution from (9.3) in (9.12) gives

$$\boldsymbol{\xi}' = CA\mathbf{x},$$

and, since by (9.10)

$$\mathbf{x} = C^{-1}\boldsymbol{\xi},$$

we get the desired relation

$$\boldsymbol{\xi}' = CAC^{-1}\boldsymbol{\xi}. \tag{9.13}$$

The transformation determined by the matrix $S = CAC^{-1}$ is called *similar* to the transformation produced by A because formulas 9.13 and 9.3 characterize the *same* deformation of space relative to two different reference frames.

If we recall the definition (9.11), we can write (9.13) in the form

$$\boldsymbol{\xi}' = (B')^{-1}AB'\boldsymbol{\xi}, \tag{9.14}$$

which brings into explicit evidence the matrices A and B characterizing, respectively, the deformation of space and the transformation of base vectors. We note that the determinants of all similar transformations are equal. An important special case of the transformation 9.2, corresponding to the rotation of the vector $\mathbf{x}$ to a new position, is discussed in the next section.

10. Orthogonal Transformation in E_3

Let us suppose that the base vectors $\mathbf{a}^{(1)}$, $\mathbf{a}^{(2)}$, $\mathbf{a}^{(3)}$ in Sec. 9 are orthogonal unit vectors, so that the measure numbers x_j in (9.1) are the physical components of $\mathbf{x}$. Then the square of the length of the vector $\mathbf{x}$ is given by the formula

$$|\mathbf{x}|^2 = x_i x_i, \qquad (i = 1, 2, 3).$$

Let us inquire about restrictions that must be imposed on the matrix A in (9.3) if the length of $\mathbf{x}$ is to be unchanged by the transformation 9.2. This restriction demands that

$$x_i' x_i' = x_i x_i. \tag{10.1}$$

Substituting in (10.1) from (9.2), we obtain

$$(a_{ij} x_j)(a_{ik} x_k) = x_i x_i, \qquad (i, j, k = 1, 2, 3),$$

or

$$a_{ij} a_{ik} x_j x_k = \delta_{jk} x_j x_k, \tag{10.2}$$

since

$$\delta_{jk} x_j x_k = x_k x_k = x_i x_i.$$

Equating the coefficients of like products in (10.2), we obtain six equations

$$\begin{aligned}
a_{11}^2 + a_{21}^2 + a_{31}^2 &= 1, \\
a_{12}^2 + a_{22}^2 + a_{32}^2 &= 1, \\
a_{13}^2 + a_{23}^2 + a_{33}^2 &= 1, \\
a_{12}a_{13} + a_{22}a_{23} + a_{32}a_{33} &= 0, \\
a_{13}a_{11} + a_{23}a_{21} + a_{33}a_{31} &= 0, \\
a_{11}a_{12} + a_{21}a_{22} + a_{31}a_{32} &= 0,
\end{aligned}$$

or

$$a_{ij} a_{ik} = \delta_{jk}. \tag{10.3}$$

These equations are consequences of the hypothesis that the length of $\mathbf{x}$ remains invariant. The determinant of the matrix in (10.3) has the value

$$|a_{ij} a_{ik}| = 1. \tag{10.4}$$

Since the value of the determinant $|a_{ij}|$ is unchanged when its rows and columns are interchanged, we see from the rule for multiplication of determinants (Sec. 7) that

$$|a_{ij} a_{ik}| = |a_{ij}| \cdot |a_{ij}| = |A|^2.$$

Thus (10.4) yields the result that the square of the determinant $|a_{ij}|$ in (9.2) has the value 1 whenever the length of the vector is unchanged by the transformation. We conclude that $|A| = \pm 1$. The case when $|A| = +1$ corresponds to the transformation of rotation of space relative to fixed axes. The circumstance when $|A| = -1$ corresponds to the transformation of reflection (say, $x_1' = -x_1$, $x_2' = -x_2$, $x_3' = -x_3$) or a reflection followed by a rotation.

A linear transformation

$$x_i' = a_{ij}x_j, \tag{10.5}$$

in which $a_{ij}a_{ik} = \delta_{jk}$ is called an *orthogonal* transformation. It is called the transformation of *rotation* when $|a_{ij}| = +1$. If we denote by A' the transpose of A in (10.5), we can write the orthogonality conditions (10.3) in the form

$$A'A = I.$$

Multiplying this equation on the right by A^{-1}, we get

$$A' = A^{-1}, \tag{10.6}$$

so that *in an orthogonal transformation the inverse matrix A^{-1} is equal to the transpose A' of A when the base vectors are orthonormal.*

It follows that, if we write equations 10.5 in the form

$$\mathbf{x}' = A\mathbf{x},$$

then

$$\mathbf{x} = A^{-1}\mathbf{x}',$$

and by virtue of (10.6)

$$\mathbf{x} = A'\mathbf{x}'$$

or

$$x_i = a_{ji}x_j'. \tag{10.7}$$

11. Linear Transformations in *n*-Dimensional Euclidean Spaces

Our discussion of linear transformations in Euclidean 3-space can be immediately extended to n-dimensional manifolds E_n referred to a coordinate system such that the length of the vector $\mathbf{x}$ is determined from formula 5.6.

We introduce n-orthonormal vectors,

$$\begin{aligned} \mathbf{a}^{(1)}&: \quad (1, 0, 0, \ldots, 0), \\ \mathbf{a}^{(2)}&: \quad (0, 1, 0, \ldots, 0), \\ &\cdots\cdots\cdots\cdots\cdots\cdots, \\ \mathbf{a}^{(n)}&: \quad (0, 0, 0, \ldots, 1), \end{aligned}$$

and represent any vector $\mathbf{x}$: $(x_1, x_2, \ldots, x_n)$ in the form (cf. equation 9.1)

$$\mathbf{x} = x_j \mathbf{a}^{(j)}, \qquad (j = 1, \ldots, n). \tag{11.1}$$

A linear transformation of components, corresponding to equation 9.2, is

$$x_i' = a_{ij} x_j, \qquad (i, j = 1, \ldots, n). \tag{11.2}$$

We can write it in matrix notation as

$$\mathbf{x}' = A\mathbf{x}, \tag{11.3}$$

where $A = (a_{ij})$.

We suppose that $|A| \neq 0$, and denote the solution of (11.3) by

$$\mathbf{x} = A^{-1}\mathbf{x}',$$

where

$$A^{-1} = \frac{(A_{ji})}{|A|}.$$

The A_{ij}'s denote the cofactors of the elements a_{ij} in $|A|$.

Just as was done in the three-dimensional case, we can show that the product of transformations $\mathbf{x}' = A\mathbf{x}$ and $\mathbf{x}'' = B\mathbf{x}'$ is $\mathbf{x}'' = BA\mathbf{x}$. We can still use the suggestive language of geometry and speak of the set of equations 11.3 as representing the deformation of space E_n and consider that the transformation of the form

$$\mathbf{x}' = CAC^{-1}\mathbf{x} \tag{11.4}$$

represents the same deformation of space as that characterized by the matrix A in (11.3). The matrices A and CAC^{-1} are still termed *similar*.

By analogy with the three-dimensional case, a real linear transformation that leaves the length of every real vector $\mathbf{x}$: $(x_1, \ldots, x_n)$ invariant is called *orthogonal*. From computations of Sec. 10 it is obvious that the coefficients a_{ij} in an orthogonal transformation (11.2) satisfy the relations

$$a_{ij} a_{ik} = \delta_{jk}, \tag{11.5}$$

and that the matrix $A = (a_{ij})$ of an orthogonal transformation is related to its inverse by the formula $A' = A^{-1}$. The condition (11.5) is both necessary and sufficient for a transformation to be orthogonal. Since the transpose of the matrix of an orthogonal transformation is equal to its inverse, we deduce that $a_{ji} a_{ki} = \delta_{jk}$.

Any matrix satisfying the *orthogonality conditions* (11.5) is called *orthogonal*. The square of the determinant of such a matrix has the value 1.

As in the three-dimensional case we introduce a matrix $B = (b_{ij})$ defining a transformation of the base vectors $\mathbf{a}^{(i)}$ into a new set of base vectors $\boldsymbol{\alpha}^{(i)}$ in accordance with the formula

$$\boldsymbol{\alpha}^{(i)} = b_{ij}\mathbf{a}^{(j)}, \qquad (i, j = 1, \ldots, n); \tag{11.6}$$

then $C = (B')^{-1}$.

If the vectors $\mathbf{a}^{(i)}$ are orthonormal and the matrix B orthogonal, the new set of vectors $\boldsymbol{\alpha}^{(i)}$ will obviously be orthonormal. Whenever $|b_{ij}| = 1$, we shall speak of (11.6) as representing a rotation of base vectors in E_n.

We now raise the question: Is it possible to find a matrix C such that the matrix CAC^{-1} has the diagonal form

$$\Lambda = \begin{bmatrix} \lambda_1 & 0 & & 0 \\ 0 & \lambda_2 & \cdots & 0 \\ \cdots & \cdots & \cdots & \cdots \\ 0 & 0 & \cdots & \lambda_n \end{bmatrix}?$$

This means that relative to a suitable reference frame the deformation of space, characterized by (11.2), assumes the form

$$\xi_1' = \lambda_1\xi_1, \qquad \xi_2' = \lambda_2\xi_2, \qquad \ldots, \qquad \xi_n' = \lambda_n\xi_n, \tag{11.7}$$

the ξ_i''s being the components of $\mathbf{x}'$ and the ξ_i's of $\mathbf{x}$ in the new coordinate system.

In the language of transformations in E_3, equations 11.7 state that for a suitably chosen reference frame the linear deformation of space is equivalent to simple extensions or contractions along the coordinate axes. Clearly the possibility of such reduction depends on the nature of coefficients a_{ij} in (11.2).

A detailed discussion of the problem of reduction of matrices to various canonical forms is involved. In the following sections we treat only those cases that occur most frequently in applications, referring the reader for an exhaustive treatment to standard treatises on higher algebra.

12. Reduction of Matrices to the Diagonal Form

We return to the problem posed in Sec. 11, concerning the possibility of finding a nonsingular matrix C such that an arbitrary matrix A can be reduced to the diagonal form Λ by means of the similitude transformation CAC^{-1}. From the point of view of linear transformation of space, this problem is equivalent to determining the base system $\boldsymbol{\alpha}^{(i)}$, $(i = 1, \ldots, n)$, relative to which the transformation

$$x_i' = a_{ij}x_j$$

assumes the form [see (11.7)]

$$\xi_1' = \lambda_1\xi_1, \qquad \xi_2' = \lambda_2\xi_2, \qquad \ldots, \qquad \xi_n' = \lambda_n\xi_n.$$

We write $C^{-1} \equiv S$, and seek a solution of the matric equation

$$S^{-1}AS = \Lambda, \tag{12.1}$$

or

$$AS = S\Lambda, \tag{12.2}$$

where $A = (a_{ij})$ and

$$\Lambda = \begin{bmatrix} \lambda_1 & 0 & \cdots & 0 \\ 0 & \lambda_2 & \cdots & 0 \\ \cdots & \cdots & \cdots & \cdots \\ 0 & 0 & \cdots & \lambda_n \end{bmatrix}.$$

The matric equation 12.2 is equivalent to the system of linear equations

$$a_{ij}s_{jk} = s_{ik}\lambda_k, \qquad \text{(no sum on } k), \qquad (i, j, k = 1, 2, \ldots, n), \tag{12.3}$$

where

$$S = \begin{bmatrix} s_{11} & \cdots & s_{1k} & \cdots & s_{1n} \\ s_{21} & \cdots & s_{2k} & \cdots & s_{2n} \\ \cdots & \cdots & \cdots & \cdots & \cdots \\ s_{n1} & \cdots & s_{nk} & \cdots & s_{nn} \end{bmatrix}.$$

If in (12.3) we set $i = 1, 2, \ldots, n$, and fix k, we obtain a system of n equations containing the elements $(s_{1k}, s_{2k}, \ldots, s_{nk})$ appearing in the kth column of S. The elements $(s_{1k}, s_{2k}, \ldots, s_{nk})$ can be viewed as components of the vector $\mathbf{s}^{(k)}$, so that the determination of the matrix S is equivalent to finding a set of n vectors $\mathbf{s}^{(k)}$, $(k = 1, \ldots, n)$, whose components satisfy equations 12.3. Accordingly we write equation 12.3 in the form

$$A\mathbf{s}^{(k)} = \mathbf{s}^{(k)}\lambda_k, \qquad \text{(no sum on } k), \tag{12.4}$$

and note that (12.3) is equivalent to

$$(a_{ij} - \delta_{ij}\lambda_k)s_{jk} = 0, \qquad (k \text{ not summed}). \tag{12.5}$$

If this system of linear homogeneous equations is to have a nontrivial solution for the s_{jk}, then λ_k must be a root of the determinantal equation

$$|a_{ij} - \delta_{ij}\lambda| = 0,$$

which, when written out in full, is

$$\begin{vmatrix} a_{11} - \lambda & a_{12} & \cdots & a_{1n} \\ a_{21} & a_{22} - \lambda & \cdots & a_{2n} \\ \cdots & \cdots & \cdots & \cdots \\ a_{n1} & a_{n2} & & a_{nn} - \lambda \end{vmatrix} = 0. \tag{12.6}$$

This nth order algebraic equation in λ has n roots, which are known as the *characteristic values*[4] of the matrix A. If these n roots are distinct, we can readily show that the system of equations 12.4 yields a set of n *linearly independent* vectors $\mathbf{s}^{(k)}$, and hence a *nonsingular* matrix S, as required by (12.1), exists. If the roots are not distinct, it *may not* be possible to determine the desired matrix S.

We consider the case when the roots are distinct, and denote them by $\lambda_1, \lambda_2, \ldots, \lambda_n$. If we set λ_1 for λ_k in (12.5) we obtain a system of n homogeneous equations. This system will have a *nontrivial* solution $s_{11}, s_{21}, \ldots, s_{n1}$. Setting $\lambda_k = \lambda_2$ in (12.5) we get the system yielding a solution $s_{12}, s_{22}, \ldots, s_{n2}$. This gives the second column of S. Proceeding in this fashion we can determine the remaining columns and hence the matrix S, which satisfies the equation 12.2. To show that the transformation 12.1 is possible, we must demonstrate that the vectors $\mathbf{s}^{(k)}$ so calculated are linearly independent, so that S possesses an inverse S^{-1}. We shall prove this by supposing that the matrix S is singular and reaching a contradiction.

If $|S| = 0$, the vectors $\mathbf{s}^{(k)}$ appearing in the columns of S are linearly dependent, and hence there exists a set of constants c_i, not all zero, such that

$$c_1\mathbf{s}^{(1)} + c_2\mathbf{s}^{(2)} + \cdots + c_n\mathbf{s}^{(n)} = \mathbf{0}.$$

In this expression some c's may be zero. We may suppose, without loss of generality, that the first r c's do not vanish, so that we have the relation

$$c_1\mathbf{s}^{(1)} + c_2\mathbf{s}^{(2)} + \cdots + c_r\mathbf{s}^{(r)} = \mathbf{0}, \qquad r \leq n, \tag{12.7}$$

in which none of the c's (or $\mathbf{s}^{(i)}$'s) vanishes.

[4] They are also called *eigenvalues* and the corresponding vectors $\mathbf{s}^{(k)}$ are termed *characteristic vectors* or *eigenvectors*.

From (12.4) we deduce the relations

$$A\mathbf{s}^{(k)} = \mathbf{s}^{(k)}\lambda_k, \qquad A(A\mathbf{s}^{(k)}) = A\mathbf{s}^{(k)}\lambda_k = \mathbf{s}^{(k)}\lambda_k{}^2,$$

$$A[A(A\mathbf{s}^{(k)})] = \mathbf{s}^{(k)}\lambda_k{}^3, \qquad \ldots, \qquad (A)^{r-1}\mathbf{s}^{(k)} = \mathbf{s}^{(k)}\lambda_k{}^{r-1}.$$

If we multiply (12.7) by A successively $r - 1$ times and take account of the chain of relations just written, we obtain a system of equations

$$\begin{aligned} c_1\mathbf{s}^{(1)} + c_2\mathbf{s}^{(2)} + \cdots + c_r\mathbf{s}^{(r)} &= \mathbf{0}, \\ c_1\mathbf{s}^{(1)}\lambda_1 + c_2\mathbf{s}^{(2)}\lambda_2 + \cdots + c_r\mathbf{s}^{(r)}\lambda_r &= \mathbf{0}, \\ \cdots\cdots\cdots\cdots\cdots\cdots\cdots\cdots\cdots\cdots&, \\ c_1\mathbf{s}^{(1)}\lambda_1^{r-1} + c_2\mathbf{s}^{(2)}\lambda_2^{r-1} + \cdots + c_r\mathbf{s}^{(r)}\lambda_r^{r-1} &= \mathbf{0}. \end{aligned}$$

Since none of the c's or $\mathbf{s}^{(k)}$'s vanishes, this system can be satisfied only if

$$\Delta \equiv \begin{vmatrix} 1 & 1 & \cdots & 1 \\ \lambda_1 & \lambda_2 & \cdots & \lambda_r \\ \cdots & \cdots & \cdots & \cdots \\ \lambda_1^{r-1} & \lambda_2^{r-1} & \cdots & \lambda_r^{r-1} \end{vmatrix} = 0.$$

The determinant Δ, however, is a Vandermondian,[5] and its value is known to be

$$\begin{aligned} \Delta = (\lambda_2 - \lambda_1)(\lambda_3 - \lambda_1) & \cdots (\lambda_r - \lambda_1) \\ (\lambda_3 - \lambda_2) & \cdots (\lambda_r - \lambda_2) \\ & \cdots\cdots \\ & (\lambda_r - \lambda_{r-1}) = \Pi(\lambda_i - \lambda_j), \qquad i > j. \end{aligned}$$

This is never zero if the λ's are distinct. Thus the assumption that the matrix S is singular is incorrect, and hence the matrix can always be reduced to the diagonal form whenever the characteristic values of the matrix A are distinct.

If the roots of the equation $|A - \lambda I| = 0$ are not distinct, the reduction of A to the diagonal form by the transformation $S^{-1}AS$ may be impossible. In this event there are other canonical representations which are discussed in books on higher algebra.[6] In several special cases, however, the

[5] See L. J. Paige and J. D. Swift, *Elements of Linear Algebra*, Ginn and Co., Boston (1961).

[6] See F. D. Murnaghan, *Applied Mathematics*, John Wiley and Sons, New York, (1948); G. Birkhoff and S. MacLane, *A Survey of Modern Algebra*, The Macmillan Co., New York (1941).

reduction of the matrix A to the diagonal form, even when the characteristic equation $|A - \lambda I| = 0$ has multiple roots, can be achieved. We turn to the consideration of these cases in the following sections.

13. Real Symmetric Matrices and Quadratic Forms

Let us assume that the matrix $A = (a_{ij})$ in a linear transformation

$$x_i' = a_{ij}x_j, \qquad (i, j = 1, 2, \ldots, n), \tag{13.1}$$

is *real* and *symmetric*, so that $a_{ij} = a_{ji}$ (or $A' = A$) for all values of i and j. We will show that the matrix A can be reduced to the diagonal form by the transformation $S^{-1}AS$. Moreover, S can be an *orthogonal* matrix.

Linear transformations with real symmetric matrices occur commonly in the study of deformations taking place in elastic media. Real symmetric matrices also enter prominently in the study of real quadratic forms

$$Q(x_1, x_2, \ldots, x_n) \equiv a_{ij}x_ix_j, \qquad (i, j = 1, \ldots, n) \tag{13.2}$$

which arise in many problems in dynamics and geometry. We can assume without loss of generality that the coefficients a_{ij} in (13.2) are symmetric, since (13.2) can always be written

$$Q(x_1, x_2, \ldots, x_n) = \frac{a_{ij} + a_{ji}}{2} x_ix_j,$$

in which the coefficients are obviously symmetric. In dealing with quadratic forms we shall always suppose that they have been symmetrized.

It will follow from the developments in this section that the problems of reduction of the set of linear forms (13.1) to the form

$$\xi_1' = \lambda_1\xi_1, \qquad \xi_2' = \lambda_2\xi_2, \qquad \ldots, \qquad \xi_n' = \lambda_n\xi_n$$

and of the quadratic form (13.2) to the form

$$Q = \lambda_1\xi_1^2 + \lambda_2\xi_2^2 + \cdots + \lambda_n\xi_n^2 \tag{13.3}$$

are mathematically identical.

We note first several properties of quadratic forms (13.2). If we introduce a linear transformation

$$x_i = s_{ik}\xi_k \qquad \text{or} \qquad \mathbf{x} = S\boldsymbol{\xi}, \tag{13.4}$$

the form Q in (13.2) becomes

$$\begin{aligned} Q &= a_{ij}(s_{ik}\xi_k)(s_{jl}\xi_l) \\ &= a_{ij}s_{ik}s_{jl}\xi_k\xi_l. \end{aligned}$$

We denote the coefficients of $\xi_k \xi_l$ by c_{kl}, so that

$$Q = c_{kl}\xi_k\xi_l,$$

where

$$c_{kl} = a_{ij}s_{ik}s_{jl}. \tag{13.5}$$

Since $a_{ij} = a_{ji}$, and i and j in (13.5) are the summation indices, an interchange of k and l does not alter the value of (13.5). Thus $c_{kl} = c_{lk}$, and hence the matrix $C = (c_{ij})$ is symmetric. We thus have the result that *the symmetry of quadratic form* (13.2) *is not destroyed by subjecting the variables* x_i *to a linear transformation.*

Let us write (13.5) in the form

$$c_{kl} = s_{ik}(a_{ij}s_{jl}),$$

and observe that $a_{ij}s_{jl}$ is an element in the ith row and lth column of the matrix

$$AS \equiv B.$$

Thus

$$c_{kl} = s_{ik}b_{il} \tag{13.6}$$

can be regarded as the element in the kth row and the lth column of the matrix $S'B$, and

$$C = S'AS. \tag{13.7}$$

We have established a

THEOREM. *If the variables* x_i *in the quadratic form* $Q = a_{ij}x_ix_j$, *with a matrix* A, *are subjected to a linear transformation* $x_i = s_{ij}\xi_j$, *with a matrix* S, *the resulting quadratic form has the matrix* $S'AS$.

We note, as a corollary of this theorem, that the determinant of the resulting quadratic form has the value $|A|\,|S|^2$.

If the transformation 13.4 is orthogonal, then $S' = S^{-1}$ and we can write (13.7) as

$$C = S^{-1}AS.$$

It follows from this result that the determination of an *orthogonal transformation* which reduces the form 13.2 to the sum of the squares 13.3 reduces to the solution of the matrix equation

$$S^{-1}AS = \Lambda. \tag{13.8}$$

This is precisely the problem we considered in Sec. 12. It follows from the discussion of that section that the system of homogeneous equations

$$a_{ij}s_{jk} = s_{ik}\lambda_k, \qquad \text{(no sum on } k\text{)}, \tag{13.9}$$

obtained from

$$AS = S\Lambda$$

(see equations 12.3) will have a nontrivial solution for the vectors $\mathbf{s}^{(k)}$: $(s_{1k}, s_{2k}, \ldots, s_{nk})$ if, and only if, the λ's in (13.9) satisfy the equation $|a_{ij} - \delta_{ij}\lambda| = 0$, or

$$|A - \lambda I| = 0. \tag{13.10}$$

If the matrix A is arbitrary, the characteristic equation 13.10, in general, has complex roots; and if these roots are distinct, the methods discussed in Sec. 12 permit us to calculate a set of n linearly independent vectors $\mathbf{s}^{(k)}$ composing the matrix S. In the present case, however, the matrix S has to be orthogonal and hence real. Now, if the roots of the characteristic equation 13.10 are real, then it follows at once from equation 13.9 that the solutions $\mathbf{s}^{(k)}$: $(s_{1k}, s_{2k}, \ldots, s_{nk})$ can be taken to be real since the a_{ij}'s are real. We prove a

THEOREM. *If the matrix A is real and symmetric, then the roots of the characteristic equation* $|A - \lambda I| = 0$ *are all real.*

The system of equations 13.9 can be written compactly as

$$A\mathbf{s}^{(k)} = \mathbf{s}^{(k)}\lambda_k, \qquad \text{(no sum on } k). \tag{13.11}$$

We can regard $A\mathbf{s}^{(k)}$ as a vector with components

$$a_{i1}s_{1k} + a_{i2}s_{2k} + \cdots + a_{in}s_{nk} \qquad (i = 1, 2, \ldots, n).$$

Let λ_k be a root of (13.10), real or complex, and $\mathbf{s}^{(k)}$ a vector, real or complex, satisfying the system 13.11. We multiply (13.11) scalarly by $\bar{\mathbf{s}}^{(k)}$ and get

$$\bar{\mathbf{s}}^{(k)} \cdot A\mathbf{s}^{(k)} = |\mathbf{s}^{(k)}|^2 \, \lambda_k. \tag{13.12}$$

Now, the left-hand member in this product (recall definition 6.1)

$$\bar{\mathbf{s}}^{(k)} \cdot A\mathbf{s}^{(k)} = a_{ij}\bar{s}_{ik}s_{jk}, \qquad \text{(no sum on } k)$$

is real if $a_{ij} = a_{ji}$. To prove this, note that the conjugate of $a_{ij}\bar{s}_{ik}s_{jk}$ is equal to the original expression,

$$a_{ij}s_{ik}\bar{s}_{jk} = a_{ji}s_{ik}\bar{s}_{jk} = a_{ij}\bar{s}_{ik}s_{jk}.$$

Since the left-hand member of (13.12) is real, and $|\mathbf{s}^{(k)}|^2$ is real, it follows that λ_k is real. This completes the proof of the theorem.

We prove next that, if λ_i and λ_j are two distinct roots of (13.10), then the vectors $\mathbf{s}^{(i)}$ and $\mathbf{s}^{(j)}$, corresponding to these roots, are orthogonal.

Since $\mathbf{s}^{(i)}$ and $\mathbf{s}^{(j)}$ satisfy (13.11), we have the identities

$$A\mathbf{s}^{(i)} = \mathbf{s}^{(i)}\lambda_i, \qquad \text{(no sum)},$$
$$A\mathbf{s}^{(j)} = \mathbf{s}^{(j)}\lambda_j, \qquad \text{(no sum)},$$

where all the vectors involved are real. If we multiply the first of these scalarly by $\mathbf{s}^{(j)}$ on the right and the second by $\mathbf{s}^{(i)}$ on the left and subtract, we get

$$A\mathbf{s}^{(i)} \cdot \mathbf{s}^{(j)} - \mathbf{s}^{(i)} \cdot A\mathbf{s}^{(j)} = (\lambda_i - \lambda_j)\mathbf{s}^{(i)} \cdot \mathbf{s}^{(j)},$$

and the left-hand member vanishes since $\mathbf{s}^{(i)} \cdot A\mathbf{s}^{(j)} = A\mathbf{s}^{(i)} \cdot \mathbf{s}^{(j)}$ because of symmetry of A. This establishes the orthogonality of $\mathbf{s}^{(i)}$ and $\mathbf{s}^{(j)}$, whenever the roots λ_i and λ_j are unequal. Since equation 13.11 is homogeneous, we can multiply it by a suitable constant making the length of $\mathbf{s}^{(k)}$ equal to 1. We shall suppose that this has been done.

We recall that a set of orthogonal vectors is necessarily linearly independent. Hence, if all roots of $|A - \lambda I| = 0$ are distinct, the vectors $\mathbf{s}^{(k)}$ will be orthonormal, and, accordingly, the matrix S, accomplishing the transformation $S^{-1}AS = \Lambda$, will be orthogonal.

It remains to consider the case of reduction of real quadratic forms 13.2 to the diagonal form 13.3 when the equation

$$|A - \lambda I| = 0 \tag{13.10}$$

has multiple roots. The demonstration that the reduction is possible in this case hinges on one important property of all similar matrices, namely: *the characteristic roots of all similar matrices are equal.* The proof of this is easy. We replace A in the left-hand member of (13.10) by some similar matrix $S^{-1}AS$ and obtain the polynomial in λ,

$$\begin{aligned} |S^{-1}AS - \lambda I| &= |S^{-1}(A - \lambda I)S| \\ &= |S^{-1}| \cdot |A - \lambda I| \cdot |S| \\ &= |A - \lambda I|. \end{aligned}$$

It follows that the characteristic equations associated with $S^{-1}AS$ and A are identical, and hence their roots are equal.

Now let us suppose that (13.10) has multiple roots. Let $\lambda = \lambda_1$ be some root of (13.10), and let us determine the solution of (13.11) $\mathbf{s}^{(1)}$: $(s_{11}, s_{21}, \ldots, s_{n1})$ corresponding to $\lambda = \lambda_1$, which is such that $\mathbf{s}^{(1)} \cdot \mathbf{s}^{(1)} = 1$. This can be done whether λ_1 is a multiple root or not. We can adjoin to the vector $\mathbf{s}^{(1)}$ a set of $n - 1$ orthonormal vectors forming a complete system of vectors in our n-dimensional manifold. These vectors can be used as a basis for our space instead of the original set of orthonormal base vectors $\mathbf{a}^{(1)}, \ldots, \mathbf{a}^{(n)}$, and we can pass from the reference frame determined by the $\mathbf{a}^{(i)}$'s to the new frame by an orthogonal transformation. Hence the matrix of the quadratic form 13.2, when referred to the new frame, will assume the form $A_1 = S_1^{-1}AS_1$, where S_1 is orthogonal.

Moreover,

$$|A_1 - \lambda I| = 0 \tag{13.13}$$

has the same characteristic roots as (13.10). The equation [cf. (13.11)]

$$A_1 \mathbf{s} = \mathbf{s}\lambda \tag{13.14}$$

for $\lambda = \lambda_1$ has the solutions $\mathbf{s}^{(1)}$: $(1, 0, 0, \ldots, 0)$, since we chose it to be a unit vector, and $\mathbf{s}^{(1)}$ is one of the base vectors of the new reference frame. If we insert this solution in (13.14), we get an identity

$$A_1 \cdot \begin{bmatrix} 1 \\ 0 \\ \cdot \\ \cdot \\ \cdot \\ 0 \end{bmatrix} = \begin{bmatrix} \lambda_1 \\ 0 \\ \cdot \\ \cdot \\ \cdot \\ 0 \end{bmatrix},$$

from which it follows that the matrix A_1 has the following elements:

$$a_{11}^{(1)} = \lambda_1, \qquad a_{21}^{(1)} = a_{31}^{(1)} = \cdots = a_{n1}^{(1)} = 0. \tag{13.15}$$

The original matrix A is symmetric, and, since orthogonal transformations do not destroy the symmetry, the matrix A_1 is also symmetric.[7] Thus

$$A_1' = A_1,$$

and we can write instead of (13.15)

$$a_{11}^{(1)} = \lambda_1, \qquad a_{12}^{(1)} = a_{21}^{(1)} = a_{31}^{(1)} = a_{13}^{(1)} = \cdots = a_{n1}^{(1)} = a_{1n}^{(1)} = 0,$$

so that

$$A_1 = \begin{bmatrix} \lambda_1 & 0 & \cdots & 0 \\ 0 & a_{22}^{(1)} & \cdots & a_{2n}^{(1)} \\ \cdots & \cdots & \cdots & \cdots \\ 0 & a_{n2}^{(1)} & \cdots & a_{nn}^{(1)} \end{bmatrix},$$

Thus the quadratic form 13.2, when referred to our new frame, has the structure

$$Q = \lambda_1 \xi_1^2 + a_{ij}^{(1)} \xi_i \xi_j, \qquad (i, j = 2, 3, \ldots, n).$$

We succeeded in separating one square and reduced the problem to a consideration of the form $a_{ij}^{(1)} \xi_i \xi_j$ in $n - 1$ variables. We can apply

[7] For $A_1' = (S_1^{-1} A S_1)' = S_1' A' (S_1^{-1})' = S_1^{-1} A S_1$, since $S^{-1} = S'$ for orthogonal matrices.

similar reasoning to the $(n-1) \times (n-1)$ matrix $A_2 = (a_{ij}^{(1)})$ and consider the form $a_{ij}^{(1)}\xi_i\xi_j$, $(i, j = 2, 3, \ldots, n)$, in the $n-1$ dimensional subspace E_{n-1} of E_n, determined by the base vectors other than $\mathbf{s}^{(1)}$. In E_{n-1}, we can calculate a unit vector $\mathbf{s}^{(2)}$ satisfying the equation

$$A_2\mathbf{s} = \mathbf{s}\lambda,$$

corresponding to $\lambda = \lambda_2$, and construct a new base system by an orthogonal transformation in which $\mathbf{s}^{(2)}$ is a base vector. This will yield a matrix

$$A_2 = \begin{bmatrix} \lambda_2 & 0 & \cdots & 0 \\ 0 & a_{33}^{(2)} & \cdots & a_{3n}^{(2)} \\ \cdots & \cdots & \cdots & \cdots \\ 0 & a_{n3}^{(2)} & \cdots & a_{nn}^{(2)} \end{bmatrix},$$

and hence Q of the form

$$Q = \lambda_1\xi_1^2 + \lambda_2\xi_2^2 + a_{ij}^{(2)}\xi_i\xi_j, \qquad (i, j = 3, \ldots, n).$$

The continuation of this process will reduce the original quadratic form 13.2 to the form

$$Q = \lambda_1\xi_1^2 + \lambda_2\xi_2^2 + \cdots + \lambda_n\xi_n^2.$$

Since each successive reduction is performed by an orthogonal transformation, the product of orthogonal transformations is equivalent to a single orthogonal transformation S. The resulting diagonal matrix Λ,

$$S^{-1}AS = \Lambda = \begin{bmatrix} \lambda_1 & 0 & \cdots & 0 \\ 0 & \lambda_2 & \cdots & 0 \\ \cdot & & \cdot & \cdot \\ \cdot & & \cdot & \cdot \\ \cdot & & \cdot & \cdot \\ 0 & 0 & \cdots & \lambda_n \end{bmatrix},$$

contains the number of like roots λ equal to the multiplicity of the roots in $|A - \lambda I| = 0$. Since the matrix $S^{-1}AS$ is similar to A, the characteristic roots $\lambda_1, \lambda_2, \ldots, \lambda_n$, of $|\Lambda - \lambda I| = 0$, are identical with those of $|A - \lambda I| = 0$.

The directions determined by the characteristic vectors $\mathbf{s}^{(k)}$ associated with the matrix A are called the *principal directions* of the matrix A.

Problem

If $\mathbf{x}$: $(x_1, x_2, \ldots, x_n)$ is a unit vector and $Q = a_{ij}x_ix_j$ is a real symmetric quadratic form with nonsingular matrix A, then the extreme values of Q are the characteristic values of A. Prove it. *Hint:* Maximize Q subject to the constraining condition $x_ix_i = 1$ and deduce the system of equations $(a_{ij} - \delta_{ij}\lambda)x_i = 0$, where λ is the Lagrange multiplier.

14. Illustrations of Reduction of Quadratic Forms

We shall interpret the results of Sec. 13 in the language of analytic geometry and give two examples providing concrete illustrations of reduction of quadratic forms to the canonical form by means of orthogonal transformations.

If we suppose that the dimensionality of space $n = 3$, and set $a_{ij}x_ix_j$ equal to a constant c, then the equation

$$a_{ij}x_ix_j = c, \qquad (i, j = 1, 2, 3) \tag{14.1}$$

represents a quadratic surface Q referred to a reference frame with base vectors $\mathbf{a}^i$. An orthogonal transformation $S^{-1}AS = \Lambda$, leading to the quadratic form

$$\lambda_1\xi_1^2 + \lambda_2\xi_2^2 + \lambda_3\xi_3^2 = c, \tag{14.2}$$

can be interpreted as a transformation of coordinate axes yielding a frame with base vectors directed along the principal axes of the quadratic.

Let the quadratic exemplifying (14.1) be

$$Q \equiv 2x_1^2 + 2x_2^2 - 15x_3^2 + 8x_1x_2 - 12x_2x_3 - 12x_1x_3 = c.$$

In order to determine the coefficients λ_i in (14.2) for this particular case, we symmetrize Q and obtain

$$\begin{aligned} Q = 2x_1^2 &+ 4x_1x_2 - 6x_1x_3 \\ &+ 4x_2x_1 + 2x_2^2 - 6x_2x_3 \\ &- 6x_3x_1 - 6x_3x_2 - 15x_3^2, \end{aligned}$$

from which the characteristic equation $|A - \lambda I| = 0$ can be written down at once. We have

$$|A - \lambda I| = \begin{vmatrix} 2-\lambda & 4 & -6 \\ 4 & 2-\lambda & -6 \\ -6 & -6 & -15-\lambda \end{vmatrix} = 0.$$

Expanding this determinant leads to a cubic

$$\lambda^3 + 11\lambda^2 - 144\lambda - 324 = 0,$$

which has the roots

$$\lambda_1 = -2, \qquad \lambda_2 = -18, \qquad \lambda_3 = 9.$$

Thus, relative to a new reference frame, Q assumes the form

$$-2\xi_1^2 - 18\xi_2^2 + 9\xi_3^2 = c,$$

representing an hyperboloid.

For the determination of the new base vectors $\mathbf{s}^{(i)}$, we have the system of equations 13.9,

$$a_{ij}s_{jk} = s_{ik}\lambda_k, \qquad \text{(no sum on } k\text{)},$$

or

$$(a_{ij} - \delta_{ij}\lambda_k)s_{jk} = 0.$$

Writing these out, we have

$$\begin{aligned} (2 - \lambda_k)s_{1k} + 4s_{2k} - 6s_{3k} &= 0, \\ 4s_{1k} + (2 - \lambda_k)s_{2k} - 6s_{3k} &= 0, \\ -6s_{1k} - 6s_{2k} - (15 + \lambda_k)s_{3k} &= 0. \end{aligned} \tag{14.3}$$

Substituting $\lambda_1 = -2$ gives three equations, two of which are identical. The linearly independent equations are

$$\begin{aligned} 4s_{11} + 4s_{21} - 6s_{31} &= 0, \\ -6s_{11} - 6s_{21} - 13s_{31} &= 0. \end{aligned}$$

Solving these yields the components of $\mathbf{s}^{(1)}$,

$$s_{11} = c, \qquad s_{21} = -c, \qquad s_{31} = 0,$$

where c is arbitrary. We determine the constant c so that the length of $\mathbf{s}^{(1)}$ is unity. Thus

$$s_{11}^2 + s_{21}^2 + s_{31}^2 = 1,$$

and hence $c = 1/\sqrt{2}$ and our normalized components are

$$s_{11} = \frac{1}{\sqrt{2}}, \qquad s_{21} = -\frac{1}{\sqrt{2}}, \qquad s_{31} = 0.$$

These determine the first column of the matrix S.

The substitution of $\lambda_2 = -18$ in (14.3) leads to three homogeneous equations

$$\begin{aligned} 20s_{12} + 4s_{22} - 6s_{32} &= 0, \\ 4s_{12} + 20s_{22} - 6s_{32} &= 0, \\ -6s_{12} - 6s_{22} + 3s_{32} &= 0, \end{aligned}$$

the solution of which is readily found to be

$$s_{12} = \tfrac{1}{4}c, \qquad s_{22} = \tfrac{1}{4}c, \qquad s_{32} = c.$$

The normalized solution is

$$s_{12} = \frac{1}{3\sqrt{2}}, \qquad s_{22} = \frac{1}{3\sqrt{2}}, \qquad s_{32} = \frac{4}{3\sqrt{2}}.$$

The elements entering in the third column of S are determined from the system 14.3 by setting $\lambda_3 = 9$. This yields the equations

$$\begin{aligned} -7s_{13} + 4s_{23} - 6s_{33} &= 0, \\ 4s_{13} - 7s_{23} - 6s_{33} &= 0, \\ -6s_{13} - 6s_{23} - 24s_{33} &= 0, \end{aligned}$$

which are satisfied by

$$s_{13} = c, \qquad s_{23} = c, \qquad s_{33} = -\tfrac{1}{2}c.$$

Normalizing to unity, we obtain $\mathbf{s}^{(3)}$ in the form

$$s_{13} = \tfrac{2}{3}, \qquad s_{23} = \tfrac{2}{3}, \qquad s_{33} = -\tfrac{1}{3}.$$

Accordingly, the orthogonal transformation yielding the desired canonical form is

$$\begin{cases} \xi_1 = \dfrac{1}{\sqrt{2}}\, x_1 - \dfrac{1}{\sqrt{2}}\, x_2 + 0 \cdot x_3, \\ \xi_2 = \dfrac{1}{3\sqrt{2}}\, x_1 + \dfrac{1}{3\sqrt{2}}\, x_2 + \dfrac{4}{3\sqrt{2}}\, x_3, \\ \xi_3 = \tfrac{2}{3}x_1 + \tfrac{2}{3}x_2 - \tfrac{1}{3}x_3. \end{cases}$$

To illustrate reduction in the event the characteristic equation has multiple roots, we take

$$Q = 3x_1^2 + 2x_2^2 + 3x_3^2 + 2x_1x_3 = c.$$

Here the characteristic equation of the matrix of Q is

$$-\begin{vmatrix} 3-\lambda & 0 & 1 \\ 0 & 2-\lambda & 0 \\ 1 & 0 & 3-\lambda \end{vmatrix} = \lambda^3 - 8\lambda^2 + 20\lambda - 16 = 0,$$

whose roots are $\lambda_1 = \lambda_2 = 2$, $\lambda_3 = 4$. Hence the quadric surface is an ellipsoid of revolution whose equation can be taken in the form

$$2(\xi_1^2 + \xi_2^2) + 4\xi_3^2 = c.$$

The equations for the determination of the new base vectors are

$$\begin{aligned} (3-\lambda)s_{1k} + 0s_{2k} + s_{3k} &= 0, \\ 0s_{1k} + (2-\lambda)s_{2k} + 0s_{3k} &= 0, \\ s_{1k} + 0s_{2k} + (3-\lambda)s_{3k} &= 0. \end{aligned}$$

Setting $\lambda_1 = 2$ yields only one equation

$$s_{11} + s_{31} = 0$$

for the determination of $\mathbf{s}^{(1)}$, so that the normalized solution can be taken as

$$s_{11} = \frac{1}{\sqrt{2}}, \qquad s_{21} = 0, \qquad s_{31} = -\frac{1}{\sqrt{2}}$$

The second characteristic root $\lambda_2 = 2$ gives the equation

$$s_{12} + s_{32} = 0, \tag{14.4}$$

and since $\mathbf{s}^{(2)}$ must be normal to $\mathbf{s}^{(1)}$, we have the orthogonality condition

$$s_{11}s_{12} + s_{21}s_{22} + s_{31}s_{32} = 0,$$

or

$$\frac{1}{\sqrt{2}} s_{12} - \frac{1}{\sqrt{2}} s_{32} = 0. \tag{14.5}$$

Equations 14.4 and 14.5 state that $s_{12} = 0$, $s_{22} = 1$, $s_{32} = 0$.

Finally, for the determination of the third base vector we have the system of equations

$$\begin{aligned} -s_{13} + s_{33} &= 0, \\ -2s_{23} &= 0, \\ s_{13} - s_{33} &= 0, \end{aligned}$$

obtained by setting $\lambda = 4$. The normalized solution of this system is $s_{13} = 1/\sqrt{2}$, $s_{23} = 0$, $s_{33} = 1/\sqrt{2}$. Hence the matrix S has the form

$$\begin{bmatrix} \frac{1}{\sqrt{2}} & 0 & \frac{1}{\sqrt{2}} \\ 0 & 1 & 0 \\ -\frac{1}{\sqrt{2}} & 0 & \frac{1}{\sqrt{2}} \end{bmatrix},$$

from which the equations of connection between the variables x_i and ξ_i can be written down at once.

15. Classification and Properties of Real Quadratic Forms

In this section we summarize several properties of real quadratic forms

$$Q = a_{ij}x_ix_j, \qquad (i, j = 1, \ldots, n), \tag{15.1}$$

which are of considerable importance in applications.

We have shown that the real quadratic form Q can be reduced by an orthogonal transformation

$$\xi_i = s_{ij}x_j, \tag{15.2}$$

to the canonical form

$$Q = \lambda_1\xi_1^2 + \lambda_2\xi_2^2 + \cdots + \lambda_n\xi_n^2. \tag{15.3}$$

The problem of reduction of (15.1) to the form 15.3 is equivalent to the search of an orthogonal matrix $S = (s_{ij})$ satisfying the matric equation

$$S^{-1}AS = \Lambda, \qquad (\text{or } S'AS = \Lambda), \tag{15.4}$$

where the elements along the diagonal in the Λ matrix are the roots of the determinantal equation

$$|A - \lambda I| = 0, \tag{15.5}$$

and A is a real symmetric matrix.

Since the determinant of S does not vanish, it is clear from (15.4) that the rank of A is equal to the rank of Λ. If the characteristic equation 15.5 has n nonvanishing roots, then the number of terms actually appearing in (15.3) is n. If, however, equation 15.5 has $r < n$ nonvanishing roots, then the reduced form 15.3 will have the appearance

$$Q = \lambda_1\xi_1^2 + \lambda_2\xi_2^2 + \cdots + \lambda_r\xi_r^2, \tag{15.6}$$

and we shall say that the rank of (15.1) is r. The number of *positive* λ's appearing in (15.6) is called the *index* of Q. If we have a form (15.6) with p positive and $r - p$ negative λ's, we can introduce a real transformation $\xi_i = (1/\sqrt{\lambda_i})\xi_i'$ for terms with positive λ's and $\xi_i = (1/\sqrt{-\lambda_i})\xi_i'$ for terms with negative λ's so that it assumes the form

$$Q = \xi_1'^2 + \xi_2'^2 + \cdots + \xi_p'^2 - \xi_{p+1}'^2 - \xi_{p+2}'^2 - \cdots - \xi_r'^2. \tag{15.7}$$

Thus *every real quadratic form Q can be reduced by a real linear transformation $\xi_i' = c_{ij}x_j$ to the canonical form 15.7.* The matrix (c_{ij}), of course, is not necessarily orthogonal.

The form 15.7 provides a means for the classification of quadratic forms.

We consider the following cases.

1. The index p in (15.7) is equal to n, so that equation 15.5 has n positive roots. In this case we say that the form 15.1 is *positive definite.*

2. If the index $p = 0$, so that all roots of (15.5) are negative and the rank of Q is n, the form 15.1 is *negative definite.*

3. If the index p is equal to the rank r and $r < n$, then the form is said to be *positive*. On the other hand, if the index is zero and the rank $r < n$, the form Q is *negative*.

4. The forms whose canonical representation 15.3 contains both positive and negative λ's are called *indefinite*.

We observe that positive and negative *definite* forms never vanish for real nonzero values of the variables x_i. They vanish if, and only if, all x_i's vanish. In contradistinction, the positive and negative forms may vanish for nonzero values of the arguments x_i. To see this, note that, if $r < n$, then

$$Q = \lambda_1 \xi_1^2 + \lambda_2 \xi_2^2 + \cdots + \lambda_r \xi_r^2.$$

We can make (15.1) vanish by choosing the x_j in (15.2) so that

$$\xi_1 = \xi_2 = \cdots = \xi_r = 0.$$

The nonvanishing values of x_j will surely exist, since the system of r homogeneous equations,

$$s_{ij} x_j = 0, \qquad (i = 1, \ldots, r),$$

in n unknowns x_j, has nontrivial solutions whenever $r < n$.

It follows at once from (15.4), and from the fact that in a positive definite form the λ_i's in Λ are all positive, that the determinant $|a_{ij}|$ of the positive definite form is necessarily positive. The converse of this, clearly, is not true. This can be readily seen by noting that $|A| = |\Lambda|$, and the positive value of $|\Lambda|$ admits indefinite as well as definite forms.

16. Simultaneous Reduction of Two Quadratic Forms to a Sum of Squares

We conclude our study of quadratic forms by investigating the possibility of simultaneous reduction of two real quadratic forms to the sum of squares by a single linear transformation. This problem arises, among other places, in a study of oscillations of mechanical systems about the state of equilibrium.

Consider two real quadratic forms

$$Q_1 = a_{ij} x_i x_j \quad \text{and} \quad Q_2 = b_{ij} x_i x_j, \tag{16.1}$$

each of rank n, one of which, say Q_1, is positive definite. Let it be required to find a linear transformation, not necessarily orthogonal, such that both forms reduce to the sum of squares.

If Q_1 is positive definite and of rank n, then there exists a linear transformation $x_i = c_{ij} \xi_j$, not necessarily orthogonal, under which Q_1 reduces to the form

$$Q_1 = \xi_1^2 + \xi_2^2 + \cdots + \xi_n^2. \tag{16.2}$$

Under the same transformation Q_2 will assume the form

$$Q_2 = b_{ij}'\xi_i\xi_j. \tag{16.3}$$

Now, under a suitable *orthogonal* transformation $\xi_i = d_{ij}\eta_j$ on the variables ξ_i, Q_2 can be reduced to the form

$$Q_2 = \lambda_i\eta_i^2, \tag{16.4}$$

and, since orthogonal transformations leave the scalar product $\xi_i\xi_i$ invariant, the form Q_1 will be unchanged, and we have

$$Q_1 = \eta_i\eta_i. \tag{16.5}$$

Now Q_1 and Q_2 are in the desired forms, and, since the product of successive linear transformations from x_i to η_i is a linear transformation $x_i = s_{ij}\eta_j$, it follows that the simultaneous reduction can be accomplished.

The numbers λ_i in (16.4) are called the *characteristic numbers of the form Q_2 relative to Q_1*. We proceed to derive the equation for the characteristic numbers λ_i.

We recall that if the variables x_i in a form $Q = a_{ij}x_ix_j$ with a matrix A are subjected to a linear transformation $x_i = s_{ij}\eta_j$ with the matrix S, then the matrix of the resulting quadratic form is $S'AS$. The determinant of this matrix has the value $|S|^2\,|A|$. Now let us construct the quadratic form

$$\begin{aligned} Q &= Q_2 - \lambda Q_1 \\ &= (b_{ij} - \lambda a_{ij})x_ix_j, \end{aligned} \tag{16.6}$$

where λ is an arbitrary parameter. Under successive linear transformations from the variables x_i to η_i, Q_2 and Q_1 assume the forms (16.4) and (16.5), and hence (16.6) reduces to

$$\begin{aligned} Q &= \lambda_1\eta_1^2 + \lambda_2\eta_2^2 + \cdots + \lambda_n\eta_n^2 \\ &\quad - \lambda\eta_1^2 - \lambda\eta_2^2 - \cdots - \lambda\eta_n^2 \\ &= \sum_{i=1}^{n} (\lambda_i - \lambda)\eta_i^2. \end{aligned} \tag{16.7}$$

The determinant Δ in (16.7) is

$$\Delta = (\lambda_1 - \lambda)(\lambda_2 - \lambda)\cdots(\lambda_n - \lambda), \tag{16.8}$$

whereas the determinant of Q in (16.6) is

$$D = |b_{ij} - \lambda a_{ij}|. \tag{16.9}$$

It follows from remarks just made regarding the value of the determinant in the transformed quadratic form that the determinants D and Δ can differ only by a constant multiple equal to the square of the determinant

$|S|$ of the transformation from the initial variables x_i to the final variables η_i. Since this determinant does not vanish, and since it contains no parameter λ, the roots of polynomials 16.8 and 16.9 are identical. Taking account of the structure of expression 16.8, we conclude that the coefficients λ_i in (16.4) are the roots of the determinantal equation

$$D = \begin{vmatrix} b_{11} - \lambda a_{11} & b_{12} - \lambda a_{12} & \cdots & b_{1n} - \lambda a_{1n} \\ b_{21} - \lambda a_{21} & b_{22} - \lambda a_{22} & \cdots & b_{2n} - \lambda a_{2n} \\ \cdots & \cdots & \cdots & \cdots \\ b_{n1} - \lambda a_{n1} & b_{n2} - \lambda a_{n2} & \cdots & b_{nn} - \lambda a_{nn} \end{vmatrix} = 0.$$

In application of these results to the study of small vibrations of mechanical systems about the point of equilibrium, the forms Q_1 and Q_2 are identified with the kinetic and potential energies of the system. The final coordinates η_i are termed *normal* coordinates, and the characteristic numbers λ_i are related to *normal modes of vibration* (see Sec. 89).

17. Unitary Transformations and Hermitean Matrices

In a variety of circumstances arising in applied mathematics it becomes necessary to extend the concept of orthogonal transformations to vectors defined in a complex field.

If we consider a nonsingular transformation

$$x_i' = a_{ij}x_j, \qquad (i, j = 1, \ldots, n), \tag{17.1}$$

in which the coefficients a_{ij} are complex numbers and the set of numbers $(x_1, \ldots, x_n)$ represents the components of the vector $\mathbf{x}$, the question naturally arises about restrictions that must be imposed on the matrix (a_{ij}) if the length $|\mathbf{x}|$ of the vector is to be preserved. The imposition of the condition of invariance of length, namely,

$$\bar{x}_i'x_i' = \bar{x}_ix_i,$$

leads at once to the conclusion that [cf. (11.5)]

$$\bar{a}_{ij}a_{ik} = \delta_{jk}, \tag{17.2}$$

where bars, as usual, denote conjugate complex values. We deduce from (17.2) that the absolute value of the square of the determinant $|a_{ij}|$ is 1.

Matrices $A = (a_{ij})$ whose elements satisfy the conditions 17.2 are called *unitary*, and the corresponding transformations 17.1 are *unitary transformations*. We can write (17.2) in the form

$$\bar{A}'A = I, \tag{17.3}$$

where $\bar{A}$ is the *conjugate* matrix formed by replacing every element a_{ij} in A by $\bar{a}_{ij}$. From (17.3) we conclude at once that $\bar{A}' = A^{-1}$.

A bilinear form

$$H = a_{ij}\bar{x}_i x_j, \qquad (i, j = 1, \ldots, n), \tag{17.4}$$

where $a_{ij} = \bar{a}_{ji}$, is called a *Hermitean* form, and the matrix $(a_{ij}) = A$, corresponding to it, is a *Hermitean matrix*. It follows from the definition of the Hermitean matrix that the elements along its diagonal are real and that

$$\bar{A}' = A, \qquad \text{or} \qquad A' = \bar{A}.$$

We observe that the Hermitean forms can assume, for arbitrary x_i, only real values, since

$$\begin{aligned} \bar{H} &= \bar{a}_{ij}x_i\bar{x}_j \\ &= a_{ji}x_i\bar{x}_j \\ &= a_{ij}\bar{x}_i x_j = H. \end{aligned}$$

It is clear that the Hermitean forms are a generalization of real quadratic forms.

One can raise the question of the possibility of reduction of the form 17.4 to the canonical form

$$H = \lambda_1\bar{\xi}_1\xi_1 + \lambda_2\bar{\xi}_2\xi_2 + \cdots + \lambda_n\bar{\xi}_n\xi_n, \tag{17.5}$$

with the aid of the transformation

$$x_i = u_{ij}\xi_j \qquad \text{or} \qquad \mathbf{x} = U\boldsymbol{\xi},$$

where $U \equiv (u_{ij})$ is a unitary matrix. A computation similar to that carried out in Sec. 13 leads to the solution of the matric equation

$$U^{-1}AU = \Lambda, \tag{17.6}$$

where Λ is a diagonal matrix. The procedure in this case is, in every respect, similar to the one followed in the discussion of real symmetric matrices. We multiply (17.6) by U and obtain

$$AU = U\Lambda, \tag{17.7}$$

which represents a system of linear homogeneous equations for the determination of vectors $\mathbf{u}^{(k)}$: $(u_{1k}, u_{2k}, \ldots, u_{nk})$ entering in the columns of U. A necessary and sufficient condition that the system represented by (17.7) have a solution is that

$$|A - \lambda I| = 0. \tag{17.8}$$

The possibility of constructing a unitary matrix U satisfying equation 17.6 hinges on the fact that here the roots of (17.8) are also real. The fact that the characteristic roots λ_i must necessarily be real follows from the observation that $U^{-1}AU$ is a Hermitean matrix whenever A is Hermitean and U is unitary.[7] Thus Λ in (17.6) is Hermitean, and consequently the elements along its diagonal are real.

Problems

1. Reduce the matrix

$$A = (a_{ij}) = \begin{pmatrix} 1 & -1 \\ -1 & 1 \end{pmatrix}$$

to the diagonal form S by the similitude transformation $C^{-1}AC$. Show that

$$C = \begin{pmatrix} 1 & 1 \\ 1 & -1 \end{pmatrix}, \quad C^{-1} = \begin{pmatrix} \frac{1}{2} & \frac{1}{2} \\ \frac{1}{2} & \frac{1}{2} \end{pmatrix}, \quad \text{and } S = \begin{pmatrix} 0 & 0 \\ 0 & 2 \end{pmatrix}.$$

Discuss the meaning of A when it is viewed as an operator characterizing the deformation of space.

2. Diagonalize the matrices:

$$\begin{pmatrix} 1 & -1 & -1 \\ -1 & 1 & -1 \\ -1 & -1 & 1 \end{pmatrix}, \quad \begin{pmatrix} -1 & 1 & 2 \\ 0 & -2 & 1 \\ 0 & 0 & -3 \end{pmatrix}.$$

[7] For $(U^{-1}AU)' = U'A'(U^{-1})'$ and $\overline{(U^{-1}AU)'} = \overline{U'}\,\overline{A'}\,\overline{(U^{-1})'}$. Since A is Hermitean, $\bar{A}' = A$, and since U is unitary, $\bar{U}' = U^{-1}$ and $\overline{(U^{-1})'} = U$. Thus we have $\overline{(U^{-1}AU)'} = U^{-1}AU$.

2

TENSOR THEORY

18. Scope of Tensor Analysis. Invariance

Tensor analysis is concerned with a study of abstract objects, called *tensors*, whose properties are independent of the reference frames used to describe the objects. A tensor is represented in a particular reference frame by a set of functions, termed its *components*, just as a vector is determined in a given reference frame by a set of components. Whether a given set of functions represents a tensor depends on the law of transformation of these functions from one coordinate system to another. The situation here is identical with that already encountered in Chapter 1. In a given reference frame a vector **A** is determined uniquely by a set of components A_i. If a new coordinate system is introduced, the same vector **A** is determined by a set of components B_i, and these new components are related, in a definite way, to the old ones. It is the law of transformation of components of a vector that is the essence of the vector idea, and the same is true of tensors.

Since tensor analysis deals with entities and properties that are independent of the choice of reference frames it forms an ideal tool for the study of natural laws. Indeed, whether a logical deduction based on a conglomerate of observational facts deserves the name of a natural law is often determined by the generality of such a deduction, and by its validity in a sufficiently wide class of reference systems. This is intimately bound up with the possibility of formulating the deduction in the form of a tensor equation because tensor equations are invariant with respect to a given category of coordinate transformations. The concept of invariance of mathematical objects, under coordinate transformations, permeates the structure of tensor analysis to such an extent that it is important to get at the outset a clear notion of the particular brand of invariance we have in mind. We shall suppose that a point is an invariant. In a given reference frame a point P is determined by a set of coordinates x^i. If the coordinate system is changed, the point P is described by a new set of coordinates y^i, but the transformation of coordinates does nothing to the point itself.

Again, a pair of points (P_1, P_2) determines a vector $\overrightarrow{P_1P_2}$. This vector, in a particular reference frame, is uniquely determined by a set of components A_i. A transformation of coordinates does nothing to the vector $\overrightarrow{P_1P_2}$, but in the new reference frame $\overrightarrow{P_1P_2}$ is characterized by a different set of components B_i. A set of points, such as those forming a curve or surface, is also invariant. The curve may be described in a given coordinate system by an equation which usually changes its form when the coordinates are changed, but the curve itself remains unaltered. We shall say, in general, that *an object, whatever its nature, is an invariant, provided that it is not altered by a transformation of coordinates.*

19. Transformation of Coordinates

In Chapter 1 we discussed at some length linear transformations of coordinates. Here we will deal with real, single-valued, reversible functional transformations of the form

$$(19.1) \qquad T\colon \quad y^i = y^i(x^1, x^2, \ldots, x^n), \qquad (i = 1, 2, \ldots, n),$$

where we use superscripts to identify the variables. A particular set of n real numbers $(x_0^{\,1}, x_0^{\,2}, \ldots, x_0^{\,n})$ can be thought to specify a point P_0 in the n-dimensional metric manifold covered by a coordinate system X. The set of equations 19.1 will be viewed as a transformation of coordinate systems, so that the n-tuple of numbers $(y_0^{\,1}, y_0^{\,2}, \ldots, y_0^{\,n})$ obtained by substituting in (19.1) the coordinates $(x_0^{\,1}, x_0^{\,2}, \ldots x_0^{\,n})$ represents the coordinates of P_0 in the Y-reference frame. Since the transformation T in (19.1) was assumed to be reversible and one-to-one, we can write

$$(19.2) \qquad T^{-1}\colon \quad x^i = x^i(y^1, y^2, \ldots, y^n), \quad (i = 1, 2, \ldots, n),$$

where the functions[1] $x^i(y)$ are single-valued. To ensure the satisfaction of restrictions we have just imposed on the transformation of coordinates, it will suffice to suppose that the functions $y^i(x)$ in (19.1) are continuous together with their first partial derivatives in some region R of the manifold V_n, and that the Jacobian determinant $J \equiv \left|\dfrac{\partial y^i}{\partial x^j}\right|$ does not vanish at any point of the region R. It would follow then[2] that not only a single-valued inverse (19.2) exists, but the functions $x^i(y)$ in (19.2) are also of class C^1 in some neighborhood of the point under consideration.

[1] We will often use the notation $x^i(y)$ and $f(x)$ to mean $x^i(y^1, \ldots, y^n)$ and $f(x^1, x^2, \ldots, x^n)$, respectively.

[2] See, for example, I. S. Sokolnikoff, *Advanced Calculus*, pp. 433–438. We use the symbol C^n to denote the class of functions which are continuous together with their first n partial derivatives.

We observe that, if the functions $y^i(x)$ in (19.1) are of class C^1, then, by Taylor's formula,

$$y^i = a_0^{\ i} + a_j^{\ i} x^j,$$

where $a_j^{\ i}$ is the value of $\partial y^i/\partial x^j$ evaluated at some point P' of the region R. The point P' depends, of course, on the choice of values $(x^1, x^2, \ldots, x^n)$. Thus the transformation 19.1, with stated properties, is *locally linear*. The nonvanishing of the Jacobian guarantees that this system of linear equations has a unique solution. Throughout the rest of this book we shall suppose that all encountered transformations of coordinates are of the form 19.1, in which the functions $y^i(x)$ are at least of class C^1 in some region R, and that $\left|\dfrac{\partial y^i}{\partial x^j}\right| \neq 0$ at any point of R. For brevity we shall refer to a class of coordinate transformations with these properties as *admissible transformations.*

As an example of an admissible transformation consider a system of equations specifying the relation between the spherical polar coordinates x^i and the rectangular cartesian coordinates y^i,

$$T: \begin{cases} y^1 = x^1 \sin x^2 \cos x^3, \\ y^2 = x^1 \sin x^2 \sin x^3, \\ y^3 = x^1 \cos x^2. \end{cases}$$

If we suppose that $x^1 > 0$, $0 < x^2 < \pi$, and $0 \leq x^3 < 2\pi$, then $J \neq 0$ and the inverse transformation is given by

$$T^{-1}: \begin{cases} x^1 = +\sqrt{(y^1)^2 + (y^2)^2 + (y^3)^2}, \\[2ex] x^2 = \tan^{-1} \dfrac{\sqrt{(y^1)^2 + (y^2)^2}}{y^3}, \\[2ex] x^3 = \tan^{-1} \dfrac{y^2}{y^1}. \end{cases}$$

Problem

Discuss the transformations in which the coordinates y^i are rectangular cartesian:

(*a*)

$$\begin{aligned} y^1 &= \frac{1}{\sqrt{6}} x^1 + \frac{2}{\sqrt{6}} x^2 + \frac{1}{\sqrt{6}} x^3, \\ y^2 &= \frac{1}{\sqrt{2}} x^1 - \frac{1}{\sqrt{3}} x^2 + \frac{1}{\sqrt{3}} x^3, \\ y^3 &= \frac{1}{\sqrt{2}} x^1 \qquad\qquad\;\; - \frac{1}{\sqrt{2}} x^3. \end{aligned}$$

(*b*)

$$\begin{aligned} y^1 &= x^1 \cos x^2, \\ y^2 &= x^1 \sin x^2, \\ y^3 &= x^3. \end{aligned}$$

20. Properties of Admissible Transformations of Coordinates

From a summary of certain important properties of admissible coordinate transformations contained in this section, we will see that it is quite immaterial what particular reference frame is selected to describe the invariant entities. It will be shown that all admissible transformations of coordinates form a group, and hence every coordinate system in the family can be obtained from the particular one by an admissible transformation. This fact is of great moment in the construction of a theory that lays claim to its independence of the accidental choice of reference systems.

THEOREM I. *If a transformation of coordinates T possesses an inverse T^{-1} and if J and K are the Jacobians of T and T^{-1}, respectively, then* $JK = 1$.

The proof is easy. We insert the values of x^i from (19.2) in (19.1) and obtain a set of identities in y^i,

$$y^i \underset{y}{\equiv} y^i[x^1(y^1, \ldots, y^n), \ldots, x^n(y^1, \ldots, y^n)].$$

The differentiation with respect to y^j yields

$$\frac{\partial y^i}{\partial y^j} = \delta_j^{\,i} = \frac{\partial y^i}{\partial x^\alpha}\frac{\partial x^\alpha}{\partial y^j}, \qquad (\alpha = 1, 2, \ldots, n).$$

But

$$\left|\frac{\partial y^i}{\partial x^\alpha}\frac{\partial x^\alpha}{\partial y^j}\right| = \left|\frac{\partial y^i}{\partial x^k}\right| \cdot \left|\frac{\partial x^i}{\partial y^k}\right| = J \cdot K.$$

Since $|\delta_j^{\,i}| = 1$, we see that $J \cdot K = 1$. Incidentally, it follows from this result that $J \neq 0$ in R.

Consider now any two admissible transformations

$$T_1\colon \quad y^i = y^i(x^1, \ldots, x^n),$$

and

$$T_2\colon \quad z^i = z^i(y^1, \ldots, y^n), \qquad (i = 1, 2, \ldots, n).$$

The transformation

$$T_3\colon \quad z^i = z^i[y^1(x^1, \ldots, x^n), \ldots, y^n(x^1, \ldots, x^n)]$$

is called the product of T_2 and T_1, and we write $T_3 = T_2T_1$. If the Jacobian of T_3 is denoted by J_3, it follows that

$$J_3 = \left|\frac{\partial z^i}{\partial y^\alpha}\frac{\partial y^\alpha}{\partial x^j}\right| = \left|\frac{\partial z^i}{\partial y^j}\right|\left|\frac{\partial y^i}{\partial x^j}\right| = J_2J_1,$$

where J_2 and J_1 are the Jacobians of T_2 and T_1, respectively.

We can state this result as a

THEOREM II. *The Jacobian of the product transformation is equal to the product of the Jacobians of transformations entering in the product.*

These theorems enable us to establish an important

THEOREM III. *The set of all admissible transformations of coordinates forms a group.*

The truth of the theorem becomes obvious if one notes that

(*a*) The fundamental group property, namely, the product of two admissible transformations is a transformation belonging to the set of admissible transformations, is clearly satisfied. This property is known as the property of *closure*.

(*b*) The product transformation possesses an inverse, since the transformations appearing in the product have inverses.

(*c*) The identity transformation ($x^i = y^i$) obviously exists.

(*d*) The associative law $T_3(T_2T_1) = (T_3T_2)T_1$ obviously holds.

These properties are precisely the ones entering in the definition of an abstract group.

As noted in the beginning of this section, the fact that admissible transformations form a group justifies us in choosing as a point of departure *any* convenient coordinate system, as long as it is one of those admitted in the set.

21. Transformation by Invariance

Let $F(P)$ be a function of the point P in the n-dimensional manifold V_n. We will suppose that $F(P)$ is a continuous function in some region R of V_n and that V_n is covered by some convenient coordinate system X. The values of $F(P)$ depend on the point P, but not on the coordinate system used to represent P. We call $F(P)$ a *scalar point function* or simply *a scalar*. In the reference frame X, $F(P)$ may assume the form $f(x^1, \ldots, x^n)$, and, if we introduce a new reference system Y by means of a transformation

$$T\colon\ x^i = x^i(y^1, \ldots, y^n), \tag{21.1}$$

the functional form of $F(P)$ in the Y-frame is

$$f[x^1(y^1, \ldots, y^n), \ldots, x^n(y^1, \ldots, y^n)] \equiv g(y^1, \ldots, y^n), \tag{21.2}$$

since the value of $f(x^1, \ldots, x^n)$ at $P(x^1, \ldots, x^n)$ is the same[3] as that of $g(y^1, \ldots, y^n)$ at $P(y^1, \ldots, y^n)$.

[3] In a specific case, $F(P)$ may represent the temperature of some region of space and $f(x)$ is the form which the temperature function assumes in the X-reference frame; $g(y)$ is the representation of $F(P)$ in the Y-reference frame.

We can speak of $f(x)$ as being the component of the scalar function $F(P)$ in the X-coordinate system, while $g(y)$ is the component of the same scalar function in the Y-coordinate system. Alternatively, we can regard the scalar function $F(P)$ as being defined *by the totality of components* $f(x)$, $g(y)$, $h(z)$, *etc., each of which is related to one another by the substitution law typified by formula* 21.2. In other words, once the representation of the scalar $F(P)$ in one coordinate system is known, then the form of $F(P)$ in any other coordinate system Y is determined by formula 21.2. We call this substitution transformation $G^0: f[x(y)] = g(y)$ the *transformation by invariance*.

We observe that, if we have three transformations T_1, T_2, and T_3, where

$$T_1^{-1}: \quad x = x(y),$$

$$T_2^{-1}: \quad y = y(z),$$

with $T_3 = T_2T_1$, so that

$$T_3^{-1}: \quad x = x[y(z)],$$

and a scalar $F(P)$ whose component in the X-frame is $f(x)$, we can compute the transforms of $f(x)$. Indeed, the component $g(y)$ of $F(P)$ in the Y-frame is determined by the law

$$G_1^0: \quad g(y) = f[x(y)],$$

whereas the component $h(z)$ of $F(P)$ in the Z-frame is given by

$$G_2^0: \quad h(z) = g[y(z)].$$

On the other hand, using the product transformation $T_3 = T_2T_1$, we gct

$$G_3^0: \quad h(z) = f\{x[y(z)]\},$$

from which it is clear that $G_3^0 = G_2^0G_1^0$.

We can represent these transformations of coordinates and the corresponding transformation of components of $F(P)$ diagrammatically as in Fig. 7. Thus, as coordinates are subjected to a group T of admissible transformations, the components of a scalar undergo a certain transformation G^0. The relation between the successive transformations T and G^0 is such that the product of two transformations T_2T_1 corresponds to the

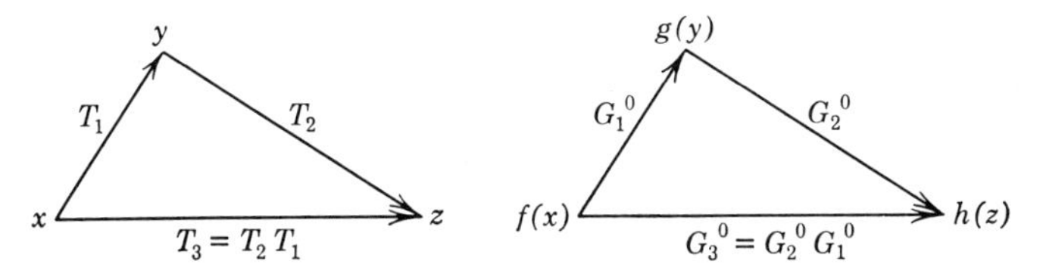

Fig. 7

product of two *corresponding* transformations $G_2^0 G_1^0$. When such a relation obtains between any two groups of transformations T and G, the groups are said to be *isomorphic*. The isomorphism between the transformations of coordinates and the transformations of functions *induced* by the transformation of coordinates is an important characteristic of a class of invariants called tensors.

22. Transformation by Covariance and Contravariance

In the preceding section we discussed the transformation of components of a scalar $F(P)$ when the coordinates of P undergo a transformation. In this section we discuss the law of transformation of entities determined by the sets of partial derivatives of a scalar. Sets of partial derivatives of the component $f(x^1, \ldots, x^n)$ of a scalar $F(P)$ are of interest in physics in connection with the notion of a gradient of potential functions.

We consider a continuously differentiable function $f(x^1, \ldots, x^n)$, representing the scalar $f(P)$, and a transformation of coordinates

$$T: \quad x^i = x^i(y^1, \ldots, y^n). \tag{22.1}$$

If we form a set of n partial derivatives

$$\frac{\partial f}{\partial x^1}, \frac{\partial f}{\partial x^2}, \ldots, \frac{\partial f}{\partial x^n}, \quad \text{or} \quad \{f_{x^i}\}, \tag{22.2}$$

the question arises: What does the set $\{f_{x^i}\}$ become when the coordinates x^i are subjected to a transformation 22.1? This question is quite without meaning unless one specifies precisely what is to be done with the set 22.2. These fractions do not automatically "become" anything until one states what law he is to use in calculating the "corresponding functions" in the Y-frame. In other words, it is necessary to agree on what the term "corresponding function" is to mean in a given situation.

For example, we might calculate the corresponding functions by the transformation of invariance G^0 of Sec. 21; that is, we can insert in each function $f_{x^i}(x^1, \ldots, x^n)$, the values of the x's from (22.1). This will yield a set of n functions

$$g_1(y^1, \ldots, y^n), g_2(y^1, \ldots, y^n), \ldots, g_n(y^1, \ldots, y^n). \tag{22.3}$$

On the other hand, if one has in mind the notion of a gradient of $f(P)$, it is necessary to say that the set of functions corresponding to (22.2) is not (22.3), but the set of n partial derivatives,

$$\frac{\partial f}{\partial y^1}, \frac{\partial f}{\partial y^2}, \ldots, \frac{\partial f}{\partial y^n}, \tag{22.4}$$

computed by the rule for differentiation of composite functions, namely,

$$G^1: \quad \frac{\partial f}{\partial y^i} = \frac{\partial f}{\partial x^\alpha}\frac{\partial x^\alpha}{\partial y^i}, \qquad (i, \alpha = 1, 2, \ldots, n). \tag{22.5}$$

If we have a function $f(x^1, \ldots, x^n)$ and a transformation

$$T_1: \quad x^i(z^1, \ldots, z^n),$$

the set of functions corresponding to (22.2), determined by the law G^1 (equation 22.5), is

$$\frac{\partial f}{\partial z^i} = \frac{\partial f}{\partial x^\alpha}\frac{\partial x^\alpha}{\partial z^i}$$

We can think of the sets of functions $\{\partial f/\partial x^i\}$, $\{\partial f/\partial y^i\}$, $\{\partial f/\partial z^i\}$, etc., as representing the *same* entity in different reference frames. At any particular point $P_0(x_0^{\ 1}, \ldots, x_0^{\ n})$ the set 22.2 determines n numbers, which can be regarded as the components of the gradient vector, and the set 22.4 represents the same vector in the Y-coordinate system.

If we have a set of n functions $A_1(x), \ldots, A_n(x)$ associated with the X-coordinate system, and if we agree to calculate the corresponding quantities $B_1(y), \ldots, B_n(y)$ in the Y-system by means of the *covariant law* G^1, namely,

$$B_i(y) = \frac{\partial x^\alpha}{\partial y^i} A_\alpha(x), \tag{22.6}$$

we say that the set $\{A_i(x)\}$ represents the components of a *covariant vector* in the X-coordinate system. The set $\{B_i(y)\}$ represents the *same* covariant vector in the Y-system, and the covariant vector itself is the totality of sets of such quantities each related to one another by the covariant law G^1.

As an illustration of the law of transformation of vectors, which is quite different from the law G^1, consider a set of n differentials

$$dx^1, dx^2, \ldots, dx^n, \tag{22.7}$$

where the x^i's are related to the variables y^i by the formula 22.1. If we have two points $P_1(x^1, \ldots, x^n)$ and $P_2(x^1 + dx^1, \ldots, x^n + dx^n)$, then the set of n numbers 22.7 determines the displacement vector from P_1 to P_2.

The same displacement vector when referred to the Y-coordinate system has for its components

$$dy^1, dy^2, \ldots, dy^n, \tag{22.8}$$

where

$$G^2: \quad dy^i = \frac{\partial y^i}{\partial x^\alpha} dx^\alpha, \qquad (i, \alpha = 1, 2, \ldots, n).$$

Note that the law G^2, for the determination of the quantities 22.8, is different from G^1. If we have a set of quantities $A_1(x)$, $A_2(x)$, . . . , $A_n(x)$, then the law G^2, determining the corresponding quantities $B_1(y)$, $B_2(y)$, . . . , $B_n(y)$, is

$$B_i = \frac{\partial y^i}{\partial x^\alpha} A_\alpha. \tag{22.9}$$

The law G^2 is the *contravariant law*, and we call the sets of quantities transforming in accordance with it the *components of a contravariant vector*.

The laws G^0, G^1, and G^2 play a fundamental role in the development of tensor analysis.

Problems

1. Show that if the transformation T: $y^i = a_j{}^i x^j$ is orthogonal, then the distinction between the covariant and contravariant laws disappears.

2. Prove the theorem: If $f(x^1, x^2, \ldots, x^n)$ is a homogeneous function of degree m, then $\dfrac{\partial f}{\partial x^i} x^i = mf$.

3. Given $f(x^1, x^2, \ldots, x^n)$ and a set of equations of transformation $x^i = x^i(y^1, y^2, \ldots, y^n)$, where each $y^i = y^i(t)$. If the transform of f by invariance is $g(y^1, y^2, \ldots, y^n)$, show that $df/dt = dg/dt$. *Hint:* $(\partial f/\partial x^\alpha)(dx^\alpha/dt) = df/dt$ and $dx^\alpha/dt = (\partial x^\alpha/\partial y^j)(dy^j/dt)$.

4. Write out the laws of transformation of components of covariant and contravariant vectors when T is the transformation from rectangular cartesian to spherical polar coordinates given in Sec. 19.

23. The Tensor Concept. Contravariant and Covariant Tensors

Consider an admissible transformation

$$T\colon\quad y^i = y^i(x^1, x^2, \ldots, x^n), \qquad (i = 1, 2, \ldots, n),$$

and a set $\{f_i\}$ of m continuous functions

$$f_i(x^1, x^2, \ldots, x^n), \qquad (i = 1, 2, \ldots, m),$$

defined in some region R of the n-dimensional space referred to the X-system of coordinates.

We associate with the given transformation T a transformation G which transforms each $f_i(x^1, x^2, \ldots, x^n)$ into a function

$$g_i(y^1, y^2, \ldots, y^n).$$

Examples of the transformation G are the transformation of invariance and the contravariant and covariant laws introduced in preceding sections. But, whatever the nature of the transformation G, it will always depend on T, and to emphasize this fact we shall say that G is a function of T. We shall call G an *induced transformation* on the set of functions f_i.

Suppose further that G, regarded as a function of T, satisfies the following conditions:

(*a*) *When T is an identity transformation, then G is an identity transformation.* This means that, if $y^i = x^i$, then

$$f_i(x^1, \ldots, x^n) = f_i(y^1, y^2, \ldots, y^n).$$

(*b*) *If T_1, T_2, T_3 are three transformations of the type T, and G_1, G_2, G_3 are the corresponding induced transformations G, and if $T_3 = T_2T_1$, then* $G_3 = G_2G_1$. In other words, the sets of transformations T and G are isomorphic. If the given set of functions $\{f_i\}$ satisfies conditions (*a*) and (*b*), we shall say that the set $\{f_i\}$ represents the components f_i of a tensor f in the X-coordinate system, the tensor f itself being the totality of sets of functions $\{f_i(x)\}$, $\{g_i(y)\}$, etc.

It should be remarked that the term *tensor* was used by A. Einstein[4] only in connection with the sets of quantities transforming in accordance with the contravariant and covariant laws. The formulation of contravariant and covariant laws, as well as an outline of the essential features of the algebra and calculus of contravariant and covariant tensors, is due to G. Ricci.[5] The much broader characterization of tensors by the isomorphism of transformations of coordinates and induced transformations is essentially due to H. Weyl and O. Veblen.[6] Because of the usefulness and commonness of covariant and contravariant laws of transformation in applications of analysis to geometry and physics, the term *tensor* is generally used in the sense contemplated by Einstein. This usage is followed in the sequel. However, the isomorphism between the laws of transformation of coordinates and the induced transformations is so fundamental to the idea of a tensor and to the invariant nature of tensor calculus that it justifies the degree of emphasis placed on it in the foregoing.

We now turn to a consideration of covariant, contravariant, and mixed tensors. It will be convenient to introduce (with Ricci) different notations for each type of such tensors so that they can be recognized at a glance. Let us consider first a set of n functions of the variables $(x^1, \ldots, x^n)$,

$$\{A(i; x)\} \qquad \text{or} \qquad A(1; x), A(2; x), \ldots, A(n; x).$$

Previously we wrote the identifying index i either as a subscript or superscript, but now we agree to use superscripts to denote the set of functions

[4] A. Einstein, *Annalen der Physik*, **49** (1916).

[5] G. Ricci, *Atti della reale accademia nazionale dei Lincei*, **5** (1889).

[6] H. Weyl, *Mathematische Zeitschrift*, **23, 24** (1925–1926). O. Veblen, *Invariants of Quadratic Differential Forms*, Cambridge Tract No. 24 (1927), pp. 19–20.

that transform in accordance with the contravariant law and subscripts for sets that transform in the covariant manner.[7] Whenever the law of transformation is neither covariant nor contravariant, or when its nature is in doubt, we write $\{A(i; x)\}$, $\{B(i; y)\}$, etc. We now lay down the following definitions.

DEFINITION 1. *A covariant tensor of rank one is the entire class of sets of quantities* $\{A(i; x)\}$, $\{B(i; y)\}$, $\{C(i; z)\}$, . . . *related to one another by the transformation of the form*

$$B(i; y) = \frac{\partial x^\alpha}{\partial y^i} A(\alpha; x), \qquad (i, \alpha = 1, 2, \ldots, n),$$

where $\{A(i, x)\}$ *is the representation of the tensor in the X-coordinate system, and* $\{B(i; y)\}$ *is its representation in any coordinate system Y related to the X-system by the transformation T.*

Frequently, we speak loosely of the given set $\{A(i; x)\}$ as being a tensor, but this usage should not conceal the fact that the tensor is the *totality* of sets of quantities typified by $\{A(i; x)\}$. The last set refers to the representation of the tensor in a particular reference frame and can be spoken of as the component of the tensor in the X-coordinate system. However, we shall use the term *component of a tensor* to mean the individual elements $A(i; x)$ in the set $\{A(i; x)\}$.

We denote components of covariant tensors by subscripts and often suppress the variables x and y entering as the arguments of A's and B's. Thus

$$B_i = \frac{\partial x^\alpha}{\partial y^i} A_\alpha \qquad \text{(covariant law).}$$

DEFINITION 2. *A contravariant tensor of rank one is the entire class of quantities such as* $\{A(i; x)\}$, $\{B(i; y)\}$, . . . *related to one another by the transformation of the form*

$$B(i; y) = \frac{\partial y^i}{\partial x^\alpha} A(\alpha; x),$$

where $\{A(i; x)\}$ *represents the tensor in the X-coordinate system and* $\{B(i; y)\}$ *in the Y-coordinate system.*

We denote components of contravariant tensors by superscripts. Thus

$$B^i = \frac{\partial y^i}{\partial x^\alpha} A^\alpha \qquad \text{(contravariant law).}$$

[7] The only exception to this convention is in the use of superscripts to identify the variables x^i, y^i, etc. These quantities do not transform according to a covariant or contravariant law unless the transformation T is affine.

The definitions of contravariant and covariant tensors of rank one are identical with those of contravariant and covariant vectors given in Sec. 22.

We speak of scalars, defined in Sec. 21, *as tensors of rank zero.*

We can generalize the definitions of tensors of rank one to include tensors of any rank as follows.

DEFINITION 3. *A set of n^r quantities $A_{i_1 i_2 \cdots i_r}(x)$, associated with the X-coordinate system, represents the components of a covariant tensor of rank r if the corresponding set of n^r quantities $B_{i_1 i_2 \cdots i_r}(y)$, associated with the Y-coordinate system, is given by*

$$B_{i_1 i_2 \cdots i_r} = \frac{\partial x^{\alpha_1}}{\partial y^{i_1}} \frac{\partial x^{\alpha_2}}{\partial y^{i_2}} \cdots \frac{\partial x^{\alpha_r}}{\partial y^{i_r}} A_{\alpha_1 \alpha_2 \cdots \alpha_r}.$$

The tensor itself is the totality of sets of such quantities as $\{A_{i_1 i_2 \cdots i_r}(x)\}$.

DEFINITION 4. *A set of n^r quantities $A^{i_1 i_2 \cdots i_r}(x)$ represents the components of a contravariant tensor of rank r in the X-coordinate system whenever the corresponding set $B^{i_1 i_2 \cdots i_r}(y)$ of n^r quantities in the Y-system is given by*

$$B^{i_1 i_2 \cdots i_r} = \frac{\partial y^{i_1}}{\partial x^{\alpha_1}} \frac{\partial y^{i_2}}{\partial x^{\alpha_2}} \cdots \frac{\partial y^{i_r}}{\partial x^{\alpha_r}} A^{\alpha_1 \alpha_2 \cdots \alpha_r}.$$

As an illustration we note that the components of the covariant tensor of rank two transform according to the law

$$B_{ij}(y) = \frac{\partial x^\alpha}{\partial y^i} \frac{\partial x^\beta}{\partial y^j} A_{\alpha\beta}(x),$$

whereas the components of the contravariant tensor are given by

$$B^{ij}(y) = \frac{\partial y^i}{\partial x^\alpha} \frac{\partial y^j}{\partial x^\beta} A^{\alpha\beta}(x).$$

There are n^2 components in each set.

We define next the *mixed* tensor.

DEFINITION 5. *The totality of sets of n^{r+s} quantities, typified in the X-coordinate system by the expressions $A^{j_1 j_2 \cdots j_s}_{i_1 i_2 \cdots i_r}(x)$, is a mixed tensor, covariant of rank r and contravariant of rank s, provided that the corresponding quantities $B^{j_1 j_2 \cdots j_s}_{i_1 i_2 \cdots i_r}(y)$ in the Y-coordinate system are given by the law*

$$B^{j_1 j_2 \cdots j_s}_{i_1 i_2 \cdots i_r} = \frac{\partial x^{\alpha_1}}{\partial y^{i_1}} \frac{\partial x^{\alpha_2}}{\partial y^{i_2}} \cdots \frac{\partial x^{\alpha_r}}{\partial y^{i_r}} \cdot \frac{\partial y^{j_1}}{\partial x^{\beta_1}} \frac{\partial y^{j_2}}{\partial x^{\beta_2}} \cdots \frac{\partial y^{j_s}}{\partial x^{\beta_s}} A^{\beta_1 \beta_2 \cdots \beta_s}_{\alpha_1 \alpha_2 \cdots \alpha_r}.$$

We note that this law for the transformation of components A^j_i of the mixed tensor gives $B^j_i(y) = \dfrac{\partial x^\alpha}{\partial y^i} \dfrac{\partial y^j}{\partial x^\beta} A^\beta_\alpha(x)$. As a simple example of a mixed tensor that already has occurred in our discussion, we cite the Kronecker delta δ^j_i. Thus $\dfrac{\partial x^\alpha}{\partial y^i}\dfrac{\partial y^j}{\partial x^\beta}\delta^\beta_\alpha = \dfrac{\partial y^j}{\partial y^i} = \delta^j_i$. The verification of the fact that the definition of covariant, contravariant, and mixed tensors satisfies properties (*a*) and (*b*), stated in the beginning of this section, is given in Sec. 24.

To distinguish tensors defined over a region of space from tensors whose domain of definition is a single point, one occasionally speaks of the former as constituting a *tensor field.*

24. Tensor Character of Covariant and Contravariant Laws

We will verify that the induced transformations defined in the preceding section satisfy the isomorphism conditions stated in Sec. 21. The fact that the transformation of invariance (leading to tensors of rank zero) fulfills these conditions was noted in Sec. 21. The proofs for contravariant and covariant tensors are special cases of the proof for a mixed tensor. Accordingly we consider a mixed tensor typified by the set of functions $A^{j_1 j_2 \cdots j_s}_{i_1 i_2 \cdots i_r}(x)$

The law G for the transformation of mixed tensors is

$$(24.1) \qquad B^{j_1 \cdots j_s}_{i_1 \cdots i_r}(y) = \frac{\partial x^{\alpha_1}}{\partial y^{i_1}} \cdots \frac{\partial x^{\alpha_r}}{\partial y^{i_r}} \cdot \frac{\partial y^{j_1}}{\partial x^{\beta_1}} \cdots \frac{\partial y^{j_s}}{\partial x^{\beta_s}} A^{\beta_1 \cdots \beta_s}_{\alpha_1 \cdots \alpha_r}(x),$$

and we must show that

(a) if $T = I$, then $G = 1$,

(b) if $T = T_2 T_1$, then $G = G_2 G_1$.

Now, if $T = I$, then

$$x^{\alpha_1} = y^{\alpha_1}, \quad x^{\alpha_2} = y^{\alpha_2}, \ldots$$

and hence

$$\frac{\partial x^{\alpha_1}}{\partial y^{i_1}} = \delta^{\alpha_1}_{i_1}, \ldots, \frac{\partial x^{\alpha_r}}{\partial y^{i_r}} = \delta^{\alpha_r}_{i_r}.$$

Moreover, $T^{-1} = I$, so that

$$y^{\alpha_1} = x^{\alpha_1}, \quad y^{\alpha_2} = x^{\alpha_2}, \ldots,$$

so that

$$\frac{\partial y^{j_1}}{\partial x^{\beta_1}} = \delta^{j_1}_{\beta_1}, \ldots, \frac{\partial y^{j_s}}{\partial x^{\beta_s}} = \delta^{j_s}_{\beta_s}.$$

Inserting these values of partial derivatives in (24.1) gives

$$B^{j_1 \cdots j_s}_{i_1 \cdots i_r}(y) = \delta^{\alpha_1}_{i_1} \cdots \delta^{\alpha_r}_{i_r} \cdot \delta^{j_1}_{\beta_1} \cdots \delta^{j_s}_{\beta_s} A^{\beta_1 \cdots \beta_s}_{\alpha_1 \cdots \alpha_r}(x)$$
$$= A^{j_1 \cdots j_s}_{i_1 \cdots i_r}(x).$$

Hence $G = I$ if $T = I$.

Suppose now that, under a transformation T_1, the variables x^i transform into y^i, and the variables y^i transform into z^i by the transformation T_2. The corresponding induced transformations G_1 and G_2 yield

$$(24.2)\quad G_1\colon\ B^{j_1 \cdots j_s}_{i_1 \cdots i_r}(y) = \frac{\partial x^{\alpha_1}}{\partial y^{i_1}} \cdots \frac{\partial x^{\alpha_r}}{\partial y^{i_r}} \cdot \frac{\partial y^{j_1}}{\partial x^{\beta_1}} \cdots \frac{\partial y^{j_s}}{\partial x^{\beta_s}} A^{\beta_1 \cdots \beta_s}_{\alpha_1 \cdots \alpha_r}(x),$$

and

$$(24.3)\quad G_2\colon\ C^{j_1 \cdots j_s}_{i_1 \cdots i_r}(z) = \frac{\partial y^{\alpha_1}}{\partial z^{i_1}} \cdots \frac{\partial y^{\alpha_r}}{\partial z^{i_r}} \cdot \frac{\partial z^{j_1}}{\partial y^{\beta_1}} \cdots \frac{\partial z^{j_s}}{\partial y^{\beta_s}} B^{\beta_1 \cdots \beta_s}_{\alpha_1 \cdots \alpha_r}(y).$$

Now, under the product transformation $T_3 = T_2 T_1$, the variables x^i go into y^i and the y^i into z^i, so that T_3 carries the x^i into the z^i. Inserting the values of the B's from (24.2) into (24.3) gives

$$G_2 G_1\colon\ C^{j_1 \cdots j_s}_{i_1 \cdots i_r}(z) = \left(\frac{\partial y^{\alpha_1}}{\partial z^{i_1}} \cdots \frac{\partial y^{\alpha_r}}{\partial z^{i_r}}\right)\left(\frac{\partial z^{j_1}}{\partial y^{\beta_1}} \cdots \frac{\partial z^{j_s}}{\partial y^{\beta_s}}\right)$$
$$\times \left(\frac{\partial x^{\gamma_1}}{\partial y^{\alpha_1}} \cdots \frac{\partial x^{\gamma_r}}{\partial y^{\alpha_r}}\right)\left(\frac{\partial y^{\beta_1}}{\partial x^{\delta_1}} \cdots \frac{\partial y^{\beta_s}}{\partial x^{\delta_s}}\right) A^{\delta_1 \cdots \delta_s}_{\gamma_1 \cdots \gamma_r}(x).$$

Performing the summation on α's and β's yields

$$G_3\colon\ C^{j_1 \cdots j_s}_{i_1 \cdots i_r}(z) = \frac{\partial x^{\gamma_1}}{\partial z^{i_1}} \cdots \frac{\partial x^{\gamma_r}}{\partial z^{i_r}} \cdot \frac{\partial z^{j_1}}{\partial x^{\delta_1}} \cdots \frac{\partial z^{j_s}}{\partial x^{\delta_s}} A^{\delta_1 \cdots \delta_s}_{\gamma_1 \cdots \gamma_r}(x).$$

The resulting law G_3 is precisely the law of transformation of the components of a mixed tensor when the variables x^i are transformed into the z^i by the transformation T_3. Thus the law of transformation G is *transitive*, and this completes the proof.

The results for covariant and contravariant tensors appear as special cases obtained by suppressing the superscripts or subscripts.

The only types of tensors with which we will deal in this book are scalars, covariant, contravariant, mixed, and relative tensors. The last are defined in Sec. 28.

We establish next a useful property of the law of the transformation of tensors, which is frequently used in the sequel.

Let the components of a mixed tensor in the X-coordinate system be denoted by $A^{j_1 \cdots j_s}_{i_1 \cdots i_r}(x)$ and its components in the Y-system by $B^{j_1 \cdots j_s}_{i_1 \cdots i_r}(y)$.

Then, from the law of transformation of mixed tensors we can write

$$B_{i_1 \cdots i_r}^{j_1 \cdots j_s}(y) = \frac{\partial x^{\alpha_1}}{\partial y^{i_1}} \cdots \frac{\partial x^{\alpha_r}}{\partial y^{i_r}} \cdot \frac{\partial y^{j_1}}{\partial x^{\beta_1}} \cdots \frac{\partial y^{j_s}}{\partial x^{\beta_s}} A_{\alpha_1 \cdots \alpha_r}^{\beta_1 \cdots \beta_s}(x). \tag{24.4}$$

On the other hand, if we are given the components $B_{i_1 \cdots i_r}^{j_1 \cdots j_s}(y)$, the components $A_{\alpha_1 \cdots \alpha_r}^{\beta_1 \cdots \beta_s}(x)$ of the same tensor in the X-reference frame are determined by the formula

$$A_{\alpha_1 \cdots \alpha_r}^{\beta_1 \cdots \beta_s}(x) = \frac{\partial y^{i_1}}{\partial x^{\alpha_1}} \cdots \frac{\partial y^{i_r}}{\partial x^{\alpha_r}} \cdot \frac{\partial x^{\beta_1}}{\partial y^{j_1}} \cdots \frac{\partial x^{\beta_s}}{\partial y^{j_s}} B_{i_1 \cdots i_r}^{j_1 \cdots j_s}(y). \tag{24.5}$$

We note that we can obtain (24.5) from (24.4) formally by treating the partial derivatives and sums in (24.4) as though they were fractions and products appearing in simple algebraic expressions.

From the structure of formulas (24.4) and (24.5) we deduce an important

THEOREM. *If all components of a tensor vanish in one coordinate system, then they necessarily vanish in all other admissible coordinate systems.*

This particular theorem is of profound significance in the formulation of physical laws. It states, in effect, that, if a certain law is implied by the vanishing of components of a tensor in one particular coordinate system, then the rules for transformation of the tensor components guarantee that they will vanish in all admissible coordinate systems. A physicist has little interest in the formulation of a law that might be valid only in some special reference frame. Indeed the notion of invariance and the universality of physical laws is the cornerstone about which mathematical physics is built.

25. Algebra of Tensors

In this section we establish several rules of operation with tensors, which are algebraic in character.

THEOREM I. *The sum (or difference) of two tensors which have the same number of covariant and the same number of contravariant indices is again a tensor of the same type and rank as the given tensors.*

Proof. Consider two tensors $A(x)$ and $\bar{A}(x)$ of the same type and rank defined at the same point P, and the corresponding laws of transformation:

$$B_{i_1 \cdots i_r}^{j_1 \cdots j_s}(y) = \frac{\partial x^{\alpha_1}}{\partial y^{i_1}} \cdots \frac{\partial x^{\alpha_r}}{\partial y^{i_r}} \cdot \frac{\partial y^{j_1}}{\partial x^{\beta_1}} \cdots \frac{\partial y^{j_s}}{\partial x^{\beta_s}} A_{\alpha_1 \cdots \alpha_r}^{\beta_1 \cdots \beta_s}(x),$$

$$\bar{B}_{i_1 \cdots i_r}^{j_1 \cdots j_s}(y) = \frac{\partial x^{\alpha_1}}{\partial y^{i_1}} \cdots \frac{\partial x^{\alpha_r}}{\partial y^{i_r}} \cdot \frac{\partial y^{j_1}}{\partial x^{\beta_1}} \cdots \frac{\partial y^{j_s}}{\partial x^{\beta_s}} \bar{A}_{\alpha_1 \cdots \alpha_r}^{\beta_1 \cdots \beta_s}(x).$$

Then

$$B_{i_1 \cdots i_r}^{j_1 \cdots j_s} \pm \bar{B}_{i_1 \cdots i_r}^{j_1 \cdots j_s}$$
$$= \left(\frac{\partial x^{\alpha_1}}{\partial y^{i_1}} \cdots \frac{\partial x^{\alpha_r}}{\partial y^{i_r}}\right) \cdot \left(\frac{\partial y^{j_1}}{\partial x^{\beta_1}} \cdots \frac{\partial y^{j_s}}{\partial x^{\beta_s}}\right) \cdot (A_{\alpha_1 \cdots \alpha_r}^{\beta_1 \cdots \beta_s} \pm \bar{A}_{\alpha_1 \cdots \alpha_r}^{\beta_1 \cdots \beta_s}).$$

It follows from this that $A + \bar{A}$ is a tensor, and we write

$$A_{\alpha_1 \cdots \alpha_r}^{\beta_1 \cdots \beta_s}(x) \pm \bar{A}_{\alpha_1 \cdots \alpha_r}^{\beta_1 \cdots \beta_s}(x) \equiv \mathscr{A}_{\alpha_1 \cdots \alpha_r}^{\beta_1 \cdots \beta_s}(x),$$

which is a tensor of the same type and rank as the given tensors.

It is clear from the laws of transformation of tensors that, if each component of a tensor is multiplied by a constant, the resulting set of functions is a tensor. This fact, in conjunction with Theorem I, permits us to state a

COROLLARY. *Any linear combination of tensors of the same type and rank is again a tensor of the same type and rank.*

THEOREM II. *The equation* $A_{\alpha_1 \cdots \alpha_r}^{\beta_1 \cdots \beta_s}(x) = \bar{A}_{\alpha_1 \cdots \alpha_r}^{\beta_1 \cdots \beta_s}(x)$ *is a tensor equation; that is, if this equation is true in some coordinate system, then it is true in all admissible systems.*

Proof. It follows from Theorem I that the difference of two tensors is a tensor. Hence

$$A_{\alpha_1 \cdots \alpha_r}^{\beta_1 \cdots \beta_s} - \bar{A}_{\alpha_1 \cdots \alpha_r}^{\beta_1 \cdots \beta_s} = 0.$$

However, we proved in Sec. 24 that, if all components of a tensor vanish in one coordinate system, they vanish in all admissible coordinate systems. We shall call the tensor all whose components vanish the *zero tensor*.

THEOREM III. *The set of quantities consisting of the product of each element of the set* $A_{i_1 \cdots i_p}^{j_1 \cdots j_q}(x)$, *representing a tensor* A, *by each element of the set* $\bar{A}_{k_1 \cdots k_r}^{l_1 \cdots l_s}(x)$, *representing a tensor* $\bar{A}$, *defines the tensor* $\mathscr{A}$, *called the outer product. This tensor is contravariant of rank* $q + s$ *and covariant of rank* $p + r$.

From the definition of outer product, the components of $\mathscr{A}$ in the X-reference frame are given by the formula

$$\mathscr{A}_{i_1 \cdots i_p k_1 \cdots k_r}^{j_1 \cdots j_q l_1 \cdots l_s} \equiv A_{i_1 \cdots i_p}^{j_1 \cdots j_q} \bar{A}_{k_1 \cdots k_r}^{l_1 \cdots l_s}.$$

The fact that the set of functions $\mathscr{A}_{i_1 \cdots k_r}^{j_1 \cdots l_s}$ defines a tensor follows directly from the law of transformation of components $A_{i_1 \cdots i_p}^{j_1 \cdots j_q}$ and $\bar{A}_{k_1 \cdots k_r}^{l_1 \cdots l_s}$.

We will denote the outer product $\mathscr{A}$ of A and $\bar{A}$ by writing the symbols in juxtaposition. Thus $\mathscr{A} = A\bar{A}$. It is obvious that *the outer product is distributive with respect to addition*, so that

$$(A + B)C = AC + BC.$$

We introduce next the operation of *contraction* which yields tensors.

THEOREM IV. *If, in a mixed tensor, contravariant of rank s and covariant of rank r, we equate a covariant and a contravariant index and sum with respect to that index, then the resulting set of n^{r+s-2} sums is a mixed tensor, covariant of rank $r - 1$ and contravariant of ranks $s - 1$.*

To avoid complications in writing we illustrate the procedure used in the proof by considering a mixed tensor A^i_{jkl}. We have

$$B^i_{jkl} = \frac{\partial y^i}{\partial x^\alpha}\frac{\partial x^\beta}{\partial y^j}\frac{\partial x^\gamma}{\partial y^k}\frac{\partial x^\delta}{\partial y^l} A^\alpha_{\beta\gamma\delta}.$$

If we equate the indices i and k and sum, we obtain the set of n^2 quantities

$$\begin{aligned} B^i_{jil} &= \frac{\partial y^i}{\partial x^\alpha}\frac{\partial x^\beta}{\partial y^j}\frac{\partial x^\gamma}{\partial y^i}\frac{\partial x^\delta}{\partial y^l} A^\alpha_{\beta\gamma\delta} \\ &= \frac{\partial x^\beta}{\partial y^j}\frac{\partial x^\delta}{\partial y^l}\delta^\gamma_\alpha A^\alpha_{\beta\gamma\delta} \\ &= \frac{\partial x^\beta}{\partial y^j}\frac{\partial x^\delta}{\partial y^l} A^\alpha_{\beta\alpha\delta} \equiv \frac{\partial x^\beta}{\partial y^j}\frac{\partial x^\delta}{\partial y^l}\bar{A}_{\beta\delta}. \end{aligned}$$

Thus $B^i_{jil} \equiv \bar{B}_{jl}$ is a covariant tensor of rank two.

In this case we can obtain three different covariant tensors of rank two by performing the operation of contraction on other covariant indices. We observe that, when as a result of contraction of one or more pairs of indices there remain no free indices, the resulting quantity is a scalar.

If it is possible to apply the operation of contraction to the outer product of two tensors A and $\bar{A}$, the result is a tensor called the *inner product* of A and $\bar{A}$. We denote the inner product by the symbol $A \cdot \bar{A}$. The proof that $A \cdot \bar{A}$ is a tensor is immediate, for the outer product of two tensors is a tensor, and the operation of contraction yields a tensor.

Example. Consider the tensors $A_{ij}(x)$, $A_k(x)$, and $A^k(x)$. If we form the outer product $A_{ij}A_k \equiv A_{ijk}$, we obtain a covariant tensor of rank three, and hence no contraction is possible here. On the other hand, the outer product of A_{ij} and A^k gives a mixed tensor $A_{ij}A^k \equiv A^k_{ij}$, and in this case we can contract to get a covariant tensor $A^\alpha_{i\alpha}$ or $A^\alpha_{\alpha j}$. As already remarked, the tensor A^i_{jkl} may be contracted in three different ways to yield $A^\alpha_{\alpha kl}$, $A^\alpha_{j\alpha l}$, and $A^\alpha_{jk\alpha}$. The tensor A^{ij}_{klm} can be contracted twice in several ways. The contraction of A^i_j yields a scalar.

26. Quotient Laws

In this section we give two useful theorems which will enable us to establish the tensor character of sets of functions without going to the trouble of determining the law of transformation directly.

We use the term *inner product* for sums of the type $A(\alpha, i_2, \ldots, i_r)A_\alpha$ (or $A(\alpha, i_2, \ldots, i_r)A^\alpha$) whether the set of functions $A(i_1, \ldots, i_r)$ represents a tensor or not. We also speak of tensors of rank one as *vectors.*

THEOREM I. *Let $\{A(i_1, i_2, \ldots, i_r)\}$ be a set of functions of the variables x^i, and let the inner product $A(\alpha, i_2, \ldots, i_r)\xi^\alpha$, with an arbitrary vector $\boldsymbol{\xi}$, be a tensor of the type $A^{j_1 \cdots j_q}_{k_1 \cdots k_p}(x)$; then the set $A(i_1, \ldots, i_r)$ represents a tensor of the type $A^{j_1 \cdots j_q}_{\alpha k_1 \cdots k_p}(x)$.*

In order to avoid writing out long formulas for the transformation of tensors with many covariant and contravariant indices, we will establish this theorem for the set of n^3 functions $A(i, j, k)$, which has all features of the more involved cases. Let us suppose that the inner product $A(\alpha, j, k)\xi^\alpha$ for an arbitrary vector $\xi^\alpha(x)$ yields a tensor of the type $A^j_k(x)$. We will prove that the set $A(i, j, k)$ is a tensor of the type A^j_{ik}. By hypothesis $A(\alpha, j, k)\xi^\alpha$ is a tensor of the type A^j_k; hence its transform $B(\alpha, j, k)\eta^\alpha$ is given by the rule

$$B(\alpha, j, k)\eta^\alpha = \frac{\partial x^\gamma}{\partial y^k}\frac{\partial y^j}{\partial x^\beta}\{A(\lambda, \beta, \gamma)\xi^\lambda\},$$

where

$$\xi^\lambda(x) = \frac{\partial x^\lambda}{\partial y^\alpha}\eta^\alpha(y).$$

Inserting this expression for ξ^λ in the right-hand member of the above formula and transposing all terms on one side of the equation yields

$$\left[B(\alpha, j, k) - \frac{\partial x^\lambda}{\partial y^\alpha}\frac{\partial x^\gamma}{\partial y^k}\frac{\partial y^j}{\partial x^\beta}A(\lambda, \beta, \gamma)\right]\eta^\alpha = 0.$$

However, $\eta^\alpha(y)$ is an arbitrary vector; hence the bracket must vanish, and we obtain

$$B(\alpha, j, k) = \frac{\partial x^\lambda}{\partial y^\alpha}\frac{\partial x^\gamma}{\partial y^k}\frac{\partial y^j}{\partial x^\beta}A(\lambda, \beta, \gamma).$$

This is precisely the law of transformation of the tensor of the type A^j_{ik}.

Clearly, we can state an analogous theorem in which the vector $\boldsymbol{\xi}$ is a covariant vector. For example, if $A(i, j, k, \alpha)\xi_\alpha$ is known to be a tensor of the type A^i_{jk}, for an arbitrary vector ξ_α, then $A(i, j, k, \alpha) = A^{i\alpha}_{jk}$. On the other hand, if $A(i, j, k, \alpha)\xi_\alpha = A^{ijk}$, then $A(i, j, k, \alpha) = A^{ijk\alpha}$. These expressions suggest that an algorithm of division can be employed to determine the tensor character. Thus let $A(i, j, k, \alpha)\xi_\alpha = A^i_{jk}$, and write symbolically.

$$A(i, j, k, \alpha) = \frac{A^i_{jk}}{\xi_\alpha}.$$

Now, if we should regard the covariant quantities appearing below the division line as contravariant when written above the line, we have

$$A(i, j, k, \alpha) = A^i_{jk}\xi^\alpha,$$

where ξ^α is the symbolic reciprocal of ξ_α. From the product $A^i_{jk}\xi^\alpha$ we see that $A(i, j, k, \alpha) = A^{i\alpha}_{jk\cdot}$. Similarly, if $A(i, j, k, \alpha)\xi_\alpha = A^{ijk}$, then

$$A(i, j, k, \alpha) = \frac{A^{ijk}}{\xi_\alpha} = A^{ijk}\xi^\alpha = A^{ijk\alpha}.$$

On the other hand, if $A(\alpha, j, k)\xi^\alpha = A^j_k$,

$$A(\alpha, j, k) = \frac{A^j_k}{\xi^\alpha} = A^j_k\xi_\alpha = A^j_{k\alpha}.$$

In the division algorithm the contravariant quantities appearing below the division line are to be regarded as covariant when written above the line.

THEOREM II. *Let $\{A(i_1, \ldots, i_r)\}$ be a set of n^r functions defined in the X-coordinate system, and let $\{B(i_1, \ldots, i_r)\}$ be the corresponding quantities in the Y-system. If, for every set of vectors with components ξ_{α_i} relative to the X-coordinates and η_{α_i} relative to the Y-coordinates, we have the equality*

$$B(\beta_1, \ldots, \beta_r)\eta^{(1)}_{\beta_1} \cdots \eta^{(r)}_{\beta_r} = A(\alpha_1, \ldots, \alpha_r)\xi^{(1)}_{\alpha_1} \cdots \xi^{(r)}_{\alpha_r}$$

(that is, the inner product is a scalar), then the set of functions $A(i_1, \ldots, i_r)$ represents a contravariant tensor of rank r in the X-coordinate system.

Proof. Since the ξ_{α_i} are the components of a covariant vector,

$$\xi^{(j)}_{\alpha_i} = \frac{\partial y^{\beta_i}}{\partial x^{\alpha_i}}\eta^{(j)}_{\beta_i}.$$

Therefore

$$\left[B(\beta_1, \ldots, \beta_r) - A(\alpha_1, \ldots, \alpha_r)\frac{\partial y^{\beta_1}}{\partial x^{\alpha_1}} \cdots \frac{\partial y^{\beta_r}}{\partial x^{\alpha_r}}\right]\eta^{(1)}_{\beta_1} \cdots \eta^{(r)}_{\beta_r} = 0.$$

However, $\eta^{(1)}_{\beta_1}, \ldots, \eta^{(r)}_{\beta_r}$ are arbitrary; hence the term in the bracket must vanish. Therefore

$$B(\beta_1, \ldots, \beta_r) = \frac{\partial y^{\beta_1}}{\partial x^{\alpha_1}} \cdots \frac{\partial y^{\beta_r}}{\partial x^{\alpha_r}} A(\alpha_1, \ldots, \alpha_r),$$

which shows that

$$A(\alpha_1, \ldots, \alpha_r) = A^{\alpha_1 \cdots \alpha_r}.$$

This particular form of the quotient law is taken by some authors as the definition of the contravariant tensor of rank r. Thus, if the multilinear form $A(\alpha_1, \ldots, \alpha_r)\xi^{(1)}_{\alpha_1} \cdots \xi^{(r)}_{\alpha_r}$ is an invariant, then $A(\alpha_1, \ldots, \alpha_r) = A^{\alpha_1 \cdots \alpha_r}$, provided that the ξ_{α_i} are the components of arbitrary vectors.

On the other hand, if $A(\alpha_1, \ldots, \alpha_r)\xi^{\alpha_1}_{(1)} \cdots \xi^{\alpha_r}_{(r)}$ is an invariant, for an arbitrary choice of ξ^α's, then

$$A(\alpha_1, \ldots, \alpha_r) = A_{\alpha_1 \cdots \alpha_r}.$$

It is obvious from proofs of Theorems I and II that many other quotient laws can be stated. For example, if the inner product $A(i, \alpha)\xi_{\alpha j}$ of the set of n^2 functions $A(i, j)$ with an *arbitrary* tensor is a covariant tensor of rank two, then $A(i, j)$ represents a mixed tensor of the type A^j_i. The reader can prove this fact by following the pattern used in proving Theorem I. The tensor properties of the set $A(i, j)$ may be surmised from the division algorithm. Thus, if $A(i, \alpha)\xi_{\alpha j} = \mathscr{A}_{ij}$, then $A(i, \alpha) = \dfrac{\mathscr{A}_{ij}}{\xi_{\alpha j}}$. Now if we write the symbolic reciprocal of $\xi_{\alpha j}$ as $\xi^{\alpha j}$, we have $A(i, \alpha) = \dfrac{\mathscr{A}_{ij}}{\xi_{\alpha j}} = \mathscr{A}_{ij}\xi^{\alpha j} = A^\alpha_i$.

27. Symmetric and Skew-Symmetric Tensors

When an interchange of two covariant (or contravariant) indices in the components $A^{i_1 \cdots i_r}_{j_1 \cdots j_s}(x)$ of a tensor does not alter the value of components, the tensor A is said to be *symmetric* with respect to those indices. For example, a covariant tensor $A_{ij}(x)$ is symmetric if $A_{ij}(x) = A_{ji}(x)$. The definition of symmetry of tensors obviously would not be satisfactory if the symmetry of its components were not preserved under the coordinate transformations. To see that this is indeed so, let us suppose that $A_{i_1 i_2 \cdots i_r}(x) = A_{i_2 i_1 \cdots i_r}(x)$. Then $A_{i_1 i_2 \cdots i_r} - A_{i_2 i_1 \cdots i_r} = 0$. However, the difference of two tensors is a tensor; and if a tensor vanishes in one coordinate system, it vanishes in all admissible systems. Hence $B_{i_1 i_2 \cdots i_r}(y) = B_{i_2 i_1 \cdots i_r}(y)$.

We may say that a tensor is *skew-symmetric* (or antisymmetric) with respect to certain indices whenever an interchange of a pair of covariant (or contravariant) indices in the components merely changes the sign of the components. The skew-symmetry of tensors is likewise an invariant property. The proof of invariance of the skew-symmetry property is similar to that given for symmetry. However, as an exercise the reader may find it instructive to construct a proof based on the use of the law of transformation of components $A^{i_1 \cdots i_r}_{j_1 \cdots j_s}$.

We will extend the notions of symmetry and skew-symmetry in Sec. 40.

28. Relative Tensors

We recall that a function $f(x^1, \ldots, x^n)$ represents a scalar in the X-reference frame whenever in the Y-reference frame, determined by the transformation $x^i = x^i(y^1, \ldots, y^n)$, the scalar is given by the formula $g(y^1, \ldots, y^n) = f[x^1(y), \ldots, x^n(y)]$. We will encounter functions $f(x)$ which transform in accordance with the more general law, namely,

$$g(y^1, \ldots, y^n) = f[x^1(y), \ldots, x^n(y)] \left| \frac{\partial x^i}{\partial y^j} \right|^W, \tag{28.1}$$

where $\left| \frac{\partial x^i}{\partial y^j} \right|$ denotes the Jacobian of the transformation and W is a constant. We observe that, if the function $f(x)$ transforms in accordance with the law 28.1, then

$$\begin{aligned} h(z) = f(x) \left| \frac{\partial x^i}{\partial z^j} \right|^W &= f(x) \left| \frac{\partial x^i}{\partial y^j} \right|^W \left| \frac{\partial y^k}{\partial z^l} \right|^W \\ &= g(y) \left| \frac{\partial y^k}{\partial z^l} \right|^W, \end{aligned}$$

where we have made use of Theorem II of Sec. 20. Thus the formula 28.1 determines a class of invariant functions known as *relative scalars of weight W*.

A relative scalar of weight zero is the scalar defined in Sec. 21. Sometimes a scalar of weight zero is called an *absolute scalar*.

A relative scalar of weight 1 is called *scalar density*. The reason for this terminology may be seen from the expression for the total mass of a distribution of matter of density $\rho(x^1, x^2, x^3)$, the coordinates x^i being rectangular cartesian. The mass contained in a volume τ is given by the integral $M = \iiint_\tau \rho(x^1, x^2, x^3)\, dx^1\, dx^2\, dx^3$. If the coordinates x^i are changed with the aid of the equations of transformation $x^i = x^i(y^1, y^2, y^3)$, $(i = 1, 2, 3)$, the mass M is given by the integral

$$\begin{aligned} M &= \iiint_\tau \rho[x(y)] \left| \frac{\partial x^i}{\partial y^j} \right| dy^1\, dy^2\, dy^3 \\ &\equiv \iiint_\tau \bar{\rho}(y^1, y^2, y^3)\, dy^1\, dy^2\, dy^3. \end{aligned}$$

It is clear that the density of distribution when referred to the Y-coordinates is $\bar{\rho}(y) = \rho(x) \left| \frac{\partial x^i}{\partial y^j} \right|$.

We can also generalize the law of transformation of components of a mixed tensor by considering the sets of quantities $A^{j_1 \cdots j_s}_{i_1 \cdots i_r}(x)$ which transform according to the formula

$$B^{j_1 \cdots j_s}_{i_1 \cdots i_r}(y) = \left| \frac{\partial x^i}{\partial y^j} \right|^W \frac{\partial y^{j_1}}{\partial x^{\alpha_1}} \cdots \frac{\partial y^{j_s}}{\partial x^{\alpha_s}} \cdot \frac{\partial x^{\beta_1}}{\partial y^{i_1}} \cdots \frac{\partial x^{\beta_r}}{\partial y^{i_r}} A^{\alpha_1 \cdots \alpha_s}_{\beta_1 \cdots \beta_r}(x). \tag{28.2}$$

The sets of quantities $A^{\alpha_1 \cdots \alpha_s}_{\beta_1 \cdots \beta_r}(x)$ obeying this law of transformation are called the components of a *relative tensor of weight W*.

From the discussion in Sec. 24, and from the transitive property of Jacobians, namely,

$$\left|\frac{\partial x^i}{\partial z^j}\right| = \left|\frac{\partial x^i}{\partial y^k}\right| \left|\frac{\partial y^k}{\partial z^j}\right|,$$

if follows that the transformation 28.2 is transitive. In addition, from the linear and homogeneous character of this transformation it follows that if all components of a relative tensor vanish in one coordinate system, they vanish in every coordinate system. An immediate corollary of this is that a tensor equation involving relative tensors when true in one coordinate system is valid in all coordinate systems. In this case the relative tensors on two sides of equations must be of the same weight.

A little reflection will convince the reader that

(*a*) Relative tensors of the same type and weight may be added, and the sum is a relative tensor of the same type and weight.

(*b*) Relative tensors may be multiplied, the weight of the product being the sum of the weights of tensors entering in the product.

(*c*) The operation of contraction on a relative tensor yields a relative tensor of the same weight as the original tensor.

To distinguish mixed tensors, considered in the preceding sections, from relative tensors, the term *absolute tensor* is frequently used to designate the former. We shall encounter several relative tensors in applications of tensor theory.

Problems

1. Given the relation $A(i, j, k)B^{jk} = C^i$, where B^{jk} is an arbitrary symmetric tensor. Prove that $A(i, j, k) + A(i, k, j)$ is a tensor. Hence deduce that, if $A(i, j, k)$ is symmetric in j and k, then $A(i, j, k)$ is a tensor.

2. Given the relation $A(i, j, k)B^{jk} = C^i$, where B^{jk} is an arbitrary skew-symmetric tensor. Prove that $A(i, j, k) - A(i, k, j)$ is a tensor. Hence, if $A(i, j, k)$ is skew-symmetric in j and k, then $A(i, j, k)$ is a tensor.

3. If $a(i, j)\, dx^i\, dx^j$ is an invariant for an arbitrary vector dx^i, and $a(i, j)$ is symmetric, show that $a(i, j)$ is a tensor a_{ij}.

4. If a_{ij} is a tensor, show that A^{ij}, the cofactor of a_{ij} in $|a_{ij}|$ divided by $|a_{ij}| \neq 0$, is a tensor.

5. If $\phi(x^1, \ldots, x^n)$ is a scalar, show that $\{\partial^2\phi/\partial x^i\, \partial x^j\}$ is a tensor with respect to a set of *linear* transformations of coordinates.

6. If $|a_{ij} - \lambda b_{ij}| = 0$ for $\lambda = \lambda_1$, in one set of variables, then $|a_{ij}' - \lambda b_{ij}'| = 0$ for $\lambda = \lambda_1$, in the new set of variables. In other words, the roots of the polynomial $|a_{ij} - \lambda b_{ij}|$ are invariants.

7. Prove that a tensor with skew-symmetric components in one coordinate system has skew-symmetric components in all coordinate systems.

8. Show that every tensor can be expressed as the sum of two tensors, one of which is symmetric and the other skew-symmetric.

9. Show that the tensor equation $a_j{}^i\lambda_i = \alpha\lambda_j$, where α is an invariant and λ_j an arbitrary vector, demands that $a_j{}^i = \delta_j{}^i\alpha$.

10. Prove directly from the law of transformation of components that symmetry of a tensor is an invariant property.

11. The square of the element of arc ds appears in the form

$$ds^2 = g_{ij}\,dx^i\,dx^j.$$

Let T be an admissible transformation of coordinates $x^i = x^i(y^1, \ldots, y^n)$; then $ds^2 = h_{ij}\,dy^i\,dy^j$. Prove that $|g_{ij}|$ is a relative scalar of weight two. *Hint:* $h_{ij}(y) = \dfrac{\partial x^\alpha}{\partial y^i}\dfrac{\partial x^\beta}{\partial y^j} g_{\alpha\beta}(x)$, and recall the rule for multiplication of determinants.

12. How many independent components are there in a skew-symmetric tensor of rank two?

13. If a_{ij} is a skew-symmetric tensor and A^i is a contravariant vector, then $a_{ij}A^iA^j = 0$.

14. Prove that, if $A(i, j, k)A^iB^jC_k$ is a scalar for arbitrary vectors A^i, B^j, and C_k, then $A(i, j, k)$ is a tensor.

29. The Metric Tensor

In Sec. 5 we introduced the idea of n-dimensional space E_n by extending the concepts familiar to us from our experience with ordinary Euclidean geometry. Thus, in defining the length $|\mathbf{x}|$ of a vector $\mathbf{x}$, we used the generalized formula of Pythagoras, $|\mathbf{x}| = \sqrt{x^ix^i}$, where the x^i are the components of the vector $\mathbf{x}$ referred to a set of orthogonal cartesian axes. (See Sec. 5.) If we now consider a displacement vector dx^i, $(i = 1, \ldots, n)$, determined by a pair of points $P(x)$ and $P'(x + dx)$, wherein the coordinates x^i are orthogonal cartesian, the formula of Pythagoras gives for the square of the distance between P and P' the expression

$$ds^2 = dx^i\,dx^i, \qquad (i = 1, 2, \ldots, n). \tag{29.1}$$

We shall call *ds the element of arc* in E_n.

A change of coordinate system, determined by the transformation

$$x^i = x^i(y^1, \ldots, y^n), \tag{29.2}$$

permits us to write the formula 29.1 as

$$ds^2 = \frac{\partial x^i}{\partial y^\alpha}\frac{\partial x^i}{\partial y^\beta}\,dy^\alpha\,dy^\beta, \tag{29.3}$$

since $dx^i = (\partial x^i/\partial y^\alpha)\, dy^\alpha$. We can thus write the formula for the square of the element of arc in the Y-reference frame as a quadratic form

$$ds^2 = g_{\alpha\beta}\, dy^\alpha\, dy^\beta, \tag{29.4}$$

where the coefficients $g_{\alpha\beta}(y)$ are defined by

$$g_{\alpha\beta}(y) = \frac{\partial x^i}{\partial y^\alpha} \frac{\partial x^i}{\partial y^\beta}. \tag{29.5}$$

These coefficients are functions of the variables (y^i), and they are obviously symmetric with respect to the indices α and β.

Since the square of the element of arc ds is an invariant, we conclude (see Problem 3) that the set of functions $g_{\alpha\beta}(y)$ represents a *symmetric tensor*. This tensor is called the *metric tensor*, because, as will be shown in Chapter 3, all essential metric properties of Euclidean space are completely determined by this tensor.

We have obtained the formula 29.4 by starting with expression 29.1, which is characteristic of the Euclidean space. A transformation of coordinates 29.2 clearly does not alter its metric properties, and formula 29.4 simply enables us to calculate distances in the Euclidean space when it is covered by a coordinate system Y. By starting with the form 29.1 and the transformation 29.2, we have shown that the set of n functions 29.2 satisfies a system of $\frac{1}{2}n(n + 1)$ partial differential equations 29.5, in which the $g_{\alpha\beta}(y)$ are known functions of the variables y. Now, if the functions $g_{\alpha\beta}$ are specified arbitrarily, the system of $\frac{1}{2}n(n + 1)$ partial differential equations 29.5 for n unknown functions $x^i(y)$, in general, will have no solution. In the event the $g_{\alpha\beta}$'s are such that the system 29.5 has a solution, the existence of a transformation of coordinates which reduces the quadratic form 29.4 to the sum of squares 29.1 is assured. In that event the metric tensor $g_{\alpha\beta}$ defines an Euclidean manifold. If, on the other hand, the functions $g_{\alpha\beta}(y)$ are such that the system 29.5 has no solution, then no admissible transformation of coordinates exists which reduces the expression 29.4 for the square of the arc element to the Pythagorean form 29.1. We shall say then that the manifold is non-Euclidean. A set of necessary and sufficient conditions for the integrability of equations 29.5 will be deduced in Sec. 39.

We suppose in the remainder of this chapter that our tensors are defined in metric manifolds and that the element of arc ds is given by the quadratic form $ds^2 = g_{ij}(x)\, dx^i\, dx^j$, where the g_{ij}'s are functions belonging to the class C^1. We also assume that the symmetric tensor $g_{ij}(x)$ is such that $|g_{ij}| \neq 0$ at any point of the region under discussion, but do not assume that our manifold is necessarily Euclidean.

Problems

1. Let E_3 be covered by orthogonal cartesian coordinates x^i, and consider a transformation

$$x^1 = y^1 \sin y^2 \cos y^3,$$
$$x^2 = y^1 \sin y^2 \sin y^3,$$
$$x^3 = y^1 \cos y^2,$$

where the y^i are spherical polar coordinates ($y^1 = r$, $y^2 = \theta$, $y^3 = \phi$). What are the metric coefficients $g_{ij}(y)$?

2. Let E_3 be covered by orthogonal cartesian coordinates x^i, and let

$$x^1 = y^1 \cos y^2,$$
$$x^2 = y^1 \sin y^2,$$
$$x^3 = y^3$$

represent a transformation to cylindrical coordinates y^i. Find the expression for ds^2 in cylindrical coordinates.

3. Let E_3 be covered by orthogonal cartesian coordinates x^i, and let $x^i = a_j{}^i y^j$, $|a_j{}^i| \neq 0$, $(i, j = 1, 2, 3)$, represent a linear transformation of coordinates. Determine the metric coefficients $g_{ij}(y)$. Discuss the case when the transformation is orthogonal.

30. The Fundamental and Associated Tensors

Let $g_{ij}(x)$ represent a symmetric tensor such that the $g_{ij}(x)$ belong to class C^1 and $g = |g_{ij}| \neq 0$ at any point of the region. We construct, with the aid of the set of functions $g_{ij}(x)$, a new set of functions $g^{ij}(x)$, representing a contravariant tensor, which is such that $g^{ij}g_{kj} = \delta^i_k$. The tensors $g_{ij}(x)$ and $g^{ij}(x)$ will play an essential role in all our subsequent considerations, and for that reason they will be called the *fundamental tensors*.

Let us form a set of n^2 functions

$$g(i, j) = \frac{G^{ij}}{g}, \tag{30.1}$$

where G^{ij} is the cofactor of the element g_{ij} in the determinant g. The notation used in the definition 30.1 anticipates that the $g(i, j)$ form a contravariant tensor, and, indeed, we will prove that they define a symmetric, contravariant tensor g^{ij}. The symmetry of the set of functions $g(i, j)$ follows directly from the observation that the determinant obtained by deleting the ith row and the jth column in a symmetric determinant g_{ij} has the same value as the determinant obtained by deleting the jth row and the ith column. We prove next, by means of a quotient law, that the $g(i, j)$'s transform according to a contravariant law. We first note that, if ξ^i is an arbitrary contravariant vector, then

$$\xi_i \equiv g_{\alpha i}\xi^\alpha \tag{30.2}$$

is an arbitrary covariant vector, since $|g_{ij}| \neq 0$. Now, if both sides of the formula 30.2 are multiplied by $g(\beta, i) = G^{\beta i}/g$ and summed on i, we get

$$(30.3)\qquad g(\beta, i)\xi_i = \frac{G^{\beta i}}{g}\, g_{\alpha i}\xi^\alpha.$$

However, by (7.4), $G^{\beta i}g_{\alpha i} = g\delta^\beta_\alpha$, so that (30.3) can be written as

$$g(\beta, i)\xi_i = \xi^\beta.$$

Since ξ_i is arbitrary, we conclude from Theorem I of Sec. 26 that $g(\beta, i)$ is a contravariant tensor of rank two. We can thus write (30.1) as

$$(30.4)\qquad g^{ij} \equiv \frac{G^{ij}}{g}.$$

The reciprocal relation $g^{ij}g_{kj} = \delta^i_k$ follows directly from the fact that $G^{ij}g_{kj} = \delta^i_k g$. Incidentally, we can conclude that the set of cofactors G^{ij} represents a contravariant tensor of weight two. This follows from Problem 11 of Sec. 28, where it is indicated that the determinant $|g_{ij}|$ is a relative scalar of weight two.

A tensor obtained by the process of inner multiplication of any tensor $A^{i_1 \cdots i_r}_{j_1 \cdots j_s}$ with either of the fundamental tensors g_{ij} or g^{ij} is called a tensor *associated* with the given tensor.

As an illustration of this definition consider a tensor A_{ijk} and form the following inner products: $g^{\alpha i}A_{ijk} \equiv A^{\alpha}_{\cdot jk}$, $g^{\alpha j}A_{ijk} \equiv A^{\alpha}_{i \cdot k}$, $g^{\alpha k}A_{ijk} \equiv A^{\alpha}_{ij\cdot}$. All these tensors are associated with the tensor A_{ijk}. Operating on these tensors with g^{ij} again, we can form other associated tensors. It will be observed that the operation of inner multiplication of g_{ij} with any tensor, say A^{ijk}_{lm}, lowers the index with respect to which the summation is performed. Thus $g_{\alpha\beta}A^{ij\alpha}_{lm} = A^{ij\cdot}_{lm\beta}$, while $g^{\alpha\beta}A^{ijk}_{l\alpha} = A^{ijk\beta}_{l\cdot}$. The procedure of raising and lowering indices is clearly reversible. In the foregoing formulas the position occupied by the raised (or lowered) index is indicated by a dot. In general, such systems as $g^{i\alpha}A_{j\alpha} = A_{j\cdot}^{\;i}$ and $g^{i\alpha}A_{\alpha j} = A^{i}_{\cdot j}$ are different. They are identical whenever $A_{ij} = A_{ji}$.

It is possible to interpret all tensors associated with a given tensor as representing the same tensor in different reference frames. This interpretation is particularly simple for the covariant vector A_i, and its associated vector $g^{i\alpha}A_\alpha = A^i$, whenever the space is Euclidean. We will return to this matter in Sec. 45.

31. Christoffel's Symbols

We introduce in this section certain combinations of partial derivatives of the fundamental tensor $g_{ij}(x)$, which will prove useful in the development of the calculus of tensors. Let us construct a set of functions denoted by the symbol

$$(31.1) \quad [ij, k] \equiv \frac{1}{2}\left(\frac{\partial g_{ik}}{\partial x^j} + \frac{\partial g_{jk}}{\partial x^i} - \frac{\partial g_{ij}}{\partial x^k}\right), \qquad (i, j, k = 1, \ldots, n),$$

and call them the *Christoffel* 3-*index symbols of the first kind.* The set of functions

$$(31.2) \qquad \begin{Bmatrix} k \\ ij \end{Bmatrix} \equiv g^{k\alpha}[ij, \alpha],$$

where $g^{k\alpha}$ is the contravariant tensor, constructed with the aid of the g_{ij}'s in the manner described in the preceding section, are the *Christoffel* 3-*index symbols of the second kind.*

Evidently there are n distinct Christoffel symbols of each kind for each independent g_{ij}, and, since the number of independent g_{ij}'s is $\frac{1}{2}n(n + 1)$, the number N of independent Christoffel symbols is $N = \frac{1}{2}n^2(n + 1)$. We proceed to deduce several properties and identities involving Christoffel's symbols, which will prove useful to us in the sequel.

It is clear from definitions 31.1 and 31.2 that the Christoffel symbols are symmetric with respect to the indices i and j. Thus

$$(31.3) \qquad [ij, k] = [ji, k],$$

and

$$(31.4) \qquad \begin{Bmatrix} k \\ ij \end{Bmatrix} = \begin{Bmatrix} k \\ ji \end{Bmatrix}.$$

We see from the defining formula 31.2 that we can pass from the symbol of the first kind $[ij, \alpha]$ to the symbol $\begin{Bmatrix} k \\ ij \end{Bmatrix}$ by forming the inner product $g^{k\alpha}[ij, \alpha]$. Now, if we multiply equation 31.2 through by $g_{k\beta}$, and recall that $g_{k\beta}g^{k\alpha} = \delta^\alpha_\beta$, we get

$$(31.5) \qquad g_{k\beta}\begin{Bmatrix} k \\ ij \end{Bmatrix} = \delta^\alpha_\beta[ij, \alpha] = [ij, \beta].$$

Formulas 31.2 and 31.5 are easy to remember if it is noted that the operation of inner multiplication of $[ij, \alpha]$ with $g^{k\alpha}$ raises the index and replaces the square brackets by the braces. The multiplication of $\begin{Bmatrix} k \\ ij \end{Bmatrix}$ by $g_{k\beta}$, on the other hand, lowers the index and replaces the braces by the square brackets. Formally, these operations of multiplication by $g^{k\alpha}$ and $g_{k\alpha}$ are analogous to raising and lowering the indices on tensors, but we will see that the Christoffel symbols, in general, are not tensors.

From (31.1) we readily deduce an expression for the partial derivative of the fundamental tensor g_{ij} in terms of the symbols of the first kind. It is

$$\frac{\partial g_{ij}}{\partial x^k} = [ik, j] + [jk, i], \tag{31.6}$$

which can also be written as

$$\frac{\partial g_{ij}}{\partial x^k} = g_{\alpha j}\begin{Bmatrix} \alpha \\ ik \end{Bmatrix} + g_{\alpha i}\begin{Bmatrix} \alpha \\ jk \end{Bmatrix} \tag{31.7}$$

if we note (31.5). An analogous formula for the partial derivatives of the contravariant tensor g^{ij} can be obtained by differentiating the identity $g_{i\alpha}g^{\alpha j} = \delta_i^j$ with respect to x^k. We get

$$\frac{\partial g_{i\alpha}}{\partial x^k} g^{\alpha j} + g_{i\alpha}\frac{\partial g^{\alpha j}}{\partial x^k} = 0,$$

or

$$g_{i\alpha}\frac{\partial g^{\alpha j}}{\partial x^k} = -g^{\alpha j}\frac{\partial g_{i\alpha}}{\partial x^k}.$$

To solve this system of equations for $\partial g^{\alpha j}/\partial x^k$ we multiply both sides by $g^{i\beta}$ and get

$$g^{i\beta}g_{i\alpha}\frac{\partial g^{\alpha j}}{\partial x^k} = -g^{i\beta}g^{\alpha j}\frac{\partial g_{i\alpha}}{\partial x^k}.$$

Since $g^{i\beta}g_{i\alpha} = \delta_\alpha^\beta$, we have

$$\frac{\partial g^{\beta j}}{\partial x^k} = -g^{i\beta}g^{\alpha j}([ik, \alpha] + [\alpha k, i]),$$

where we made use of the formula 31.6. Noting the definition 31.2, we have finally

$$\frac{\partial g^{\beta j}}{\partial x^k} = -g^{i\beta}\begin{Bmatrix} j \\ ik \end{Bmatrix} - g^{\alpha j}\begin{Bmatrix} \beta \\ \alpha k \end{Bmatrix},$$

which is the same as

$$\frac{\partial g^{ij}}{\partial x^k} = -g^{\alpha i}\begin{Bmatrix} j \\ \alpha k \end{Bmatrix} - g^{\alpha j}\begin{Bmatrix} i \\ \alpha k \end{Bmatrix}. \tag{31.8}$$

We conclude this section with a derivation of the formula for the derivative of the logarithm of the determinant $|g_{ij}|$; this will be useful to us in writing a compact expression for the divergence of a vector field, as well as in several other connections.

The determinant $g = |g_{ij}|$ can be expanded by minors to obtain

$$\begin{aligned} g &= g_{i1}G^{i1} + g_{i2}G^{i2} + \cdots + g_{in}G^{in}, \quad (\textit{no summation} \text{ on } i \text{ or } n), \\ &= g_{i\alpha}G^{i\alpha}, \qquad (\text{sum on } \alpha \textit{ only}, i \text{ fixed}), \end{aligned} \tag{31.9}$$

where G^{ij} is the cofactor of the element g_{ij}. Since the $g_{i\alpha}$'s are functions of $x^1, \ldots, x^n$, the $G^{i\alpha}$'s are also functions of the same variables. From (31.9) we deduce that

$$\frac{\partial g}{\partial g_{ij}} = \frac{\partial(g_{i\alpha}G^{i\alpha})}{\partial g_{ij}}$$

$$= g_{i\alpha}\frac{\partial G^{i\alpha}}{\partial g_{ij}} + G^{i\alpha}\frac{\partial g_{i\alpha}}{\partial g_{ij}}, \qquad \text{(sum on } \alpha \text{ only, } i \text{ fixed).}$$

Since $G^{i\alpha}$ contains no g_{ij}, $\partial G^{i\alpha}/\partial g_{ij} = 0$, and since the g_{ij}'s are independent variables in this formula, $\partial g_{i\alpha}/\partial g_{ij} = \delta^j_\alpha$. Thus

$$\frac{\partial g}{\partial g_{ij}} = G^{i\alpha}\delta^j_\alpha = G^{ij}.$$

But

$$\frac{\partial g}{\partial x^i} = \frac{\partial g}{\partial g_{\alpha\beta}}\frac{\partial g_{\alpha\beta}}{\partial x^i} = G^{\alpha\beta}\frac{\partial g_{\alpha\beta}}{\partial x^i},$$

and, if we recall that $g^{\alpha\beta} = G^{\alpha\beta}/g$, the foregoing formula becomes

$$\frac{\partial g}{\partial x^i} = gg^{\alpha\beta}\frac{\partial g_{\alpha\beta}}{\partial x^i}.$$

If we now insert for $\partial g_{\alpha\beta}/\partial x^i$ from (31.7), we get

$$\frac{\partial g}{\partial x^i} = gg^{\alpha\beta}\left(g_{\gamma\beta}\begin{Bmatrix}\gamma\\ \alpha i\end{Bmatrix} + g_{\gamma\alpha}\begin{Bmatrix}\gamma\\ \beta i\end{Bmatrix}\right)$$

$$= g\left(\begin{Bmatrix}\alpha\\ \alpha i\end{Bmatrix} + \begin{Bmatrix}\beta\\ \beta i\end{Bmatrix}\right)$$

$$= 2g\begin{Bmatrix}\alpha\\ \alpha i\end{Bmatrix}.$$

Therefore we can write $\dfrac{1}{2g}\dfrac{\partial g}{\partial x^i} = \begin{Bmatrix}\alpha\\ i\alpha\end{Bmatrix}$, and hence

$$\frac{\partial}{\partial x^i}\log\sqrt{g} = \begin{Bmatrix}\alpha\\ i\alpha\end{Bmatrix}. \tag{31.10}$$

We close this section with some remarks about different notations used for the Christoffel symbols by various authors. The notation $[ij, k]$ for the symbol of the first kind is fairly universal, but there are several different notations for the symbol $\begin{Bmatrix}k\\ ij\end{Bmatrix}$. Thus, many writers use the symbol $\{ij, k\}$.

P. Appell, in *Traité de mécanique rationelle*, vol. 5, uses $\begin{bmatrix} ij \\ k \end{bmatrix}$ for the symbol of the first kind and $\begin{Bmatrix} ij \\ k \end{Bmatrix}$ for the second kind. The followers of the Princeton school generally use the symbol Γ^k_{ij} for the symbol $\begin{Bmatrix} k \\ ij \end{Bmatrix}$ adopted in this book. Although the notation Γ^k_{ij} has some advantages, it suggests that the symbol of the second kind is a tensor. This, however, is not always true, as will be seen from the developments of Sec. 32.

Problems

1. Show that $\dfrac{\partial g_{ij}}{\partial x^k} - \dfrac{\partial g_{jk}}{\partial x^i} = [jk, i] - [ij, k]$.

2. Show that, if $g_{ij} = 0$ for $i \neq j$, then $\begin{Bmatrix} k \\ ij \end{Bmatrix} = 0$ whenever i, j, and k are distinct.

3. Show that, if $g_{ij} = 0$ for $i \neq j$, then

$$\begin{Bmatrix} i \\ ii \end{Bmatrix} = \frac{1}{2}\frac{\partial}{\partial x^i}\log g_{ii}, \qquad \begin{Bmatrix} i \\ ij \end{Bmatrix} = \frac{1}{2}\frac{\partial}{\partial x^j}\log g_{ii}, \qquad \begin{Bmatrix} i \\ jj \end{Bmatrix} = -\frac{1}{2g_{ii}}\frac{\partial g_{jj}}{\partial x^i},$$

where we suspend the summation convention and suppose that $i \neq j$.

4. If $|g_{ij}| \neq 0$, show that

$$g_{\alpha\beta}\frac{\partial}{\partial x^j}\begin{Bmatrix} \beta \\ ik \end{Bmatrix} = \frac{\partial}{\partial x^j}[ik, \alpha] - \begin{Bmatrix} \beta \\ ik \end{Bmatrix}([\beta j, \alpha] + [\alpha j, \beta]).$$

5. If $y^i = a_j{}^i x^j$ is a transformation from a set of orthogonal cartesian variables y^i to a set of oblique cartesian coordinates x^i covering E_3, what are the metric coefficients g_{ij} in $ds^2 = g_{ij}\,dx^i\,dx^j$?

32. Transformation of Christoffel's Symbols

We have already remarked that the Christoffel symbols do not, in general, represent tensors. In this section we deduce the laws of transformation for the sets of functions $[ij, k]$ and $\begin{Bmatrix} k \\ ij \end{Bmatrix}$, under coordinate transformations $y^i = y^i(x^1, \ldots, x^n)$, which will from now on belong to the class C^2. The functions $g_{ij}(x)$ are assumed to belong to class C^1, and their transforms to the Y-coordinate system are denoted by the symbols $h_{ij}(y)$, so that

$$h_{ij} = \frac{\partial x^\alpha}{\partial y^i}\frac{\partial x^\beta}{\partial y^j}g_{\alpha\beta}. \tag{32.1}$$

Let us construct the Christoffel symbols $_y[ij, k]$, where the index y signifies that they refer to the Y-coordinate system; then

$$
{}_y[ij, k] = \frac{1}{2}\left(\frac{\partial h_{ik}}{\partial y^j} + \frac{\partial h_{jk}}{\partial y^i} - \frac{\partial h_{ij}}{\partial y^k}\right). \tag{32.2}
$$

Differentiating (32.1) we get

$$
\frac{\partial h_{ij}}{\partial y^k} = g_{\alpha\beta}\left(\frac{\partial^2 x^\alpha}{\partial y^k\, \partial y^i}\frac{\partial x^\beta}{\partial y^j} + \frac{\partial^2 x^\beta}{\partial y^k\, \partial y^j}\frac{\partial x^\alpha}{\partial y^i}\right) + \frac{\partial x^\alpha}{\partial y^i}\frac{\partial x^\beta}{\partial y^j}\frac{\partial x^\gamma}{\partial y^k}\frac{\partial g_{\alpha\beta}}{\partial x^\gamma}.
$$

Since $g_{\alpha\beta} = g_{\beta\alpha}$, we can interchange the dummy indices α and β in the second term within parentheses and obtain

$$
\frac{\partial h_{ij}}{\partial y^k} = g_{\alpha\beta}\left(\frac{\partial^2 x^\alpha}{\partial y^k\, \partial y^i}\frac{\partial x^\beta}{\partial y^j} + \frac{\partial^2 x^\alpha}{\partial y^k\, \partial y^j}\frac{\partial x^\beta}{\partial y^i}\right) + \frac{\partial x^\alpha}{\partial y^i}\frac{\partial x^\beta}{\partial y^j}\frac{\partial x^\gamma}{\partial y^k}\frac{\partial g_{\alpha\beta}}{\partial x^\gamma}.
$$

The partial derivatives $\partial h_{jk}/\partial y^i$ and $\partial h_{ik}/\partial y^j$, entering in (32.2), can be obtained from this formula by a cyclic permutation of indices, and the substitution in (32.2) yields

$$
{}_y[ij, k] = \frac{\partial x^\alpha}{\partial y^i}\frac{\partial x^\beta}{\partial y^j}\frac{\partial x^\gamma}{\partial y^k}\,{}_x[\alpha\beta, \gamma] + \frac{\partial^2 x^\alpha}{\partial y^i\, \partial y^j}\frac{\partial x^\beta}{\partial y^k}\, g_{\alpha\beta}, \tag{32.3}
$$

which shows that $[\alpha\beta, \gamma]$ is not a tensor unless the second term on the right vanishes. The second term will vanish identically if the coordinate transformation is affine, that is, if $y^i = c_j{}^i x^j$ and the $c_j{}^i$'s are constants.

Similarly, we can easily show that the Christoffel symbols of the second kind, are not tensors in general. Indeed, we note from formula 31.2 that

$$
{}_y\begin{Bmatrix} k \\ ij \end{Bmatrix} = h^{k\mu}\,{}_y[ij, \mu],
$$

where

$$
h^{k\mu} = \frac{\partial y^k}{\partial x^\rho}\frac{\partial y^\mu}{\partial x^\sigma}\, g^{\rho\sigma}.
$$

If we multiply (32.3) (with k replaced by μ) on the left by $h^{k\mu}$ and on the right by its equal from the formula written just above, and simplify, we get

$$
{}_y\begin{Bmatrix} k \\ ij \end{Bmatrix} = \frac{\partial y^k}{\partial x^\rho}\frac{\partial x^\alpha}{\partial y^i}\frac{\partial x^\beta}{\partial y^j}\, g^{\rho\gamma}\,{}_x[\alpha\beta, \gamma] + \frac{\partial y^k}{\partial x^\rho}\frac{\partial^2 x^\alpha}{\partial y^i\, \partial y^j}\, g^{\rho\beta} g_{\alpha\beta}.
$$

Thus

$$
{}_y\begin{Bmatrix} k \\ ij \end{Bmatrix} = \frac{\partial y^k}{\partial x^\rho}\frac{\partial x^\alpha}{\partial y^i}\frac{\partial x^\beta}{\partial y^j}\,{}_x\begin{Bmatrix} \rho \\ \alpha\beta \end{Bmatrix} + \frac{\partial^2 x^\alpha}{\partial y^i\, \partial y^j}\frac{\partial y^k}{\partial x^\alpha}, \tag{32.4}
$$

which shows that *the symbols of the second kind are not tensors unless the coordinate transformation is affine.*

The system of equations 32.4 can be solved for $\partial^2 x^\alpha/\partial y^i\,\partial y^j$ as follows. Multiply (32.4) by $\partial x^m/\partial y^k$, sum with respect to the common value $k = \gamma$, and obtain

$$ {}_y\begin{Bmatrix}\gamma\\ ij\end{Bmatrix}\frac{\partial x^m}{\partial y^\gamma} = \frac{\partial x^m}{\partial y^\gamma}\frac{\partial y^\gamma}{\partial x^\rho}\frac{\partial x^\alpha}{\partial y^i}\frac{\partial x^\beta}{\partial y^j}\,{}_x\begin{Bmatrix}\rho\\ \alpha\beta\end{Bmatrix} + \frac{\partial^2 x^\alpha}{\partial y^i\,\partial y^j}\frac{\partial x^m}{\partial y^\gamma}\frac{\partial y^\gamma}{\partial x^\alpha}. $$

Since $\partial x^m/\partial x^\rho = \delta^m_\rho$ and $\partial x^m/\partial x^\alpha = \delta^m_\alpha$, this expression yields

$$ \frac{\partial^2 x^m}{\partial y^i\,\partial y^j} = {}_y\begin{Bmatrix}\gamma\\ ij\end{Bmatrix}\frac{\partial x^m}{\partial y^\gamma} - {}_x\begin{Bmatrix}m\\ \alpha\beta\end{Bmatrix}\frac{\partial x^\alpha}{\partial y^i}\frac{\partial x^\beta}{\partial y^j}. \tag{32.5} $$

Obviously y and x can be interchanged, and it follows from (32.5) that

$$ \frac{\partial^2 y^m}{\partial x^i\,\partial x^j} = {}_x\begin{Bmatrix}\gamma\\ ij\end{Bmatrix}\frac{\partial y^m}{\partial x^\gamma} - {}_y\begin{Bmatrix}m\\ \alpha\beta\end{Bmatrix}\frac{\partial y^\alpha}{\partial x^i}\frac{\partial y^\beta}{\partial x^j}. \tag{32.6} $$

The important formulas 32.5 and 32.6 were first deduced in an entirely different way by E. B. Christoffel in a memoir concerned with a study of equivalence of quadratic differential forms.[8] We will make use of these formulas to define the operations of tensorial differentiation.

33. Covariant Differentiation of Tensors

We have observed, in Sec. 22, that the set of partial derivatives $\dfrac{\partial f}{\partial x^i}$, of a scalar function $f(x^1, \ldots, x^n)$, represents a covariant vector, since $\dfrac{\partial f}{\partial y^i} = \dfrac{\partial f}{\partial x^\alpha}\dfrac{\partial x^\alpha}{\partial y^i}$. But if we form the set of partial derivatives $\dfrac{\partial}{\partial y^j}\left(\dfrac{\partial f}{\partial y^i}\right)$ of the covariant vector $\dfrac{\partial f}{\partial y^i}$, we get

$$ \frac{\partial^2 f}{\partial y^j\,\partial y^i} = \frac{\partial}{\partial y^j}\left(\frac{\partial f}{\partial x^\alpha}\frac{\partial x^\alpha}{\partial y^i}\right) $$

$$ = \frac{\partial^2 f}{\partial x^\alpha\,\partial x^\beta}\frac{\partial x^\beta}{\partial y^j}\frac{\partial x^\alpha}{\partial y^i} + \frac{\partial f}{\partial x^\alpha}\frac{\partial^2 x^\alpha}{\partial y^i\,\partial y^j}, $$

which, because of the presence of the term $\dfrac{\partial f}{\partial x^\alpha}\dfrac{\partial^2 x^\alpha}{\partial y^i\,\partial y^j}$, shows that the set of second derivatives $\left\{\dfrac{\partial^2 f}{\partial y^i\,\partial y^j}\right\}$ does not transform according to a tensorial law. It follows from this example that the set of partial derivatives

[8] E. B. Christoffel, *Crelle Journal*, **70** (1869).

of a covariant vector, in general, is not a tensor. Indeed, if we have a covariant vector $A_\alpha(x)$, then

$$B_i(y) = \frac{\partial x^\alpha}{\partial y^i} A_\alpha,$$

and

$$\frac{\partial B_i}{\partial y^j} = \frac{\partial x^\alpha}{\partial y^i}\frac{\partial x^\beta}{\partial y^j}\frac{\partial A_\alpha}{\partial x^\beta} + \frac{\partial^2 x^\alpha}{\partial y^j\,\partial y^i} A_\alpha, \tag{33.1}$$

so that the derivatives of a vector do not form a tensor unless the coordinate transformation $x^i = x^i(y)$ is affine. If we insert in (33.1) for $\dfrac{\partial^2 x^\alpha}{\partial y^j\,\partial y^i}$ from the Christoffel formula 32.5, we get

$$\frac{\partial B_i}{\partial y^j} = \frac{\partial x^\alpha}{\partial y^i}\frac{\partial x^\beta}{\partial y^j}\frac{\partial A_\alpha}{\partial x^\beta} + {}_y\begin{Bmatrix}\gamma\\ ij\end{Bmatrix}\frac{\partial x^\alpha}{\partial y^\gamma} A_\alpha - {}_x\begin{Bmatrix}\alpha\\ \gamma\beta\end{Bmatrix}\frac{\partial x^\gamma}{\partial y^i}\frac{\partial x^\beta}{\partial y^j} A_\alpha.$$

Since $\dfrac{\partial x^\alpha}{\partial y^\gamma} A_\alpha = B_\gamma$, we have on rearranging

$$\frac{\partial B_i}{\partial y^j} - {}_y\begin{Bmatrix}\gamma\\ ij\end{Bmatrix} B_\gamma = \left(\frac{\partial A_\alpha}{\partial x^\beta} - {}_x\begin{Bmatrix}\gamma\\ \alpha\beta\end{Bmatrix} A_\gamma\right)\frac{\partial x^\alpha}{\partial y^i}\frac{\partial x^\beta}{\partial y^j}, \tag{33.2}$$

from which it is clear that the set of n^2 functions $\dfrac{\partial A_i}{\partial x^j} - {}_x\begin{Bmatrix}\alpha\\ ij\end{Bmatrix} A_\alpha$ obeys the law of transformation for a covariant tensor of rank two. This leads us to formulate a

DEFINITION 1. *The set of n^2 functions $\dfrac{\partial A_i}{\partial x^j} - \begin{Bmatrix}\alpha\\ ij\end{Bmatrix} A_\alpha$ defines the covariant x^j derivative (with respect to g_{ij}) of the covariant tensor A_i.*

We denote the covariant x^j derivative of A_i by the symbol $A_{i,j}$. Thus

$$A_{i,j} \equiv \frac{\partial A_i}{\partial x^j} - \begin{Bmatrix}\alpha\\ ij\end{Bmatrix} A_\alpha. \tag{33.3}$$

It should be noted that in order to compute the covariant derivative it is necessary to have the set of Christoffel symbols; that is, the fundamental tensor g_{ij} must be given in advance.

Similarly, if we start with a contravariant vector A^α, and differentiate the relation $B^i(y) = \dfrac{\partial y^i}{\partial x^\alpha} A^\alpha(x)$, we obtain

$$\frac{\partial B^i}{\partial y^j} = \frac{\partial A^\alpha}{\partial x^\beta}\frac{\partial x^\beta}{\partial y^j}\frac{\partial y^i}{\partial x^\alpha} + A^\alpha \frac{\partial^2 y^i}{\partial x^\alpha\,\partial x^\beta}\frac{\partial x^\beta}{\partial y^j},$$

and making use of the formula 32.6, we find

$$\frac{\partial B^i}{\partial y^j} + {}_y\begin{Bmatrix} i \\ \gamma j \end{Bmatrix} B^\gamma = \left(\frac{\partial A^\alpha}{\partial x^\beta} + {}_x\begin{Bmatrix} \alpha \\ \gamma\beta \end{Bmatrix} A^\gamma\right)\frac{\partial x^\beta}{\partial y^j}\frac{\partial y^i}{\partial x^\alpha}.$$

Thus the set of n^2 quantities $A(i, j) \equiv \dfrac{\partial A^i}{\partial x^j} + \begin{Bmatrix} i \\ \alpha j \end{Bmatrix} A^\alpha$ forms a mixed tensor of rank two. Accordingly, we introduce a

DEFINITION 2. *The set of n^2 functions $\dfrac{\partial A^i}{\partial x^j} + \begin{Bmatrix} i \\ \alpha j \end{Bmatrix} A^\alpha$ represents the covariant x^j derivative (with respect to g_{ij}) of the contravariant tensor A^i.*

We denote the covariant x^j derivative of the contravariant tensor A^i by the symbol $A^i_{,j}$. Thus

$$A^i_{,j} \equiv \frac{\partial A^i}{\partial x^j} + \begin{Bmatrix} i \\ \alpha j \end{Bmatrix} A^\alpha. \tag{33.4}$$

The definitions 33.3 and 33.4 can be extended, in an obvious way, to mixed tensors. Thus we define the covariant x^l derivative (with respect to a given tensor g_{ij}) of the mixed tensor $A^{j_1 \cdots j_s}_{i_1 \cdots i_r}$ by the formula

$$\begin{aligned} A^{j_1 \cdots j_s}_{i_1 \cdots i_r, l} \equiv{} & \frac{\partial A^{j_1 \cdots j_s}_{i_1 \cdots i_r}}{\partial x^l} \\ & - \begin{Bmatrix} \alpha \\ i_1 l \end{Bmatrix} A^{j_1 \cdots j_s}_{\alpha i_2 \cdots i_r} - \begin{Bmatrix} \alpha \\ i_2 l \end{Bmatrix} A^{j_1 \cdots j_s}_{i_1 \alpha i_3 \cdots i_r} - \cdots - \begin{Bmatrix} \alpha \\ i_r l \end{Bmatrix} A^{j_1 \cdots j_s}_{i_1 \cdots \alpha} \\ & + \begin{Bmatrix} j_1 \\ \alpha l \end{Bmatrix} A^{\alpha j_2 \cdots j_s}_{i_1 \cdots i_r} + \begin{Bmatrix} j_2 \\ \alpha l \end{Bmatrix} A^{j_1 \alpha j_3 \cdots j_s}_{i_1 \cdots i_r} + \cdots + \begin{Bmatrix} j_s \\ \alpha l \end{Bmatrix} A^{j_1 \cdots \alpha}_{i_1 \cdots i_r}. \end{aligned} \tag{33.5}$$

A verification of the fact that the set of functions $A^{j_1 \cdots j_s}_{i_1 \cdots i_r, l}(x)$ forms a tensor of the type indicated by the indices presents no difficulty.

If A is a tensor of rank zero, we define its covariant derivative to be the ordinary derivative. Thus $A_{,l} = \partial A/\partial x^l$. This definition is consistent with the formula 33.5. We also note that, if the g_{ij}'s are constants, the Christoffel symbols vanish identically, and hence the covariant derivatives reduce to the ordinary derivatives. This will surely be true if the g_{ij}'s are the metric coefficients of an Euclidean space covered by a cartesian reference system.

We remark in conclusion that the covariant x^l derivatives of relative tensors are defined as follows. If $f(x)$ is a relative scalar of weight W, so that $g(y) = f(x)\left|\dfrac{\partial x^i}{\partial y^j}\right|^W$, then

$$f_{,l} \equiv \frac{\partial f}{\partial x^l} - Wf\begin{Bmatrix} \alpha \\ l\alpha \end{Bmatrix}. \tag{33.6}$$

This set of functions represents a relative vector of weight W. If $A^{j_1\cdots j_s}_{i_1\cdots i_r}$ is a relative tensor of weight W, then its covariant x^l derivative is a relative tensor of weight W, determined by the formula

$$A^{j_1\cdots j_s}_{i_1\cdots i_r,l} \equiv \frac{\partial A^{j_1\cdots j_s}_{i_1\cdots i_r}}{\partial x^l} - WA^{j_1\cdots j_s}_{i_1\cdots i_r}\begin{Bmatrix}\alpha\\ l\alpha\end{Bmatrix}$$

$$-\begin{Bmatrix}\alpha\\ i_1 l\end{Bmatrix}A^{j_1\cdots j_s}_{\alpha i_2\cdots i_r} - \cdots - \begin{Bmatrix}\alpha\\ i_r l\end{Bmatrix}A^{j_1\cdots j_s}_{i_1\cdots \alpha}$$

$$+\begin{Bmatrix}j_1\\ \alpha l\end{Bmatrix}A^{\alpha j_2\cdots j_s}_{i_1\cdots i_r} + \cdots + \begin{Bmatrix}j_s\\ \alpha l\end{Bmatrix}A^{j_1\cdots \alpha}_{i_1\cdots i_r}.$$

Problems

1. Prove that the following expressions are tensors.

(*a*) $$A^{ij}_{,l} = \frac{\partial A^{ij}}{\partial x^l} + \begin{Bmatrix}i\\ \alpha l\end{Bmatrix}A^{\alpha j} + \begin{Bmatrix}j\\ \alpha l\end{Bmatrix}A^{i\alpha}.$$

(*b*) $$A^i_{j,l} = \frac{\partial A^i_j}{\partial x^l} - \begin{Bmatrix}\alpha\\ jl\end{Bmatrix}A^i_\alpha + \begin{Bmatrix}i\\ \alpha l\end{Bmatrix}A^\alpha_j.$$

(*c*) $$A_{ij,l} = \frac{\partial A_{ij}}{\partial x^l} - \begin{Bmatrix}\alpha\\ il\end{Bmatrix}A_{\alpha j} - \begin{Bmatrix}\alpha\\ jl\end{Bmatrix}A_{i\alpha}.$$

(*d*) $$A^r_{ijk,l} = \frac{\partial A^r_{ijk}}{\partial x^l} - \begin{Bmatrix}\alpha\\ il\end{Bmatrix}A^r_{\alpha jk} - \begin{Bmatrix}\alpha\\ jl\end{Bmatrix}A^r_{i\alpha k} - \begin{Bmatrix}\alpha\\ kl\end{Bmatrix}A^r_{ij\alpha} + \begin{Bmatrix}r\\ \alpha l\end{Bmatrix}A^\alpha_{ijk}.$$

2. Prove that ${}_a\begin{Bmatrix}k\\ ij\end{Bmatrix} - {}_b\begin{Bmatrix}k\\ ij\end{Bmatrix}$ are components of a tensor of rank three, where ${}_a\begin{Bmatrix}k\\ ij\end{Bmatrix}$ and ${}_b\begin{Bmatrix}k\\ ij\end{Bmatrix}$ are the Christoffel symbols formed from the symmetric tensors $a_{ij}(x)$ and $b_{ij}(x)$.

3. Use the formula $\frac{\partial}{\partial x^l}\left|\frac{\partial y^i}{\partial x^j}\right| = \frac{\partial^2 y^\alpha}{\partial x^l\,\partial x^\beta}\frac{\partial x^\beta}{\partial y^\alpha}\left|\frac{\partial y^i}{\partial x^j}\right|$ and the law of transformation of relative scalars of weight W to deduce formula 33.6.

34. Formulas for Covariant Differentiation

It is easy to deduce from the structure of formula 33.5 that the rules for covariant differentiation of sums and products of tensors are identical with those used in the ordinary differentiation. Indeed, if $A^{j_1\cdots j_s}_{i_1\cdots i_r}(x)$ and $\mathscr{A}^{j_1\cdots j_s}_{i_1\cdots i_r}(x)$ are two tensors, then the formula

$$(A^{j_1\cdots j_s}_{i_1\cdots i_r} + \mathscr{A}^{j_1\cdots j_s}_{i_1\cdots i_r})_{,l} = A^{j_1\cdots j_s}_{i_1\cdots i_r,l} + \mathscr{A}^{j_1\cdots j_s}_{i_1\cdots i_r,l}$$

follows directly from inspection of (33.5). To prove that the derivatives of the outer and inner products are given by the familiar rules,

$$(A^{j_1\cdots j_s}_{i_1\cdots i_r}\mathscr{A}^{j_{s+1}\cdots j_v}_{i_{r+1}\cdots i_w})_{,l} = A^{j_1\cdots j_s}_{i_1\cdots i_r}\mathscr{A}^{j_{s+1}\cdots j_v}_{i_{r+1}\cdots i_w,l} + A^{j_1\cdots j_s}_{i_1\cdots i_r,l}\mathscr{A}^{j_{s+1}\cdots j_v}_{i_{r+1}\cdots i_w},$$

$$(A^{j_1\cdots j_{s-1}\alpha}_{i_1\cdots i_r}\mathscr{A}^{j_{s+1}\cdots j_v}_{i_{r+1}\cdots i_{w-1}\alpha})_{,l} = A^{j_1\cdots j_{s-1}\alpha}_{i_1\cdots i_r,l}\mathscr{A}^{j_{s+1}\cdots j_v}_{i_{r+1}\cdots i_{w-1}\alpha} + A^{j_1\cdots j_{s-1}\alpha}_{i_1\cdots i_r}\mathscr{A}^{j_{s+1}\cdots j_v}_{i_{r+1}\cdots i_{w-1}\alpha,l},$$

we need only insert for A in formula 33.5 the product $A\mathscr{A}$. We illustrate the procedure by considering the product $A^{j_1j_2}\mathscr{A}_{i_1i_2} \equiv \mathfrak{A}^{j_1j_2}_{i_1i_2}$. We have

$$\begin{aligned}
\mathfrak{A}^{j_1j_2}_{i_1i_2,l} &= \frac{\partial \mathfrak{A}^{j_1j_2}_{i_1i_2}}{\partial x^l} - \begin{Bmatrix}\alpha\\ i_1 l\end{Bmatrix}\mathfrak{A}^{j_1j_2}_{\alpha i_2} - \begin{Bmatrix}\alpha\\ i_2 l\end{Bmatrix}\mathfrak{A}^{j_1j_2}_{i_1\alpha} \\
&\quad + \begin{Bmatrix}j_1\\ \alpha l\end{Bmatrix}\mathfrak{A}^{\alpha j_2}_{i_1i_2} + \begin{Bmatrix}j_2\\ \alpha l\end{Bmatrix}\mathfrak{A}^{j_1\alpha}_{i_1i_2} \\
&= A^{j_1j_2}\left(\frac{\partial \mathscr{A}_{i_1i_2}}{\partial x^l} - \begin{Bmatrix}\alpha\\ i_1 l\end{Bmatrix}\mathscr{A}_{\alpha i_2} - \begin{Bmatrix}\alpha\\ i_2 l\end{Bmatrix}\mathscr{A}_{i_1\alpha}\right) \\
&\quad + \mathscr{A}_{i_1i_2}\left(\frac{\partial A^{j_1j_2}}{\partial x^l} + \begin{Bmatrix}j_1\\ \alpha l\end{Bmatrix}A^{\alpha j_2} + \begin{Bmatrix}j_2\\ \alpha l\end{Bmatrix}A^{j_1\alpha}\right) \\
&= A^{j_1j_2}\mathscr{A}_{i_1i_2,l} + \mathscr{A}_{i_1i_2}A^{j_1j_2}_{,l}.
\end{aligned}$$

This establishes the desired result. As an exercise the reader may show that

$$(A_{j\alpha}\mathscr{A}^{i\alpha})_{,l} = A_{j\alpha,l}\mathscr{A}^{i\alpha} + A_{j\alpha}\mathscr{A}^{i\alpha}_{,l}.$$

He can also show that the operations of covariant differentiation and contraction can be permuted.

We conclude this section by remarking that in covariant differentiation the Kronecker deltas behave like constants. Indeed, from (33.5) we have

$$\begin{aligned}
\delta^i_{j,l} &= \frac{\partial \delta^i_j}{\partial x^l} - \begin{Bmatrix}\alpha\\ jl\end{Bmatrix}\delta^i_\alpha + \begin{Bmatrix}i\\ \alpha l\end{Bmatrix}\delta^\alpha_j \\
&= 0 - \begin{Bmatrix}i\\ jl\end{Bmatrix} + \begin{Bmatrix}i\\ jl\end{Bmatrix} \equiv 0.
\end{aligned}$$

Problems

1. Note that the operation of contraction of indices $A^\alpha_{i\alpha}$ is equivalent to multiplying A^α_{ij} by δ^j_α. Using this, show that the operation of contraction can be performed on a tensor either before or after covariant differentiation.

2. Show that the operation of raising or lowering of indices can be performed either before or after covariant differentiation.

35. Ricci's Theorem

We will show in this section that the fundamental tensors g_{ij} and g^{ij} behave in covariant differentiation as though they were constants. This follows from

RICCI'S THEOREM. *The covariant derivative of either of the fundamental tensors is zero.*

Proof. Consider first the tensor g_{ij} and form

$$g_{ij,l} = \frac{\partial g_{ij}}{\partial x^l} - g_{\alpha j}\begin{Bmatrix}\alpha \\ il\end{Bmatrix} - g_{i\alpha}\begin{Bmatrix}\alpha \\ jl\end{Bmatrix}.$$

The right-hand member of this expression vanishes identically by virtue of (31.7), so that $g_{ij,l} = 0$.

We can perform a similar calculation for the tensor g^{ij}, but it may prove more instructive to differentiate the inner product $g^{i\alpha}g_{\alpha j} = \delta^i_j$. Thus

$$g^{i\alpha}_{,l}g_{\alpha j} + g^{i\alpha}g_{\alpha j,l} = \delta^i_{j,l};$$

since $\delta^i_{j,l} = 0$ and $g_{\alpha j,l} = 0$, we have

$$g_{\alpha j}g^{i\alpha}_{,l} = 0.$$

However, since $|g_{\alpha j}| \neq 0$, the only solution of this system of homogeneous equations is $g^{i\alpha}_{,l} = 0$.

As an immediate corollary of Ricci's theorem we note that the fundamental tensors may be taken outside the sign of covariant differentiation, and hence the operations of lowering and raising indices are permutable with covariant differentiation. Thus

$$(g_{\alpha i}A^{\alpha}_{jk})_{,l} = g_{\alpha i}A^{\alpha}_{jk,l}.$$

36. Riemann-Christoffel Tensor

We recall that a sufficient condition for the equality of mixed partial derivatives $\dfrac{\partial^2 u}{\partial x\,\partial y}$ and $\dfrac{\partial^2 u}{\partial y\,\partial x}$ of a function $u(x, y)$ is that $u(x, y)$ be of class C^2. We will assume henceforth that the tensor components under consideration belong to class C^2, but this restriction alone, as we shall see presently, is not sufficient to insure the equality of mixed covariant derivatives. Indeed, it will be shown that, if the order of covariant differentiation is to be immaterial, our tensors must be defined over a particular metric manifold X for which a certain tensor of rank four, made up entirely of the g_{ij}'s, vanishes. This tensor, known as the *Riemann-Christoffel tensor*, plays a basic role in many investigations of differential geometry, dynamics of rigid and deformable bodies, electrodynamics, and relativity.

The covariant derivative of a tensor is a tensor; hence it can be differentiated covariantly again to obtain a new tensor. This tensor is called the *second covariant derivative* of the given tensor.

Consider the covariant x^j derivative of A_i with respect to g_{ij},

$$A_{i,j} = \frac{\partial A_i}{\partial x^j} - \begin{Bmatrix} \alpha \\ ij \end{Bmatrix} A_\alpha. \tag{36.1}$$

Now, if (36.1) is differentiated covariantly with respect to x^k, there results the tensor

$$\begin{aligned} A_{i,jk} &= \frac{\partial A_{i,j}}{\partial x^k} - \begin{Bmatrix} \alpha \\ ik \end{Bmatrix} A_{\alpha,j} - \begin{Bmatrix} \alpha \\ jk \end{Bmatrix} A_{i,\alpha} \\ &= \frac{\partial}{\partial x^k}\left(\frac{\partial A_i}{\partial x^j} - \begin{Bmatrix} \alpha \\ ij \end{Bmatrix} A_\alpha\right) - \begin{Bmatrix} \alpha \\ ik \end{Bmatrix}\left(\frac{\partial A_\alpha}{\partial x^j} - \begin{Bmatrix} \beta \\ \alpha j \end{Bmatrix} A_\beta\right) \\ &\qquad - \begin{Bmatrix} \alpha \\ jk \end{Bmatrix}\left(\frac{\partial A_i}{\partial x^\alpha} - \begin{Bmatrix} \gamma \\ i\alpha \end{Bmatrix} A_\gamma\right). \end{aligned} \tag{36.2}$$

On the other hand,

$$\begin{aligned} A_{i,kj} &= \frac{\partial}{\partial x^j}\left(\frac{\partial A_i}{\partial x^k} - \begin{Bmatrix} \alpha \\ ik \end{Bmatrix} A_\alpha\right) - \begin{Bmatrix} \alpha \\ ij \end{Bmatrix}\left(\frac{\partial A_\alpha}{\partial x^k} - \begin{Bmatrix} \beta \\ \alpha k \end{Bmatrix} A_\beta\right) \\ &\qquad - \begin{Bmatrix} \alpha \\ kj \end{Bmatrix}\left(\frac{\partial A_i}{\partial x^\alpha} - \begin{Bmatrix} \gamma \\ i\alpha \end{Bmatrix} A_\gamma\right). \end{aligned} \tag{36.3}$$

Carrying out the indicated differentiation in (36.2) and (36.3) yields

$$\begin{aligned} A_{i,jk} &= \frac{\partial^2 A_i}{\partial x^k \partial x^j} - \frac{\partial \begin{Bmatrix} \alpha \\ ij \end{Bmatrix}}{\partial x^k} A_\alpha - \begin{Bmatrix} \alpha \\ ij \end{Bmatrix} \frac{\partial A_\alpha}{\partial x^k} - \begin{Bmatrix} \alpha \\ ik \end{Bmatrix} \frac{\partial A_\alpha}{\partial x^j} \\ &\qquad + \begin{Bmatrix} \alpha \\ ik \end{Bmatrix}\begin{Bmatrix} \beta \\ \alpha j \end{Bmatrix} A_\beta - \begin{Bmatrix} \alpha \\ jk \end{Bmatrix} \frac{\partial A_i}{\partial x^\alpha} + \begin{Bmatrix} \alpha \\ jk \end{Bmatrix}\begin{Bmatrix} \gamma \\ i\alpha \end{Bmatrix} A_\gamma, \end{aligned} \tag{36.4}$$

$$\begin{aligned} A_{i,kj} &= \frac{\partial^2 A_i}{\partial x^j \partial x^k} - \frac{\partial \begin{Bmatrix} \alpha \\ ik \end{Bmatrix}}{\partial x^j} A_\alpha - \begin{Bmatrix} \alpha \\ ik \end{Bmatrix} \frac{\partial A_\alpha}{\partial x^j} - \begin{Bmatrix} \alpha \\ ij \end{Bmatrix} \frac{\partial A_\alpha}{\partial x^k} \\ &\qquad + \begin{Bmatrix} \alpha \\ ij \end{Bmatrix}\begin{Bmatrix} \beta \\ \alpha k \end{Bmatrix} A_\beta - \begin{Bmatrix} \alpha \\ kj \end{Bmatrix} \frac{\partial A_i}{\partial x^\alpha} + \begin{Bmatrix} \alpha \\ kj \end{Bmatrix}\begin{Bmatrix} \gamma \\ i\alpha \end{Bmatrix} A_\gamma. \end{aligned} \tag{36.5}$$

If we subtract (36.5) from (36.4), we get

$$A_{i,jk} - A_{i,kj} = \begin{Bmatrix} \alpha \\ ik \end{Bmatrix}\begin{Bmatrix} \beta \\ \alpha j \end{Bmatrix} A_\beta - \frac{\partial \begin{Bmatrix} \alpha \\ ij \end{Bmatrix}}{\partial x^k} A_\alpha - \begin{Bmatrix} \alpha \\ ij \end{Bmatrix}\begin{Bmatrix} \beta \\ \alpha k \end{Bmatrix} A_\beta + \frac{\partial \begin{Bmatrix} \alpha \\ ik \end{Bmatrix}}{\partial x^j} A_\alpha,$$

and an interchange of α and β in the first terms of each preceding line gives

$$(36.6)\quad A_{i,jk} - A_{i,kj} = \left[\frac{\partial \left\{ \begin{matrix} \alpha \\ ik \end{matrix} \right\}}{\partial x^j} - \frac{\partial \left\{ \begin{matrix} \alpha \\ ij \end{matrix} \right\}}{\partial x^k} + \left\{ \begin{matrix} \beta \\ ik \end{matrix} \right\} \left\{ \begin{matrix} \alpha \\ \beta j \end{matrix} \right\} - \left\{ \begin{matrix} \beta \\ ij \end{matrix} \right\} \left\{ \begin{matrix} \alpha \\ \beta k \end{matrix} \right\} \right] A_\alpha.$$

Since A_i is an arbitrary covariant tensor of rank one, and since the difference of two tensors $A_{i,kj} - A_{i,kj}$ is a covariant tensor of rank three, we know by the Quotient Theorem I of Sec. 26 that the expression in the bracket of (36.6) is a mixed tensor of rank four; that is,

$$\frac{\partial \left\{ \begin{matrix} \alpha \\ ik \end{matrix} \right\}}{\partial x^j} - \frac{\partial \left\{ \begin{matrix} \alpha \\ ij \end{matrix} \right\}}{\partial x^k} + \left\{ \begin{matrix} \beta \\ ik \end{matrix} \right\} \left\{ \begin{matrix} \alpha \\ \beta j \end{matrix} \right\} - \left\{ \begin{matrix} \beta \\ ij \end{matrix} \right\} \left\{ \begin{matrix} \alpha \\ \beta k \end{matrix} \right\} = R^\alpha_{ijk}.$$

Furthermore, if the left-hand member of (36.6) is to vanish, that is, if the order of covariant differentiation is to be immaterial, then

$$R^\alpha_{ijk} = 0$$

since A_α is arbitrary. In general, however, $R^\alpha_{ijk} \neq 0$, so that the order of covariant differentiation is not immaterial. It is clear from (36.6) that *a necessary and sufficient condition for the validity of inversion of the order of covariant differentiation is that the tensor R^α_{ijk} vanishes identically.*

The tensor

$$(36.7)\qquad R^i_{jkl} = \begin{vmatrix} \dfrac{\partial}{\partial x^k} & \dfrac{\partial}{\partial x^l} \\ \left\{ \begin{matrix} i \\ jk \end{matrix} \right\} & \left\{ \begin{matrix} i \\ jl \end{matrix} \right\} \end{vmatrix} + \begin{vmatrix} \left\{ \begin{matrix} i \\ \alpha k \end{matrix} \right\} & \left\{ \begin{matrix} i \\ \alpha l \end{matrix} \right\} \\ \left\{ \begin{matrix} \alpha \\ jk \end{matrix} \right\} & \left\{ \begin{matrix} \alpha \\ jl \end{matrix} \right\} \end{vmatrix}$$

is called the *mixed Riemann-Christoffel tensor* or the *Riemann-Christoffel tensor of the second kind.*

The associated tensor

$$(36.8)\qquad R_{ijkl} \equiv g_{i\alpha} R^\alpha_{jkl}$$

is known as the *covariant Riemann-Christoffel tensor*, or the *Riemann Christoffel tensor of the first kind.*

It is not difficult to verify that the defining formula 36.8 for R_{ijkl} can be written in the convenient determinantal form

$$(36.9)\qquad R_{ijkl} = \begin{vmatrix} \dfrac{\partial}{\partial x^k} & \dfrac{\partial}{\partial x^l} \\ [jk, i] & [jl, i] \end{vmatrix} + \begin{vmatrix} \left\{ \begin{matrix} \alpha \\ jk \end{matrix} \right\} & \left\{ \begin{matrix} \alpha \\ jl \end{matrix} \right\} \\ [ik, \alpha] & [il, \alpha] \end{vmatrix},$$

which will be found useful in listing properties of this tensor in Sec. 37.

We remark in conclusion that formula 36.6 is a special case of an identity, established by Ricci, which we record here without proof, although the nature of proof is quite clear from the proof of the case treated previously. This identity reads

$$A_{i_1\cdots i_m,jk} - A_{i_1\cdots i_m,kj} = \sum_{\alpha=1}^{m} A_{i_1\cdots i_{\alpha-1}hi_{\alpha+1}\cdots i_m} R^h_{i_\alpha jk}.$$

In the special case of a tensor of rank two it assumes the form

$$A_{ij,kl} - A_{ij,lk} = A_{i\alpha} R^\alpha_{jkl} + A_{\alpha j} R^\alpha_{ikl}.$$

Problems

1. Show that

$$R_{ijkl} = \frac{\partial}{\partial x^k}[jl, i] - \frac{\partial}{\partial x^l}[jk, i] + \begin{Bmatrix} \alpha \\ jk \end{Bmatrix}[il, \alpha] - \begin{Bmatrix} \alpha \\ jl \end{Bmatrix}[ik, \alpha].$$

2. Show that

$$R_{ijkl} = \frac{1}{2}\left(\frac{\partial^2 g_{il}}{\partial x^j\,\partial x^k} + \frac{\partial^2 g_{jk}}{\partial x^i\,\partial x^l} - \frac{\partial^2 g_{ik}}{\partial x^j\,\partial x^l} - \frac{\partial^2 g_{jl}}{\partial x^i\,\partial x^k}\right) + g^{\alpha\beta}([jk, \beta][il, \alpha] - [jl, \beta][ik, \alpha]).$$

3. Using the formula of Problem 2 show that

$$R_{ijkl} = -R_{jikl} = -R_{ijlk} = R_{klij}$$

and

$$R_{ijkl} + R_{iklj} + R_{iljk} = 0.$$

4. If ϕ is a scalar, then $g^{ij}\phi_{,ij}$ is a scalar and is equal to

$$\frac{1}{\sqrt{g}}\frac{\partial}{\partial x^i}\left(\sqrt{g}\, g^{ij}\frac{\partial \phi}{\partial x^j}\right).$$

5. Referring to Problem 4, show that $g^{ij}\phi_{,ij} = 0$ reduces to $\partial^2\phi/\partial x^i\,\partial x^i = 0$ when the g_{ij} are the metric coefficients of E_3 referred to a cartesian frame. This implies that Laplace's equation in general curvilinear coordinates has the form $g^{ij}\phi_{,ij} = 0$, since this is a tensor equation.

6. Referring to Problem 5, show that Laplace's equation in polar coordinates has the form

$$\frac{\partial^2\phi}{(\partial y^1)^2} + \frac{1}{(y^1)^2}\frac{\partial^2\phi}{(\partial y^2)^2} + \frac{1}{(y^1 \sin y^2)^2}\frac{\partial^2\phi}{(\partial y^3)^2} + \frac{2}{y^1}\frac{\partial\phi}{\partial y^1} + \frac{1}{(y^1)^2}\cot y^2\frac{\partial\phi}{\partial y^2} = 0.$$

37. Properties of Riemann-Christoffel Tensors

From defining formula 36.7 for a mixed tensor R^i_{jkl}, we see immediately that the set of functions R^i_{jkl} is skew-symmetric with respect to the last two covariant indices. Thus

$$R^i_{jkl} = -R^i_{jlk}, \tag{37.1}$$

and hence $R^i_{j(\alpha)(\alpha)} = 0$.

We have defined the covariant tensor R_{ijkl} by the formula

$$R_{ijkl} = g_{i\alpha} R^{\alpha}_{jkl},$$

and, if we multiply this equation through by $g^{i\beta}$ and sum, we get

$$R^{\beta}_{jkl} = g^{i\beta} R_{ijkl}, \tag{37.2}$$

so that the Riemann-Christoffel tensor of the second kind is obtained by raising the first covariant index in the tensor R_{ijkl}. To determine the properties of the set of functions defining the Riemann-Christoffel tensor of the first kind we expand the determinants in (36.9) and insert for Christoffel's symbols in the first determinant the definitions 31.1. We get after a simple calculation the formula

$$R_{ijkl} = \frac{1}{2}\left(\frac{\partial^2 g_{il}}{\partial x^j \partial x^k} - \frac{\partial^2 g_{jl}}{\partial x^i \partial x^k} - \frac{\partial^2 g_{ik}}{\partial x^j \partial x^l} + \frac{\partial^2 g_{jk}}{\partial x^i \partial x^l}\right) + g^{\alpha\beta}([jk, \beta][il, \alpha] - (jl, \beta][ik, \alpha]), \tag{37.3}$$

from which it is obvious that

(*a*) $\qquad R_{jikl} = -R_{ijkl}.$

(*b*) $\qquad R_{ijlk} = -R_{ijkl}.$

(*c*) $\qquad R_{klij} = R_{ijkl}.$

(*d*) $\qquad R_{ijkl} + R_{iklj} + R_{iljk} = 0.$

The last identity can be verified by direct substitution; by raising indices we obtain an identity analogous to (*d*) for the mixed tensor R^i_{jkl}.

(*e*) $\qquad R^i_{jkl} + R^i_{klj} + R^i_{ljk} = 0.$

(*f*) The components of a Riemann-Christoffel tensor with more than two like indices are necessarily zero. The identities (*a*) and (*b*) state that the tensor R_{ijkl} is skew-symmetric with respect to the first two and last two indices, and the identity (*c*) signifies that R_{ijkl} is symmetric with respect to groups of first two and last two indices. It follows from these identities that distinct, nonvanishing components of R_{ijkl} are of three types:

1. Symbols with two distinct indices, that is, symbols of the type R_{ijij}.
2. Symbols with only three distinct indices, which are of the type R_{ijik}.
3. Symbols R_{ijkl} with four distinct indices.

It is now an easy matter to verify[9] that the total number N of distinct nonvanishing components of R_{ijkl} is $N = n^2(n^2 - 1)/12$.

[9] There are $n_1 = n(n-1)/2$ distinct nonvanishing symbols of the type R_{ijij}, $n_2 = \dfrac{n(n-1)(n-2)}{2}$ of the type R_{ijik}, and $n_3 = \dfrac{n(n-1)(n-2)(n-3)}{12}$ of the the type R_{ijkl}.

In a three-dimensional space, distinct, nonvanishing components R_{ijkl} have the suffixes: 1212, 1313, 2323, 1213, 2123, 3132, and in two dimensions from the total of $2^4 = 16$ components there is only one distinct nonvanishing component R_{1212}. We will see that this tensor characterizes an extremely important property of surfaces.

38. Ricci Tensor. Bianchi Identities. Einstein Tensor

We define the *Ricci tensor* R_{ij} by the formula $R_{ij} = R^{\alpha}_{ij\alpha}$, which, by virtue of (36.7), can be written as

$$R_{ij} = \begin{vmatrix} \dfrac{\partial}{\partial x^j} & \dfrac{\partial}{\partial x^\alpha} \\ \begin{Bmatrix} \alpha \\ ij \end{Bmatrix} & \begin{Bmatrix} \alpha \\ i\alpha \end{Bmatrix} \end{vmatrix} + \begin{vmatrix} \begin{Bmatrix} \alpha \\ \beta j \end{Bmatrix} & \begin{Bmatrix} \alpha \\ \beta\alpha \end{Bmatrix} \\ \begin{Bmatrix} \beta \\ ij \end{Bmatrix} & \begin{Bmatrix} \beta \\ i\alpha \end{Bmatrix} \end{vmatrix}.$$

In Sec. 31 we have shown that $\dfrac{\partial}{\partial x^i} \log \sqrt{g} = \begin{Bmatrix} \alpha \\ i\alpha \end{Bmatrix}$, so that

$$R_{ij} = \frac{\partial^2 \log \sqrt{g}}{\partial x^j \, \partial x^i} - \frac{\partial \begin{Bmatrix} \alpha \\ ij \end{Bmatrix}}{\partial x^\alpha} + \begin{Bmatrix} \alpha \\ \beta j \end{Bmatrix} \begin{Bmatrix} \beta \\ i\alpha \end{Bmatrix} - \begin{Bmatrix} \beta \\ ij \end{Bmatrix} \frac{\partial \log \sqrt{g}}{\partial x^\beta}.$$

From inspection of this result we see that the tensor R_{ij} is symmetric. Since $R_{ij} = R_{ji}$, the number of distinct components of R_{ij} is $\frac{1}{2}n(n+1)$. In a four-dimensional manifold $n = 4$, so that, if we set $R_{ij} = 0$, we obtain ten partial differential equations, which Einstein has adopted as his equations of the gravitational field in free space in the general theory of relativity.[10] In the development of that theory another tensor, introduced by Einstein, plays an important role. This tensor is most readily obtained from the identity

$$R^i_{jkl,m} + R^i_{jlm,k} + R^i_{jmk,l} = 0, \tag{38.1}$$

due to Bianchi.

Since the covariant derivative of the fundamental tensor g_{ij} vanishes, the Bianchi identity can be written in the form

$$R_{ijkl,m} + R_{ijlm,k} + R_{ijmk,l} = 0. \tag{38.2}$$

If we multiply equation 38.2 by $g^{il}g^{jk}$ and make use of the skew-symmetric properties of the Riemann tensor R_{ijkl}, we get

$$g^{jk}R_{jk,m} - g^{jk}R_{jm,k} - g^{il}R_{im,l} = 0.$$

[10] See Problem 2.

This result can be written as

$$R_{,m} - 2R^k_{m,k} = 0,$$

where $R \equiv g^{ij}R_{ij}$, or in alternative form

$$(R^k_m - \tfrac{1}{2}\delta^k_m R)_{,k} = 0, \tag{38.3}$$

where $R^k_m = g^{jk}R_{jm}$. The tensor

$$R^i_j - \tfrac{1}{2}\delta^i_j R \equiv G^i_j,$$

in parentheses in equation 38.3, is known as the *Einstein tensor.*

Problems

1. Show that $R^{\alpha}_{\ \alpha jk} \equiv 0$.

2. If $R_{ij} = \rho g_{ij}$, then $\rho = R/n$, where $R = g^{ij}R_{ij}$. (The equation $R_{ij} = \rho g_{ij}$ is known as the Einstein gravitational equation at points where matter is present. It corresponds to the Poisson equation $\nabla^2 V = \rho$ in the Newtonian theory of gravitation.)

3. If $n = 2$, show that $R_{11}/g_{11} = R_{22}/g_{22} = R_{12}/g_{12} = -R_{1212}/g$.

4. If $n = 3$, the tensor R_{ijkl} has six distinct components, and there are six equations $R_{jk} = g^{il}R_{ijkl}$. Prove that the solutions of these equations for R_{ijkl} are given by

$$R_{ijkl} = g_{il}R_{jk} + g_{jk}R_{il} - g_{ik}R_{jl} - g_{jl}R_{ik} + \frac{R}{2}(g_{ik}g_{jl} - g_{il}g_{jk}),$$

where $R = g^{ij}R_{ij}$.

5. Verify Bianchi's identity 38.2.

39. Riemannian and Euclidean Spaces. Existence Theorem

Let the n-dimensional space V_n be covered by a coordinate system X. We will metrize V_n by prescribing the element of arc ds, so that

$$ds^2 = g_{ij}\, dx^i\, dx^j \tag{39.1}$$

is a positive definite quadratic form in the differentials dx^i. The functions $g_{ij}(x)$ are assumed to be of class C^1 in V_n. The space V_n so metrized is called a *Riemannian n-dimensional* space R_n.

We will now consider in some detail the following question: *What restriction must be imposed on the symmetric tensor $g_{ij}(x)$ so that there be a coordinate system Y, defined by*

$$T\colon\quad y^i = y^i(x^1, \ldots, x^n), \qquad (i = 1, \ldots, n),$$

with $y^i(x)$ of class C^2 in R_n, in which the tensor $g_{ij}(x)$ has constant components h_{ij} throughout R_n?

This is one of the basic problems of differential geometry, which occurs also under a different guise in dynamics, elasticity, relativity, and other branches of applied mathematics.

We note first that the components of $g_{ij}(x)$, when referred to the Y-frame, are given by

$$h_{ij} = \frac{\partial x^\alpha}{\partial y^i}\frac{\partial x^\beta}{\partial y^j} g_{\alpha\beta}. \tag{39.2}$$

If h_{ij}'s are constants, then the Christoffel symbols ${}_y\begin{Bmatrix} k \\ ij \end{Bmatrix}$ vanish identically. Conversely, if the ${}_y\begin{Bmatrix} k \\ ij \end{Bmatrix}$ vanish identically, $h_{ij,l} = \dfrac{\partial h_{ij}}{\partial y^l}$, and, since $h_{ij,l} = 0$ by Ricci's theorem ,we have $\partial h_{ij}/\partial y^l = 0$ in R_n. Consequently, the h_{ij} are constants throughout R_n. This permits us to state a

THEOREM I. *A necessary and sufficient condition that the metric coefficients $g_{ij}(x)$ reduce to constants h_{ij} in some reference frame Y is that the Christoffel symbols ${}_y\begin{Bmatrix} k \\ ij \end{Bmatrix}$ vanish identically.*

From this theorem we can deduce at once a system of differential equations that must be satisfied by functions $y^i(x^1, \dots x^n)$, if there is to be a coordinate system Y in which the h_{ij}'s are constants. The law of transformation 32.6 demands that

$$-{}_y\begin{Bmatrix} m \\ \alpha\beta \end{Bmatrix} \frac{\partial y^\alpha}{\partial x^i}\frac{\partial y^\beta}{\partial x^j} = \frac{\partial^2 y^m}{\partial x^i \partial x^j} - {}_x\begin{Bmatrix} \gamma \\ ij \end{Bmatrix} \frac{\partial y^m}{\partial x^\gamma},$$

and, since ${}_y\begin{Bmatrix} m \\ \alpha\beta \end{Bmatrix} = 0$, we have the system of equations

$$\frac{\partial^2 y^m}{\partial x^i \partial x^j} - \begin{Bmatrix} \gamma \\ ij \end{Bmatrix} \frac{\partial y^m}{\partial x^\gamma} = 0, \tag{39.3}$$

in which the symbols $\begin{Bmatrix} \gamma \\ ij \end{Bmatrix}$ are formed from the $g_{ij}(x)$. The system 39.3, of second-order partial differential equations, can be rewritten in an equivalent form as a system of first-order partial differential equations

$$\begin{cases} \dfrac{\partial y}{\partial x^i} = u_i, & (i = 1, 2, \dots, n), \\[2ex] \dfrac{\partial u_i}{\partial x^j} = \begin{Bmatrix} \gamma \\ ij \end{Bmatrix} u_\gamma, & (\gamma = 1, 2, \dots, n). \end{cases} \tag{39.4}$$

This system, in general, will be incompatible, and we now turn to the determination of the necessary and sufficient conditions for the existence of solution of the system 39.4.

In order to phrase these conditions in a symmetric form, we will consider the system

$$(39.5)\quad \frac{\partial f^\alpha}{\partial x^i} = F_i^\alpha(f^1, f^2, \ldots, f^m; x^1, x^2, \ldots, x^n), \qquad \begin{matrix}(\alpha = 1, 2, \ldots, m),\\ (i = 1, 2, \ldots, n),\end{matrix}$$

where the F_i^α are known functions of the f's and x's. Equations 39.5 specialize to (39.4) if we set $f^1 = y, f^2 = u_1, \ldots, f^m = u_n$. The functions F_i^α are defined over the n-dimensional region R and for arbitrary values of the functions f^i, that is, for $-\infty < f^i < \infty$. Let us refer to the region of definition of functions F_i^α as R'. This region consists of the region R of the variables x^i and the set of ranges

$$-\infty < f^i < \infty.$$

We will suppose that the functions F_i^α are of class C^1 in R'. Since the region R' is open, we will assume that the $\partial F_i^\alpha/\partial f^j$ are bounded in R'. The restrictions imposed on the choice of functions F_i^α are clearly satisfied by functions appearing in the right-hand members of equations 39.4.

Since the F_i^α are of class C^1 in R', it follows that the f^α's are of class C^2, and hence

$$(39.6)\qquad \frac{\partial^2 f^\alpha}{\partial x^i\,\partial x^j} = \frac{\partial^2 f^\alpha}{\partial x^j\,\partial x^i}.$$

This is a *necessary condition for the integrability of the system 39.5*. Differentiating equations 39.5 with respect to x^j, we obtain

$$\begin{aligned}\frac{\partial^2 f^\alpha}{\partial x^i\,\partial x^j} &= \frac{\partial F_i^\alpha}{\partial x^j} + \frac{\partial F_i^\alpha}{\partial f^\beta}\frac{\partial f^\beta}{\partial x^j}\\ &= \frac{\partial F_i^\alpha}{\partial x^j} + \frac{\partial F_i^\alpha}{\partial f^\beta} F_j^\beta,\end{aligned}$$

where the last step results from the substitution of the expression for $\partial f^\beta/\partial x^j$ from (39.5). Now, if we form (39.6), we get as a necessary condition for integrability the set of equations

$$(39.7)\quad \frac{\partial F_i^\alpha}{\partial x^j} + \frac{\partial F_i^\alpha}{\partial f^\beta} F_j^\beta = \frac{\partial F_j^\alpha}{\partial x^i} + \frac{\partial F_j^\alpha}{\partial f^\beta} F_i^\beta, \qquad \begin{cases}(\alpha, \beta = 1, \ldots, m),\\ (i, j = 1, \ldots, n).\end{cases}$$

We see that if the system 39.5 has a solution, then either (39.7) are identities in f^α and x^i or else there are certain functional relations existing between the f's and x's. If (39.7) are identities, the system of equations 39.5 is said to be *completely integrable*. It is then possible to prove that the integrability conditions (39.7) are not only necessary but also *sufficient* to guarantee the existence of solutions of the system 39.5.

There are several proofs of the existence of solution of complete systems of partial differential equations; perhaps the simplest of these was given by T. Y. Thomas in 1934 in a paper entitled "Systems of Total Differential Equations Defined over Simply Connected Domains," *Annals of Mathematics*, **35**, 730–734 (1934). An earlier proof, assuming the analyticity of functions F_i^α, was given by Bouquet[11] in 1872, and there are other proofs by G. Darboux and E. Cartan. We shall not go into a discussion of the sufficiency of conditions 39.7, but will merely state an

EXISTENCE THEOREM. *Let R be an open n-dimensional simply connected region referred to the X-system of coordinates, and R′ the region composed of R and the ranges* $-\infty < f^i < \infty$. *If the functions* $F_\alpha^{\,i}(x, f)$ *are of class* C^1 *in R′ and have bounded derivatives* $\partial F_\alpha^{\,i}/\partial f^j$ *in R′, and if furthermore the integrabiltiy conditions* 39.7 *are satisfied identically, then the system* 39.5 *has one and only one set of solutions*

$$f^\alpha(x^1, \ldots, x^n), \qquad (\alpha = 1, \ldots, m),$$

which for an arbitrary set of values $(x_0^{\,1}, \ldots, x_0^{\,n})$ *take on the arbitrarily prescribed values* $C^\alpha = f^\alpha(x_0^{\,1}, \ldots, x_0^{\,n})$.

We will now apply these results to the special case of the system 39.4 by identifying it with (39.5).

The dependent variables in (39.4) are $y, u_1, \ldots, u_n$, whereas in (39.5) they are $f^1, f^2, \ldots, f^m$. Thus we set

$$f^1 = y, \qquad f^2 = u_1, \ldots, f^{n+1} = u_n,$$

and the system 39.4 reads

$$\frac{\partial f^1}{\partial x^i} = F_i^1 = u_i, \qquad (i = 1, 2, \ldots, n),$$

and

$$\frac{\partial f^\alpha}{\partial x^i} = F_i^\alpha = \begin{Bmatrix} \gamma \\ \alpha - 1 \quad i \end{Bmatrix} u_\gamma, \qquad \begin{matrix} (\alpha = 2, 3, \ldots, n + 1), \\ (i, \gamma = 1, 2, \ldots, n). \end{matrix}$$

The substitution of the expressions for F_i^α in the integrability conditions 39.7 gives

$$(39.8) \qquad \begin{cases} \begin{Bmatrix} \gamma \\ ij \end{Bmatrix} u_\gamma = \begin{Bmatrix} \gamma \\ ji \end{Bmatrix} u_\gamma \\ R^\gamma_{kij} u_\gamma = 0. \end{cases}$$

The first of these sets of equations is satisfied identically because of the symmetry of Christoffel symbols. The second set states that the set of

[11] J. C. Bouquet, *Bull. Sci. Math. et Astron.*, **3**, (1872) p. 265, G. Darboux, *Leçons sur les systèmes othogonaux*, (1910) pp. 326–335, E. Cartan, *Géometrié des espaces de Riemann*, (1928) pp. 54–57. The proof by T. Y. Thomas is quite close in spirit to that given by Cartan.

equations 39.4 will have a solution if the Riemann-Christoffel tensor R^i_{jkl} vanishes identically. Since this tensor vanishes when metric coefficients are constants, we can enunciate a basic

THEOREM II. *A necessary and sufficient condition that a symmetric tensor g_{ij} with $|g_{ij}| \neq 0$, reduce under a suitable transformation of coordinates to a tensor h_{ij}, where the h_{ij}'s are constants, is that the Riemann-Christoffel tensor formed from the g_{ij}'s be a zero tensor.*

We note further that, if the quadratic form $Q = h_{ij}y^iy^j$ is positive definite, there exists a nonsingular *linear* transformation reducing Q to the canonical form $Q = (y^1)^2 \cdots + (y^n)^2$. Thus, if the $g_{ij}(x)$ are the coefficients in the positive definite quadratic differential form

$$ds^2 = g_{ij}\, dx^i\, dx^j, \tag{39.1}$$

characterizing metric properties of R_n, there exists a real functional transformation T: $y^i = y^i(x)$ which reduces it to the form

$$ds^2 = (dy^1)^2 + \cdots + (dy^n)^2, \tag{39.9}$$

provided that R^i_{jkl} vanishes identically in R_n.

We recall that a metric manifold R_n in which it is possible to effect the reduction of the form 39.1 to 39.9 is called an *Euclidean n-dimensional manifold E_n, and we see that R_n is Euclidean if, and only if, the Riemann tensor of the manifold is a zero tensor.*

Problems

1. Verify the substitutions in the integrability conditions 39.7 leading to equations 39.8.

2. Referring to the system 39.5, show that it is completely equivalent to the system of total differential equations

$$df^\alpha = F^\alpha_i\, dx^i.$$

3. What are the integrability conditions for the equation

$$P(x, y, z)\, dx + Q(x, y, z)\, dy + R(x, y, z)\, dz = 0?$$

Consider also the system

$$\frac{\partial F}{\partial x} = P, \qquad \frac{\partial F}{\partial y} = Q, \qquad \frac{\partial F}{\partial z} = R.$$

4. Prove a theorem: If $P\, dx + Q\, dy + R\, dz = 0$ is integrable, then

$$\lambda P\, dx + \lambda Q\, dy + \lambda R\, dz = 0$$

is also integrable for any $\lambda(x, y, z)$ of class C^1.

5. Deduce the integrability conditions for the equation

$$P_i(x^1, \ldots, x^n)\, dx^i = 0, \qquad (i = 1, \ldots, n).$$

40. The e-Systems and the Generalized Kronecker Deltas

The notions of symmetry and skew-symmetry with respect to pairs of indices (see Sec. 27) can be extended to cover the sets of quantities that are symmetric or skew-symmetric with respect to more than two indices. We will consider in this section the sets of quantities $A^{i_1 \cdots i_k}$ or $A_{i_1 \cdots i_k}$, depending on k indices, written as subscripts or superscripts, although the quantities A may not represent tensors.

DEFINITION 1. *The system of quantities $A^{i_1 \cdots i_k}$ (or $A_{i_1 \cdots i_k}$), depending on k indices, is said to be completely symmetric if the value of the symbol A is unchanged by any permutation of the indices.*

DEFINITION 2. *The system $A^{i_1 \cdots i_k}$ (or $A_{i_1 \cdots i_k}$), depending on k indices, is said to be completely skew-symmetric if the value of the symbol A is unchanged by any even permutation of the indices, and A merely changes sign after an odd permutation of the indices.*

We recall that any permutation of n distinct objects, say a permutation of n distinct integers, can be accomplished by a finite number of interchanges of pairs of these objects and that the number of interchanges required to bring about a given permutation from a prescribed order is always even or always odd.

It follows at once from definition 2 that in any skew-symmetric system the term containing two like indices is necessarily zero. Thus, if one has a skew-symmetric system of quantities A_{ijk}, where i, j, k assume values $1, 2, 3$, then $A_{122} = 0$, $A_{123} = -A_{213}$, $A_{312} = A_{123}$, etc. In general, the components A_{ijk} of a skew-symmetric system satisfy the relations $A_{ijk} = -A_{ikj} = -A_{jik} = A_{jki} = A_{kij} = -A_{kji}$.

Consider now a skew-symmetric system of quantities $A_{i_1 \cdots i_n}$ (or $A^{i_1 \cdots i_n}$), in which the indices $i_1, \ldots, i_n$ assume values $1, 2, \ldots, n$. We define the e-system as follows.

DEFINITION 3. *If the value of $A_{i_1 \cdots i_n}$ (or $A^{i_1 \cdots i_n}$) is $+1$ when $i_1 i_2 \cdots i_n$ is an even permutation of the numbers $12 \cdots n$, and -1 when $i_1 i_2 \cdots i_n$ is an odd permutation of $12 \cdots n$, and if it is zero in all other cases, then the system $A_{i_1 \cdots i_n}$ (or $A^{i_1 \cdots i_n}$) is called the e-system.*

We shall use the symbols $e_{i_1 \cdots i_n}$ or $e^{i_1 \cdots i_n}$ to denote the e-systems. It will be shown in Sec. 41 that the e-systems are relative tensors.

As an illustration we note that the components of the system e_{ij} are: $e_{11} = 0$, $e_{12} = 1$, $e_{21} = -1$, $e_{22} = 0$. If the e-system depends on three indices ijk, then $e_{ijk} = 0$ if any two indices are alike, whereas $e_{ijk} = e_{123} = 1$ if ijk is an even permutation of 123 and $e_{ijk} = -e_{123} = -1$ if ijk is an odd permutation of 123.

Closely allied to the e-systems are the *generalized Kronecker deltas*, which we proceed to define.

DEFINITION 4. *A symbol $\delta^{i_1 \cdots i_k}_{j_1 \cdots j_k}$ depending on k superscripts and k subscripts, each of which runs from* 1 *to n, is called a generalized Kronecker delta provided that:* (*a*) *it is completely skew-symmetric in superscripts and subscripts;* (*b*) *if the superscripts are distinct from each other and the subscripts are the same set of numbers as the superscripts, the value of the symbol is* $+1$ *or* -1 *according as an even or odd number of transpositions is required to arrange the superscripts in the same order as the subscripts;* (*c*) *in all other cases the value of the symbol is zero.*

As an illustration consider δ^{ij}_{kl}. It follows from definition 4 that if $i = j$ or $k = l$, or if the set ij is not the set kl, then $\delta^{ij}_{kl} = 0$. In all other cases δ^{ij}_{kl} equals $+1$ or -1 according to whether kl is an even or an odd permutation of ij. Thus

$$
\begin{aligned}
0 &= \delta^{11}_{ij} = \delta^{22}_{ij} = \delta^{12}_{13} = \cdots, \\
1 &= \delta^{12}_{12} = \delta^{13}_{13} = \delta^{21}_{21} = \cdots, \\
-1 &= \delta^{12}_{21} = \delta^{13}_{31} = \delta^{21}_{12} = \cdots.
\end{aligned}
$$

We prove in Sec. 41 that the generalized Kronecker deltas are tensors.

From definition 3, it follows that the direct product $e^{i_1 i_2 \cdots i_n} e_{j_1 j_2 \cdots j_n}$ of the two systems $e^{i_1 \cdots i_n}$ and $e_{j_1 \cdots j_n}$ is the generalized Kronecker delta. For example, $e^{\alpha\beta\gamma} e_{ijk}$ has the following values:

(*a*) Zero, if two or more subscripts or superscripts are alike.

(*b*) $+1$, if the difference in the number of transpositions of $\alpha\beta\gamma$ and ijk from 123 is an even number.

(*c*) -1, if the difference in the number of transpositions of $\alpha\beta\gamma$ and ijk from 123 is an odd number.

A little reflection will show that another way of phrasing statements (*b*) and (*c*) is the following:

(*b'*) $e^{\alpha\beta\gamma} e_{ijk} = +1$, if an even number of transposition is required to arrange the subscripts in the same order as the superscripts.

(*c'*) $e^{\alpha\beta\gamma} e_{ijk} = -1$, if an odd number of transpositions is required to arrange the subscripts in the same order as the superscripts.

We can thus write

$$e^{\alpha\beta\gamma} e_{ijk} = \delta^{\alpha\beta\gamma}_{ijk}.$$

It is clear from definitions 3 and 4 that the e-symbols can be defined in terms of the Kronecker deltas,

$$e^{i_1 i_2 \cdots i_n} = \delta^{i_1 i_2 \cdots i_n}_{12 \cdots n} \qquad \text{and} \qquad \delta^{12 \cdots n}_{i_1 i_2 \cdots i_n} = e_{i_1 i_2 \cdots i_n},$$

since $e = +1$ or -1 when the set of distinct integers $i_1 i_2 \cdots i_n$ is obtained from the set $12 \cdots n$, by an even or an odd permutation, and $e = 0$ in all other cases. The e-systems and generalized Kronecker deltas prove useful in calculations involving alternating sets of quantities.

We consider next several examples which permit us to deduce a number of identities involving operations on these symbols.

Let us contract $\delta^{ijk}_{\alpha\beta\gamma}$ on k and γ. The result for $n = 3$ is

$$\delta^{ijk}_{\alpha\beta k} = \delta^{ij1}_{\alpha\beta 1} + \delta^{ij2}_{\alpha\beta 2} + \delta^{ij3}_{\alpha\beta 3} \equiv \delta^{ij}_{\alpha\beta}.$$

We observe that this expression vanishes if i and j are equal or if α and β are equal. If we set $i = 1$ and $j = 2$, we get $\delta^{123}_{\alpha\beta 3} = \delta^{12}_{\alpha\beta}$, and hence $\delta^{12}_{\alpha\beta} = 0$, unless $\alpha\beta$ is a permutation of 12. In the latter case $\delta^{12}_{\alpha\beta} = 1$ if $\alpha\beta$ is an even permutation of 12, and $\delta^{12}_{\alpha\beta} = -1$ for an odd permutation. Similar results hold for all values of i and j selected from the set of numbers 1, 2, 3. We thus see that $\delta^{ij}_{\alpha\beta}$ is equal to

(*a*) 0, if two of the subscripts or superscripts are alike, or when the subscripts and superscripts are not formed from the same numbers.

(*b*) +1, if ij is an even permutation of $\alpha\beta$.

(*c*) −1, if ij is an odd permutation of $\alpha\beta$.

If we contract $\delta^{ij}_{\alpha\beta}$ and halve the result, we obtain a system depending on two indices

$$\delta^{i}_{\alpha} \equiv \tfrac{1}{2}\delta^{ij}_{\alpha j} = \tfrac{1}{2}(\delta^{i1}_{\alpha 1} + \delta^{i2}_{\alpha 2} + \delta^{i3}_{\alpha 3}).$$

If we set $i = 1$ in δ^{i}_{α}, we get $\delta^{1}_{\alpha} = \tfrac{1}{2}(\delta^{12}_{\alpha 2} + \delta^{13}_{\alpha 3})$. This vanishes unless $\alpha = 1$, in which event $\delta^{1}_{1} = 1$. Similar results can be obtained by setting $i = 2$ or $i = 3$. Thus δ^{i}_{α} has the values

(*a*) 0 if $i \neq \alpha$, $(\alpha, i = 1, 2, 3)$.

(*b*) 1, if $i = \alpha$.

By counting the number of terms appearing in the sums it is not difficult to show that, in general,

$$\delta^{i}_{\alpha} = \frac{1}{n-1}\,\delta^{ij}_{\alpha j} \quad \text{and} \quad \delta^{ij}_{ij} = n(n-1). \tag{40.1}$$

We can also deduce that

$$\delta^{i_1 i_2 \cdots i_r}_{j_1 j_2 \cdots j_r} = \frac{(n-k)!}{(n-r)!}\,\delta^{i_1 i_2 \cdots i_r i_{r+1} \cdots i_k}_{j_1 j_2 \cdots j_r i_{r+1} \cdots i_k}, \tag{40.2}$$

and

$$\delta^{i_1 i_2 \cdots i_r}_{i_1 i_2 \cdots i_r} = n(n-1)(n-2)\cdots(n-r+1) = \frac{n!}{(n-r)!}. \tag{40.3}$$

As a special case of (40.3) we have the formula

$$e^{i_1 i_2 \cdots i_n} e_{i_1 i_2 \cdots i_n} = n! \tag{40.4}$$

and from (40.2) we deduce the relation

$$(40.5) \qquad e^{i_1 \cdots i_r i_{r+1} \cdots i_n} e_{j_1 \cdots j_r i_{r+1} \cdots i_n} = (n-r)! \, \delta^{i_1 \cdots i_r}_{j_1 \cdots j_r}.$$

Consider next a set of n^{p+q} quantities $A^{i_1 \cdots i_p}_{j_1 \cdots j_q}$ (the i's and j's run from 1 to n), symmetric in two or more indices (which may be superscripts or subscripts). We can show that

$$\delta^{j_1 j_2 \cdots j_q}_{i_1 i_2 \cdots i_q} A^{r_1 r_2 \cdots r_p}_{j_1 j_2 \cdots j_q} = 0,$$

if $A^{i_1 \cdots i_p}_{j_1 \cdots j_q}$ is symmetric in two or more subscripts. Also

$$\delta^{s_1 s_2 \cdots s_p}_{i_1 i_2 \cdots i_p} A^{i_1 \cdots i_p}_{j_1 \cdots j_q} = 0,$$

if $A^{i_1 \cdots i_p}_{j_1 \cdots j_q}$ is symmetric in two or more superscripts.

Suppose that $A^{r_1 \cdots r_p}_{j_1 \cdots j_q}$ is symmetric in j_1 and j_2; then

$$\begin{aligned} \delta^{j_1 j_2 \cdots j_q}_{i_1 i_2 \cdots i_q} A^{r_1 \cdots r_p}_{j_1 j_2 \cdots j_q} &= \delta^{j_1 j_2 \cdots j_q}_{i_1 i_2 \cdots i_q} A^{r_1 \cdots r_p}_{j_2 j_1 \cdots j_q} \\ &= -\delta^{j_2 j_1 \cdots j_q}_{i_1 i_2 \cdots i_q} A^{r_1 \cdots r_p}_{j_2 j_1 \cdots j_q}. \end{aligned}$$

However, j_1 and j_2 are the dummy indices; hence

$$\delta^{j_1 j_2 \cdots j_q}_{i_1 i_2 \cdots i_q} A^{r_1 \cdots r_p}_{j_1 j_2 \cdots j_q} = -\delta^{j_1 j_2 \cdots j_q}_{i_1 i_2 \cdots i_q} A^{r_1 \cdots r_p}_{j_1 j_2 \cdots j_q}.$$

Thus

$$\delta^{j_1 j_2 \cdots j_q}_{i_1 i_2 \cdots i_q} A^{r_1 \cdots r_p}_{j_1 j_2 \cdots j_q} = 0.$$

Problems

1. (*a*) Show that $\delta^{ijk}_{ijk} = 3!$ if $i, j, k = 1, 2, 3$.

(*b*) Show that $\delta^{ij}_{\alpha\beta} = \begin{vmatrix} \delta^i_\alpha & \delta^i_\beta \\ \delta^j_\alpha & \delta^j_\beta \end{vmatrix}$ and $\delta^{ijk}_{\alpha\beta\gamma} = \begin{vmatrix} \delta^i_\alpha & \delta^i_\beta & \delta^i_\gamma \\ \delta^j_\alpha & \delta^j_\beta & \delta^j_\gamma \\ \delta^k_\alpha & \delta^k_\beta & \delta^k_\gamma \end{vmatrix}$.

2. Expand for $n = 3$:

(*a*) $\delta^i_\alpha \delta^\alpha_j$. (*b*) $\delta^{12}_{ij} x^i y^j$. (*c*) $\delta^{13}_{ij} x^i y^j$.

(*d*) $\delta^{\alpha\beta}_{ij} x^i y^j$. (*e*) δ^{ij}_{ij}.

3. Expand for $n = 2$:

(*a*) $e^{ij} a^1_i a^2_j$. (*b*) $e^{ij} a^2_i a^1_j$. (*c*) $e^{\alpha\beta} a^i_\alpha a^j_\beta = e^{ij} |a|$.

4. If a set of quantities $A_{i_1 \cdots i_k}$ is skew-symmetric in the subscripts (k in number), then

$$\delta^{i_1 \cdots i_k}_{j_1 \cdots j_k} A_{i_1 \cdots i_k} = k! \, A_{j_1 \cdots j_k}.$$

5. If A_{ijk} is completely symmetric and the indices run from 1 to n, show that the number of distinct terms in the set $\{A_{ijk}\}$ is

$$N = n + n(n-1) + \frac{n(n-1)(n-2)}{3!}.$$

Hint: Consider the cases where the subscripts ijk are all alike, when only two are distinct, and when all are distinct.

6. Show that the number of distinct, nonvanishing A_{ijk}'s in Problem 5 is $\frac{n(n-1)(n-2)}{3!}$ when A_{ijk} is completely skew-symmetric.

41. Application of the *e*-Systems to Determinants. Tensor Character of Generalized Kronecker Deltas

We recall that the determinant $|a^i_j|$ of nth order, with elements a^i_j, consists of the sum of products of the elements where each term in the sum contains one and only one element from each row and each column of the determinant. The sign of each term in the sum is determined by the character of permutation of the indices. Thus, if the superscripts in the product $a^1_{i_1} a^2_{i_2} \cdots a^n_{i_n}$ are arranged in the normal order $12 \cdots n$, then the product will carry the plus sign if the number of transpositions necessary to arrange the subscripts in the normal order is even. The sign is minus if the required number of transpositions is odd. Since $e^{i_1 \cdots i_n} = \delta^{i_1 i_2 \cdots i_n}_{1\,2 \cdots n}$ and $\delta^{1\,2 \cdots n}_{i_1 i_2 \cdots i_n} = e_{i_1 i_2 \cdots i_n}$, the determinant

$$|a^i_j| = \begin{vmatrix} a^1_1 & a^1_2 & \cdots & a^1_n \\ a^2_1 & a^2_2 & \cdots & a^2_n \\ \cdot & \cdot & \cdot & \cdot \\ a^n_1 & a^n_2 & \cdots & a^n_n \end{vmatrix} \equiv a \tag{41.1}$$

can be written compactly as

$$\begin{aligned} a &= e^{i_1 i_2 \cdots i_n} a^1_{i_1} a^2_{i_2} \cdots a^n_{i_n} \\ &= e_{i_1 i_2 \cdots i_n} a^{i_1}_1 a^{i_2}_2 \cdots a^{i_n}_n . \end{aligned} \tag{41.2}$$

As an example consider

$$|a^i_j| = \begin{vmatrix} a^1_1 & a^1_2 & a^1_3 \\ a^2_1 & a^2_2 & a^2_3 \\ a^3_1 & a^3_2 & a^3_3 \end{vmatrix} \equiv a.$$

If this determinant is expanded by columns we get $a = \Sigma \pm a^i_1 a^j_2 a^k_3$, where ijk is a permutation of 123. The plus or minus sign is assigned to the term $a^i_1 a^j_2 a^k_3$ according to whether this permutation is even or odd. Hence this determinant can be written $|a^i_j| = e_{ijk} a^i_1 a^j_2 a^k_3$. On the other hand, if it is expanded by rows, we can write $|a^i_j| = e^{ijk} a^1_i a^2_j a^3_k$.

Consider next the sum

$$e_{ijk} a^i_\alpha a^j_\beta a^k_\gamma, \qquad (i, j, k, \alpha, \beta, \gamma = 1, 2, 3).$$

We will show first that this system is completely skew-symmetric in $\alpha\beta\gamma$. Since the indices ijk are dummy indices, we can change them at will and write

$$e_{ijk}a^i_\alpha a^j_\beta a^k_\gamma = e_{kji}a^k_\alpha a^j_\beta a^i_\gamma = e_{kji}a^i_\gamma a^j_\beta a^k_\alpha.$$

If k and i are interchanged in e_{kji}, this e-symbol will change sign, and hence

$$e_{ijk}a^i_\alpha a^j_\beta a^k_\gamma = -e_{ijk}a^i_\gamma a^j_\beta a^k_\alpha.$$

This shows that an interchange of α and γ changes the sign, so that the system under consideration is skew-symmetric in α and γ. Similar results obviously hold for other indices. A special case of this system is the determinant $|a^i_j| = e_{ijk}a^i_1 a^j_2 a^k_3$, and it follows from the foregoing that

$$e_{ijk}a^i_\alpha a^j_\beta a^k_\gamma = |a^i_j|\, e_{\alpha\beta\gamma}.$$

Similarly, we can show that

$$e^{ijk}a^\alpha_i a^\beta_j a^\gamma_k = |a^i_j|\, e^{\alpha\beta\gamma}.$$

It follows at once from these expressions that an interchange of two columns (or two rows) of the determinant $|a_j{}^i|$ changes its sign, and if two columns in it are identical, then its value is zero.

These results can be immediatley generalized to determinants of nth order, so that for any permutation of rows we can write

$$e^{\alpha\beta\cdots\gamma}\,|a^i_j| = e^{ij\cdots k}a^\alpha_i a^\beta_j \cdots a^\gamma_k, \tag{41.3}$$

and for any permutation of columns

$$e_{ij\cdots k}|a^i_j| = e_{\alpha\beta\cdots\gamma}a^\alpha_i a^\beta_j \cdots a^\gamma_k. \tag{41.4}$$

We use formula 41.4 to establish the formula for the product of two determinants. The power and compactness of this notation are strikingly demonstrated in this derivation.

Since $|b^i_j| = e_{ij\cdots k}b^i_1 b^j_2 \cdots b^k_n$, we can write

$$\begin{aligned} |a^i_j| \cdot |b^i_j| &= |a^i_j|\, e_{ij\cdots k}b^i_1 b^j_2 \cdots b^k_n \\ &= (e_{\alpha\beta\cdots\gamma}a^\alpha_i a^\beta_j \cdots a^\gamma_k)(b^i_1 b^j_2 \cdots b^k_n), \end{aligned}$$

where we have made use of the formula 41.4. Thus

$$\begin{aligned} |a^i_j| \cdot |b^i_j| &= e_{\alpha\beta\cdots\gamma}(a^\alpha_i b^i_1)(a^\beta_j b^j_2) \cdots (a^\gamma_k b^k_n) \\ &= |c^i_j|, \end{aligned}$$

where

$$c^i_j = a^i_\alpha b^\alpha_j = a^i_1 b^1_j + a^i_2 b^2_j + \cdots + a^i_n b^n_j.$$

The expansion of the determinant in terms of the elements of the first column and their cofactors can be written

$$|a_j^i| = a_1^{i_1} e_{i_1 i_2 \cdots i_n} a_2^{i_2} \cdots a_n^{i_n} \tag{41.5}$$
$$= a_1^{\alpha} A_{\alpha}^1,$$

where $A_{\alpha}^1 = e_{\alpha i_2 \cdots i_n} a_2^{i_2} a_3^{i_3} \cdots a_n^{i_n}$ is the cofactor of the element a_1^{α}.

We derive next the formula for the partial derivatives of a determinant whose elements $a_j^{\ i}$ are functions of the variables $x^1, x^2, \ldots, x^n$. From formula 41.2 we have

$$a = e_{i_1 i_2 \cdots i_n} a_1^{i_1} a_2^{i_2} \cdots a_n^{i_n}.$$

Differentiating this expression, we get

$$\frac{\partial a}{\partial x^j} = e_{i_1 i_2 \cdots i_n} \left(\frac{\partial a_1^{i_1}}{\partial x^j} a_2^{i_2} \cdots a_n^{i_n} + a_1^{i_1} \frac{\partial a_2^{i_2}}{\partial x^j} \cdots a_n^{i_n} + \cdots + a_1^{i_1} a_2^{i_2} \cdots \frac{\partial a_n^{i_n}}{\partial x^j} \right)$$

$$= \frac{\partial a_1^{i_1}}{\partial x^j} A_{i_1}^1 + \frac{\partial a_2^{i_2}}{\partial x^j} A_{i_2}^2 + \cdots + \frac{\partial a_n^{i_n}}{\partial x^j} A_{i_n}^n$$

$$= \frac{\partial a_{\beta}^{\alpha}}{\partial x^j} A_{\alpha}^{\beta}$$

by formula of the type 41.5.

Formulas 41.3 and 41.4 permit us to establish the fact that *the permutation symbols* $e^{i_1 \cdots i_n}$ *and* $e_{i_1 \cdots i_n}$ *are relative tensors of weights* $+1$ and -1, respectively.

Consider an admissible transformation

$$T\colon\ y^i = y^i(x^1, \ldots, x^n),$$

and its Jacobian $J = \left|\dfrac{\partial y}{\partial x}\right|$. If we set $a_j^i = \dfrac{\partial y^i}{\partial x^j}$ in formula 41.3, and recall that $\dfrac{1}{J} = \left|\dfrac{\partial x}{\partial y}\right|$, we obtain at once

$$e^{i_1 \cdots i_n} = \left|\frac{\partial x}{\partial y}\right| e^{\alpha_1 \cdots \alpha_n} \frac{\partial y^{i_1}}{\partial x^{\alpha_1}} \cdots \frac{\partial y^{i_n}}{\partial x^{\alpha_n}}, \tag{41.6}$$

which is the law of transformation of relative contravariant tensors of weight $+1$. In an entirely similar way we deduce that

$$e_{i_1 i_2 \cdots i_n} = \left|\frac{\partial x}{\partial y}\right|^{-1} e_{\alpha_1 \alpha_2 \cdots \alpha_n} \frac{\partial x^{\alpha_1}}{\partial y^{i_1}} \frac{\partial x^{\alpha_2}}{\partial y^{i_2}} \cdots \frac{\partial x^{\alpha_n}}{\partial y^{i_n}}, \tag{41.7}$$

so that $e_{i_1 i_2 \cdots i_n}$ is a relative tensor of weight -1.

From formula 40.5,

$$e^{i_1 \cdots i_r i_{r+1} \cdots i_n} e_{j_1 \cdots j_r i_{r+1} \cdots i_n} = (n - r)!\, \delta^{i_1 \cdots i_r}_{j_1 \cdots j_r},$$

we see that the Kronecker delta $\delta^{i_1 \cdots i_r}_{j_1 \cdots j_r}$ is obtained by multiplying together two e-symbols, one of which is a relative tensor of weight $+1$ and the other of weight -1, and contracting with respect to a number of indices. The result is a tensor of weight zero, that is, an ordinary tensor. Thus we have proved that *the generalized Kronecker deltas are absolute tensors.*

Since $\delta^{j_1 \cdots j_q}_{i_1 \cdots i_q, s}$ reduces to $\dfrac{\partial \delta^{j_1 \cdots j_q}_{i_1 \cdots i_q}}{\partial x^s} = 0$ when the coordinate system X is cartesian, we conclude that the *covariant derivatives of generalized Kronecker deltas vanish identically.* Thus the Kronecker deltas behave as constants in a covariant differentiation.

Problems

1. Verify that $\delta^{ij}_{\alpha\beta} a^{\alpha\beta} = a^{ij} - a^{ji}$.

2. Verify that $\delta^{ijk}_{\alpha\beta\gamma} a^{\alpha\beta\gamma} = a^{ijk} - a^{ikj} + a^{jki} - a^{jik} + a^{kij} - a^{kji}$.

3. If a_{ij} satisfies the equation

$$b a_{ij} + c a_{ji} = 0,$$

then either $b = -c$ and a_{ij} is symmetric, or $b = c$ and a_{ij} is skew-symmetric. *Hint:* Since i and j take on values $1 \cdots n$, the equation can be written as

$$b a_{ji} + c a_{ij} = 0.$$

Add and obtain $(b + c)(a_{ij} + a_{ji}) = 0$.

4. Show that (a) $e_{\alpha j} e^{\alpha i} = \delta^i_j$, (b) $e_{ijk} e^{irs} = \delta^r_j \delta^s_k - \delta^r_k \delta^s_j$. *Hint:* The left-hand member in (b) is zero unless j and k are distinct and j, k is a permutation of r, s. If $j = r$ and $k = s$, then the left-hand member is $+1$; if $j = s$ and $k = r$, then its value is -1. Consider now the value of the right-hand member for the same choices of indices.

5. List the values of δ^i_i, δ^{ij}_{ji}, $\delta^i_j \delta^j_k \delta^k_i$ when the indices range from 1 to n.

3

GEOMETRY

42. Non-Euclidean Geometries

There is no branch of mathematics in which the tyranny of authority has been felt more strongly than in geometry. The traditional Euclidean geometry, based on a set of "self-evident truths" and created largely by the Alexandrian School of mathematicians (around 300 B.C.), dominated the thought and shaped the development of physics and astronomy for over 2000 years. There were a few bold souls, even among the ancient mathematicians, to whom "self-evident truths" contained in Euclid's axioms did not seem convincing, but the prestige of logical structure of Euclid's *Elements* was so high and the hand of authority so heavy that they hindered the development of mathematics for centuries.

In 1621, Sir Henry Savile raised some questions concerning what he called "two blemishes" in geometry, the theory of proportion and the theory of parallels. Euclid's axiom of parallels (Postulate V in the first book of *Elements*) is to the effect that any two given lines in a plane, when produced indefinitely, will intersect if the sum of two interior angles made by a transversal with these lines is less than two right angles. The fact that some of Euclid's propositions, dealing essentially with the converse of this postulate, can be proved without invoking Postulate V gave hope that the postulate itself might be deduced from his other axioms. However, all attempts to prove the fifth postulate proved unsuccessful, and a hope that contradictions would emerge if this postulate were abrogated while others were retained led nowhere. In 1826, a Russian mathematician, Nicolai Lobachevski, presented to the mathematicians faculty of the University of Kazan a paper based on an assumption that it is possible to draw through any point in the plane two lines parallel to a given line. The geometry developed by Lobachevski proved just as devoid of inner inconsistencies as Euclidean geometry. Indeed, it contained the latter as a special case and implied the arbitrariness of the concept of length adopted in Euclidean geometry.

In 1831, a Hungarian mathematician, John Bolyai, published results of his independent investigations which conceptually differ little from those of Lobachevski, but which perhaps contain a deeper appreciation of the metric properties of space. Bolyai pointed out, just as Lobachevski did, that his geometry in the small is approximately Euclidean and that only a physical experiment can decide whether Euclidean or non-Euclidean geometry should be adopted for the purposes of physical measurement. Thus it appears that there are no *a priori* reasons for preferring one geometry to another. However, it was only after Riemann's profound dissertation on the hypotheses underlying the foundations of geometry appeared in print (published posthumously in 1867) that the mathematical world recognized fully the role played by the metric concepts in geometry.

Riemann appears to have been unaware of the work of Lobachevski and Bolyai, although it was well known to Gauss. Later Beltrami published his classical paper on the interpretation of non-Euclidean geometries (1868) in which he analyzed the work of Lobachevski, Bolyai, and Riemann and stressed the fact that the metric properties of space are mere definitions. From these researches it appeared that three consistent geometries are possible on surfaces of constant curvature: the Lobachevskian, on a surface of constant negative curvature; the Riemannian, on a surface of constant positive curvature; and the Euclidean, on a surface of zero curvature. These geometries are also called hyperbolic, elliptic, and parabolic, respectively. We consider them briefly in the next section.

43. Length of Arc

Let the n-dimensional space R be covered by a coordinate system X, and consider a one-dimensional subspace of R determined by

$$C\colon\quad x^i = x^i(t), \qquad (i = 1, \ldots, n), \tag{43.1}$$

where t is a real parameter varying continuously in the interval $t_1 \leq t \leq t_2$. The one-dimensional manifold C is called an *arc of a curve*. In this book we deal only with those curves for which $x^i(t)$ and $\dot{x}^i(t) \equiv dx^i/dt$ are continuous functions in $t_1 \leq t \leq t_2$. The definition of the arc of a curve given here is a direct generalization of the parametric representation of curves of elementary analytic geometry.

Let $F(x^1, \ldots, x^n, \dot{x}^1, \ldots, \dot{x}^n)$, viewed as a function of t, be a prescribed continuous function in the interval $t_1 \leq t \leq t_2$. We suppose that[1] $F(x, \dot{x}) > 0$, unless every $\dot{x}^i = 0$, and that for every positive number k

$$F(x^1, \ldots, x^n, k\dot{x}^1, \ldots, k\dot{x}^n) = kF(x^1, \ldots, x^n, \dot{x}^1, \ldots, \dot{x}^n).$$

The integral

$$s = \int_{t_1}^{t_2} F(x, \dot{x})\, dt \tag{43.2}$$

is called the length of C; and the space R is said to be *metrized by formula* 43.2.

Different choices of functions $F(x, \dot{x})$ lead to different metric geometries. If one chooses to define the length of arc by the formula

$$s = \int_{t_1}^{t_2} \sqrt{g_{\alpha\beta}(x)\frac{dx^\alpha}{dt}\frac{dx^\beta}{dt}}\, dt, \qquad (\alpha, \beta = 1, \ldots, n), \tag{43.3}$$

where $g_{\alpha\beta}\dot{x}^\alpha\dot{x}^\beta$ is a positive definite quadratic form in the variables $\dot{x}^\alpha$, then the resulting geometry is the *Riemannian geometry*, and the space R metrized in this way is the *Riemannian n-dimensional space* R_n.

[1] A function $F(x, \dot{x})$ satisfying the condition $F(x, k\dot{x}) = kF(x, \dot{x})$ for every $k > 0$ is called *positively homogeneous of degree* 1 *in the* $\dot{x}^i$. This condition is both necessary and sufficient to ensure the independence of the value of the integral 43.2 of a particular mode of parametrization of C. Thus, if t in (43.1) is replaced by some function $t = \phi(s)$, and we denote $x^i[\phi(s)]$ by $\xi^i(s)$ so that $x^i(t) = \xi^i(s)$ we have the equality

$$\int_{t_1}^{t_2} F(x, \dot{x})\, dt = \int_{s_1}^{s_2} F(\xi, \xi')\, ds,$$

where $\xi'^i(s) = dx^i/ds$ and $t_1 = \phi(s_1)$ and $t_2 = \phi(s_2)$.

To prove this theorem, suppose that k is an arbitrary positive number, and set $t = ks$, so that $t_1 = ks_1$, and $t_2 = ks_2$. Then (43.1) becomes

$$C: \quad x^i(ks) = \xi^i(s)$$

and

$$\xi'^i(s) = \frac{dx^i(ks)}{ds} = k\dot{x}^i(ks).$$

If these values are inserted in (43.2), we get

$$s = \int_{ks_1}^{ks_2} F[x(ks), \dot{x}(ks)]k\, ds,$$

and if this is to equal

$$s = \int_{s_1}^{s_2} F[\xi(s), \xi'(s)]\, ds,$$

we must have the relation $F(\xi, \xi') = F(x, k\dot{x}) = kF(x, \dot{x})$. Conversely, if this relation is true for every line element of C and each $k > 0$, then the equality of integrals is assured for every choice of parameter $t = \phi(s)$, $\phi'(s) > 0$, $s_1 \leq s \leq s_2$, with $t_1 = \phi(s_1)$ and $t_2 = \phi(s_2)$.

We recall from Sec. 39 that, if there exists an admissible transformation of coordinates T: $y^i = y^i(x^1, \ldots, x^n)$, such that the square of the *element of arc ds*,

$$ds^2 = g_{\alpha\beta}\, dx^\alpha\, dx^\beta, \tag{43.4}$$

can be reduced to the form

$$ds^2 = dy^i dy^i, \tag{43.5}$$

then the Riemannian manifold R_n is said to reduce to an n-dimensional *Euclidean manifold* E_n. The reference frame Y in which the element of arc of C in E_n is given by (43.5) is called an orthogonal cartesian reference frame. Obviously, E_n is a generalization of the so-called Euclidean plane determined by the totality of pairs of real values (y^1, y^2). If these values (y^1, y^2) are associated with the points of the plane referred to a pair of orthogonal cartesian axes, then the square of the element of arc ds assumes the familiar form $ds^2 = (dy^1)^2 + (dy^2)^2$.

In what follows we find it convenient to represent pairs of real values (y^1, y^2) as points in a cartesian plane even when the metric of the y^i-manifold is not Euclidean. To illustrate what is meant, consider a sphere S of radius a, immersed in a three-dimensional Euclidean manifold E_3, with center at the origin $(0, 0, 0)$ of the set of orthogonal cartesian axes O-$X^1X^2X^3$. Let T be a plane tangent to S at $(0, 0, -a)$, and let the points of this plane be referred to a set of orthogonal cartesian axes O'-Y^1Y^2 as shown in Fig. 8. If we draw from $O(0, 0, 0)$ a radial

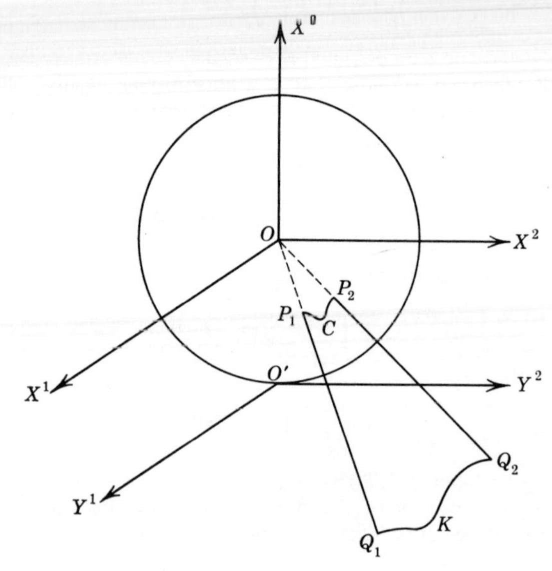

Fig. 8

line OP, intersecting the sphere S at $P(x^1, x^2, x^3)$ and the plane T at $Q(y^1, y^2, -a)$, then the points P on the lower half of the sphere S are in one-to-one correspondence with points (y^1, y^2) of the tangent plane T.

To obtain an explicit analytic form for this correspondence, we note that, if $P(x^1, x^2, x^3)$ is any point on the radial line OP, then the symmetric equations of this line furnish us with the ratios

$$\frac{x^1 - 0}{y^1 - 0} = \frac{x^2 - 0}{y^2 - 0} = \frac{x^3 - 0}{-a - 0} = \lambda,$$

or

$$x^1 = \lambda y^1, \qquad x^2 = \lambda y^2, \qquad x^3 = -\lambda a. \tag{43.6}$$

Since we are concerned with the images Q of points P lying on S, the variables x^i satisfy the equation of S,

$$(x^1)^2 + (x^2)^2 + (x^3)^2 = a^2,$$

or

$$\lambda^2[(y^1)^2 + (y^2)^2 + a^2] = a^2.$$

Solving for λ and substituting in (43.6), we get

$$x^1 = \frac{ay^1}{\sqrt{(y^1)^2 + (y^2)^2 + a^2}}, \qquad x^2 = \frac{ay^2}{\sqrt{(y^1)^2 + (y^2)^2 + a^2}}, \tag{43.7}$$
$$x^3 = \frac{-a^2}{\sqrt{(y^1)^2 + (y^2)^2 + a^2}}.$$

These are the desired equations giving the analytical one-to-one correspondence of the points Q on T and points P on the portion of S under consideration.

Let $P_1(x^1, x^2, x^3)$ and $P_2(x^1 + dx^1, x^2 + dx^2, x^3 + dx^3)$ be two nearby points on some curve C lying on S. The Euclidean distance $\overline{P_1P_2}$, along C, is given by the formula

$$ds^2 = dx^i dx^i, \qquad (i = 1, 2, 3), \tag{43.8}$$

and, since the variables x^i are related to y^i by (43.7),

$$dx^i = \frac{\partial x^i}{\partial y^\alpha} dy^\alpha, \qquad (\alpha = 1,2).$$

Thus (43.8) yields a formula

$$\begin{aligned} ds^2 &= \frac{\partial x^i}{\partial y^\alpha}\frac{\partial x^i}{\partial y^\beta} dy^\alpha\, dy^\beta \\ &= g_{\alpha\beta}(y)\, dy^\alpha\, dy^\beta, \qquad (\alpha, \beta = 1, 2), \end{aligned}$$

where the $g_{\alpha\beta}(y)$ are functions of y^i computed from (43.7) with the aid of the definition $g_{\alpha\beta} = \dfrac{\partial x^i}{\partial y^\alpha}\dfrac{\partial x^i}{\partial y^\beta}$.

If the image K of C on T is given by the equations

$$K: \begin{cases} y^1 = y^1(t), \\ y^2 = y^2(t), \qquad t_1 \le t \le t_2, \end{cases}$$

then the length of C can be computed from the integral

$$s = \int_{t_1}^{t_2} \sqrt{g_{\alpha\beta}\dot{y}^\alpha \dot{y}^\beta}\, dt.$$

A straightforward calculation gives

$$ds^2 = \frac{(dy^1)^2 + (dy^2)^2 + \dfrac{1}{a^2}(y^1\, dy^2 - y^2\, dy^1)^2}{\left\{1 + \dfrac{1}{a^2}[(y^1)^2 + (y^2)^2]\right\}^2} \tag{43.9}$$

and

$$s = \int_{t_1}^{t_2} \frac{\sqrt{(\dot{y}^1)^2 + (\dot{y}^2)^2 + \dfrac{1}{a^2}(y^1\dot{y}^2 - y^2\dot{y}^1)^2}}{1 + \dfrac{1}{a^2}[(y^1)^2 + (y^2)^2]}\, dt.$$

We see that the resulting formulas refer to a two-dimensional manifold determined by the variables (y^1, y^2) in the cartesian plane T and that the geometry of the surface of the sphere imbedded in a three-dimensional Euclidean manifold can be visualized on a two-dimensional manifold R_2 with metric determined by (43.9). If the radius of S is very large, we see from (43.9) that the terms involving $1/a^2$ can be neglected, and the geometry of the surface of the sphere is then determined approximately by the Euclidean metric

$$ds^2 = (dy^1)^2 + (dy^2)^2. \tag{43.10}$$

Thus, for large values of a, metric properties of the sphere S are indistinguishable from those of the Euclidean plane. The sum of the angles of a curvilinear triangle drawn on S will be nearly equal to 180°, since the sum of the angles of the corresponding triangle on T is 180° by Euclidean geometry. Because of the limitations of measuring devices it may be impossible to decide *a priori* whether Euclidean formula 43.10 or the more involved Riemannian formula 43.9 should be adopted as a basis for physical measurements.

The chief point of this illustration is to indicate that the geometry of a sphere, imbedded in a Euclidean 3-space with the element of arc in the form 43.8, is indistinguishable from the Riemannian geometry of a two-dimensional manifold R_2 with metric 43.9. The latter manifold, although referred to a cartesian frame Y, is not Euclidean since (43.9) cannot be reduced by an admissible transformation to (43.10).

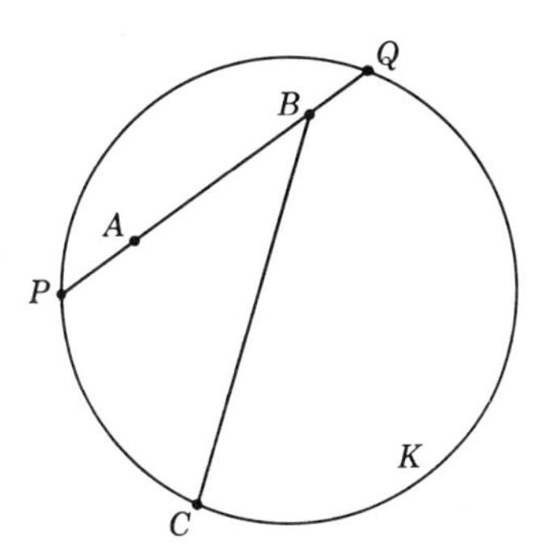

Fig. 9

Similarly, the geometry of Lobachevski can be visualized on a surface of a "pseudosphere," a surface of constant negative curvature generated by revolving a tractrix,

$$\begin{cases} x = a\left(\cos t + \log \tan \dfrac{t}{2}\right), \\ y = a \sin t, \end{cases}$$

about its asymptote. Since we will have no occasion to study the Lobachevskian or hyperbolic geometry, we will only indicate the main ideas leading to the analytical expression for the square of the element of arc

$$ds^2 = \frac{(dy^1)^2 + (dy^2)^2 - \dfrac{1}{a^2}(y^1\,dy^2 - y^2\,dy^1)^2}{\left\{1 - \dfrac{1}{a^2}[(y^1)^2 + (y^2)^2]\right\}^2}$$

which governs the study of this geometry.

Let a circle K of radius one be drawn in the plane. The universe of Lobachevskian geometry consists of points interior to K. The chords PQ of the circle are straight lines in this geometry. (See Fig. 9.) The length of the segment AB of PQ is a number given by the formula

$$\log\left(\frac{PA}{QA} : \frac{PB}{QB}\right),$$

whereas the magnitude of the angle ABC is determined as follows. Construct a sphere S of radius one tangent to K at its center. Project AB and BC on S and determine the Euclidean angle between the arcs $B'A'$ and $B'C'$ formed by the intersection of the planes passing through BC and BA perpendicular to the plane of K (Fig. 10). The Euclidean measure of $A'B'C'$ is, by definition, the measure of the angle ABC in the Lobachevski

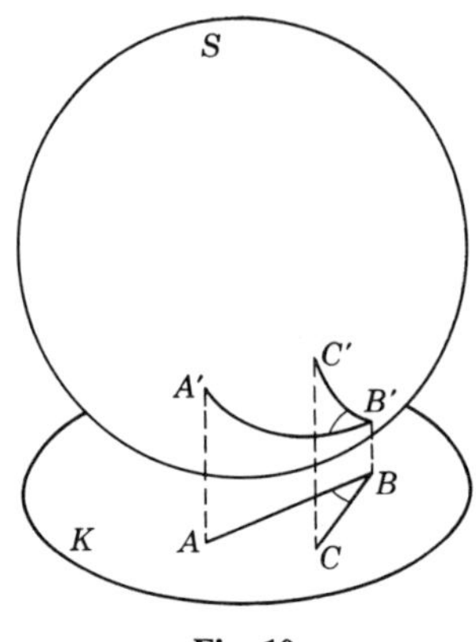

Fig. 10

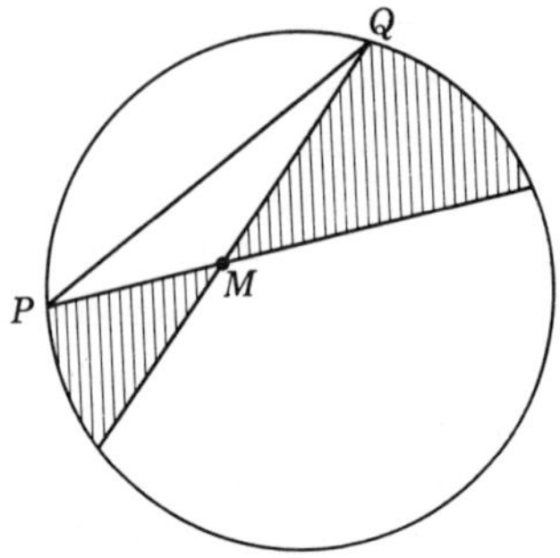

Fig. 11

plane. A pair of lines in the Lobachevski plane are considered parallel if their images on the sphere do not intersect. It can be shown that the points and lines of this geometry satisfy all postulates of Euclidean geometry except the postulate of parallels. Parallel to any given line PQ one can draw through a point M infinitely many lines which do not intersect PQ. These are the lines lying in the shaded region of Fig. 11 and passing through M. It is not difficult to prove that the sum of the angles of a triangle in this geometry is less than 180°. The consistency of Lobachevskian geometry was investigated by Cayley, Klein, and Poincaré.[2]

The discussion of this chapter is confined mainly to Euclidean geometry and those portions of Riemannian geometry that figure in applications.

44. Curvilinear Coordinates in E_3

The apparatus of tensor analysis was developed initially as a tool for the analytic study of geometries of diverse sorts. Because of its invariantive character, it was found particularly adaptable to the needs of other branches of applied mathematics. Since dynamics, mechanics of continuous media, and relativity lean rather heavily on geometrical properties of the three-dimensional space of physical experience, we devote most of this chapter to an investigation of properties of curves and surfaces imbedded in E_3.

Let the point $P(y)$, in an Euclidean 3-space E_3, be referred to a set of

[2] For details on hyperbolic geometry we refer the reader to specialized treatises on the subject, especially to F. Klein's *Nicht-Euklidische Geometrie*, **1**, pp. 161–232.

orthogonal cartesian axes Y (Fig. 12). Consider a general functional transformation

$$T: \quad x^i = x^i(y^1, y^2, y^3), \qquad (i = 1, 2, 3),$$

such that the x^i are of class C^1, and $J = \left|\frac{\partial x^i}{\partial y^j}\right| \neq 0$ in some region R of E_3. The inverse transformation,

$$T^{-1}: \quad y^i = y^i(x^1, x^2, x^3), \qquad (i = 1, 2, 3),$$

will then be single-valued, and the transformations T and T^{-1} establish one-to-one correspondence between the sets of values (x^1, x^2, x^3) and (y^1, y^2, y^3). We call the triplets of numbers (x^1, x^2, x^3) the *curvilinear coordinates* of the points P in R. The reason for this terminology is the following: if we set $x^1 =$ constant in T, then

$$x^1(y^1, y^2, y^3) = \text{constant} \tag{44.1}$$

defines a surface. If the constant is now allowed to assume different values, we get a one-parameter family of surfaces. Similarly,

$$x^2(y^1, y^2, y^3) = \text{constant}$$

and $x^3(y^1, y^2, y^3) =$ constant define two families of surfaces.

The condition that the Jacobian $J \neq 0$ in the region under consideration expresses the fact that the surfaces

$$x^1 = c_1, \qquad x^2 = c_2, \qquad x^3 = c_3 \tag{44.2}$$

intersect in one and only one point.

We call the surfaces defined by equations 44.2 the *coordinate surfaces*, and their intersections pair-by-pair are the *coordinate lines*. Thus the line of intersection of $x^1 = c_1$ and $x^2 = c_2$ is the x^3-coordinate line because

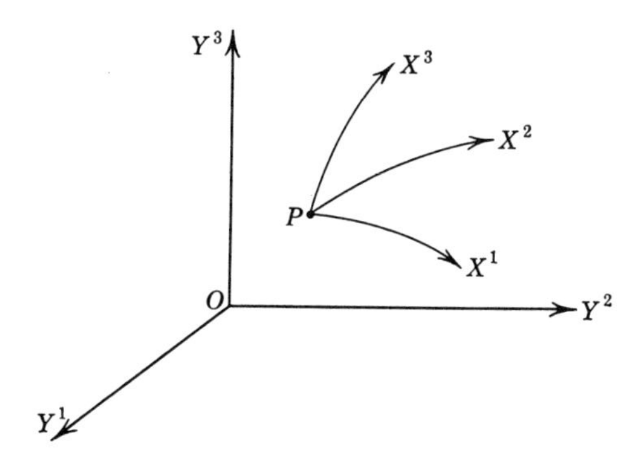

Fig. 12

along this line the variable x^3 is the only one that is changing. As an example, consider a coordinate system defined by the transformation

$$\begin{cases} y^1 = x^1 \sin x^2 \cos x^3, \\ y^2 = x^1 \sin x^2 \sin x^3, \\ y^3 = x^1 \cos x^2. \end{cases}$$

The surfaces $x^1 =$ constant are spheres, $x^2 =$ constant are circular cones, and $x^3 =$ constant are planes passing through the Y^3-axis (Fig. 13).

The inverse transformation in this case is given by

$$\begin{cases} x^1 = \sqrt{(y^1)^2 + (y^2)^2 + (y^3)^2}, \\ x^2 = \tan^{-1} \dfrac{\sqrt{(y^1)^2 + (y^2)^2}}{y^3}, \\ x^3 = \tan^{-1} \dfrac{y^2}{y^1}, \end{cases}$$

if $x^1 > 0, 0 < x^2 < \pi, 0 \leq x^3 < 2\pi$. This is the familiar spherical coordinate system.

As another illustration, the transformation

$$y^1 = x^1 \cos x^2,$$
$$y^2 = x^1 \sin x^2,$$
$$y^3 = x^3,$$

defines a cylindrical coordinate system (Fig. 14).

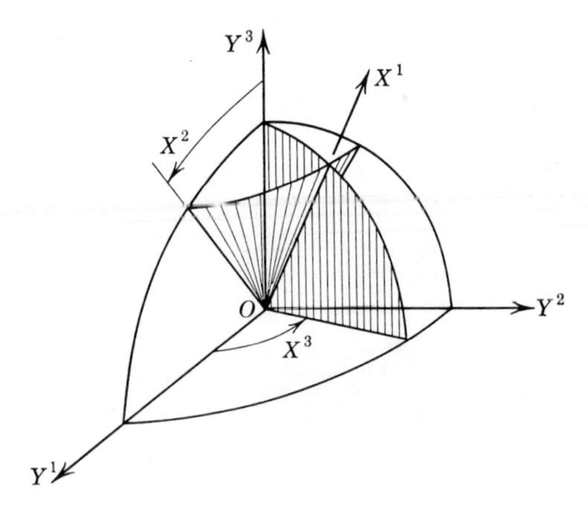

Fig. 13

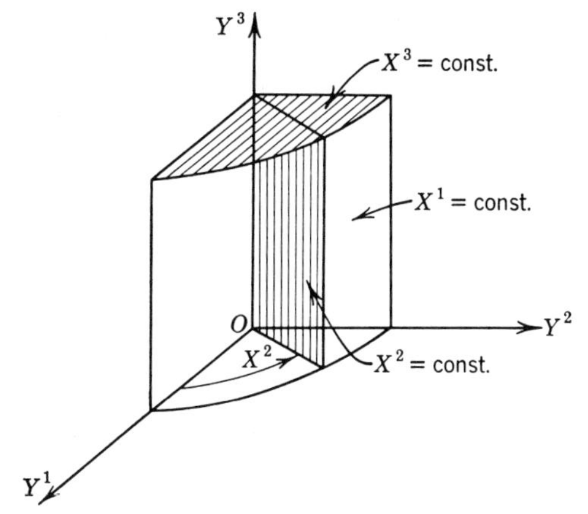

Fig. 14

Let $P(y^1, y^2, y^3)$ and $Q(y^1 + dy^1, y^2 + dy^2, y^3 + dy^3)$ be two neighboring points in R. The Euclidean distance between a pair of such points is determined by the quadratic form

$$(ds)^2 = (dy^1)^2 + (dy^2)^2 + (dy^3)^2$$
$$= dy^i\, dy^i,$$

and, since $dy^i = \dfrac{\partial y^i}{\partial x^\alpha}\, dx^\alpha$, we have

$$ds^2 = g_{ij}\, dx^i\, dx^j, \tag{44.3}$$

where

$$g_{ij} = \frac{\partial y^\alpha}{\partial x^i}\frac{\partial y^\alpha}{\partial x^j}, \qquad (\alpha = 1, 2, 3).$$

Obviously, g_{ij} is symmetric. Moreover, it is a tensor, since $(ds)^2$ is an invariant and the vector dx^i is arbitrary. Denote by g the determinant $|g_{ij}|$; this is positive in R since $g_{ij}\, dx^i\, dx^j$ is a positive definite form. Hence we can introduce the conjugate symmetric tensor g^{ij}, defined in Sec. 30 by the formula $g^{ij} = G^{ij}/g$, where G^{ij} is the cofactor of the element g_{ij} in g.

Consider now a contravariant vector $A^i(x)$, and form the invariant

$$A = (g_{ij}A^iA^j)^{1/2}. \tag{44.4}$$

Since in the orthogonal cartesian frame the invariant 44.4 assumes the form $[(A^1)^2 + (A^2)^2 + (A^3)^2]^{1/2}$, we see that A *represents the length of the vector* A^i. Similarly, the length of the covariant vector A_i is defined by the formula

$$A = (g^{ij}A_iA_j)^{1/2}. \tag{44.5}$$

In orthogonal cartesian coordinates $g^{ij} = \delta^{ij}$, and we get $A = (A_iA_i)^{1/2}$.

A vector whose length is 1 is called a *unit vector*. From formula 44.3 we see that

$$1 = g_{ij}\frac{dx^i}{ds}\frac{dx^j}{ds},$$

so that $dx^i/ds \equiv \lambda^i$ is a unit vector. If $x^i = y^i$, so that the coordinate system is cartesian, then $dx^1/ds = \lambda^1$, $dx^2/ds = \lambda^2$, $dx^3/ds = \lambda^3$ are precisely the direction cosines of the displacement vector (dx^1, dx^2, dx^3). Accordingly, we take the vector λ^i to define the direction in space relative to a curvilinear coordinate system X (Fig. 15).

Consider two directions defined by the unit vectors λ^i and μ^i at some point P (Fig. 16). Since the manifold under consideration is Euclidean, the cosine law, following from the formula of Pythagoras, gives

$$\overline{QR}^2 = \overline{PQ}^2 + \overline{PR}^2 - 2\overline{PQ}\,\overline{PR}\cos\theta,$$

and, since λ^i and μ^i are unit vectors, $\overline{PQ} = \overline{PR} = 1$, and hence

$$\overline{QR}^2 = 2(1 - \cos\theta). \tag{44.6}$$

The components of the vector joining R with Q are $\lambda^i - \mu^i$. Making use of the formula 44.4 for the length of a vector, we get

$$\begin{aligned}\overline{QR}^2 &= g_{ij}(\lambda^i - \mu^i)(\lambda^j - \mu^j) \\ &= g_{ij}\lambda^i\lambda^j + g_{ij}\mu^i\mu^j - 2g_{ij}\lambda^i\mu^j \\ &= 1 + 1 - 2g_{ij}\lambda^i\mu^j \\ &= 2(1 - g_{ij}\lambda^i\mu^j).\end{aligned} \tag{44.7}$$

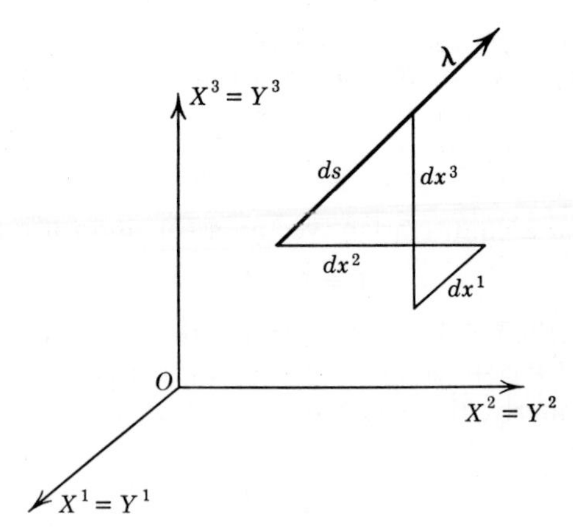

Fig. 15

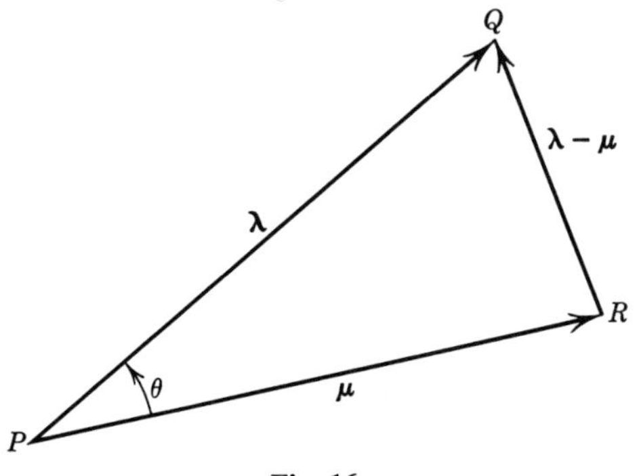

Fig. 16

It follows from (44.6) and (44.7) that the invariant $g_{ij}\lambda^i\mu^j$ is equal to $\cos\theta$, and we can write

$$\cos\theta = g_{ij}\lambda^i\mu^j. \tag{44.8}$$

We can use (44.8) to define the angle θ between two directions λ^i and μ^i if we make an unambiguous definition of $\sin\theta$.

If A^i and B^i are any two vectors, then from the definition of the length of a vector, it is clear that

$$\cos\theta = \frac{g_{ij}A^iB^j}{\sqrt{g_{ij}A^iA^j}\sqrt{g_{ij}B^iB^j}}.$$

This leads to the formula $AB\cos\theta = g_{ij}A^iB^j$, defining an invariant, which is precisely the "scalar product" $\mathbf{A}\cdot\mathbf{B}$ of elementary vector analysis.

It follows from the expression

$$ds^2 = g_{ij}\,dx^i\,dx^j,$$

for the square of the element of arc ds between $P_1(x^1, x^2, x^3)$, and $P_2(x^1 + dx^1, x^2 + dx^2, x^3 + dx^3)$, that the lengths of the elements of arc measured along the coordinate lines of our curvilinear system X are

$$ds_{(1)} = \sqrt{g_{11}}\,dx^1, \qquad ds_{(2)} = \sqrt{g_{22}}\,dx^2, \qquad ds_{(3)} = \sqrt{g_{33}}\,dx^3. \tag{44.9}$$

Thus the length of the displacement vector $(dx^1, 0, 0)$ is given by $\sqrt{g_{11}}\,dx^1$, that of $(0, dx^2, 0)$ is $\sqrt{g_{22}}\,dx^2$, and the vector $(0, 0, dx^3)$ has the length $\sqrt{g_{33}}\,dx^3$ (Fig. 17).

In addition, from (44.8) we deduce that the cosines of the angles θ_{12}, θ_{23}, θ_{13} between the coordinate lines are given by

$$\cos\theta_{12} = \frac{g_{12}}{\sqrt{g_{11}g_{22}}}, \qquad \cos\theta_{23} = \frac{g_{23}}{\sqrt{g_{22}g_{33}}}, \qquad \cos\theta_{13} = \frac{g_{13}}{\sqrt{g_{11}g_{33}}}. \tag{44.10}$$

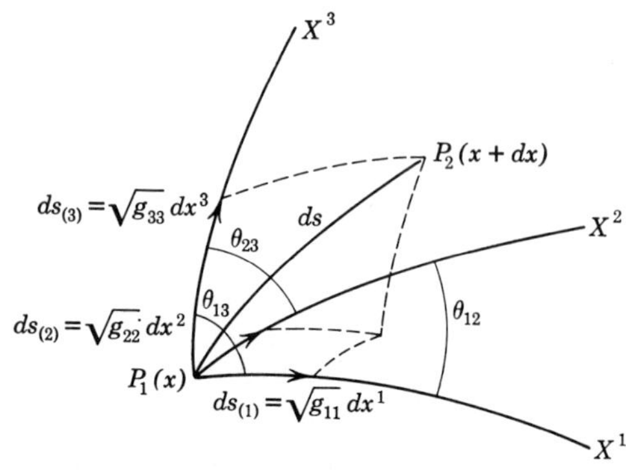

Fig. 17

For, if $\lambda^i_{(1)}$: $(dx^1/ds_{(1)}, 0, 0)$ and $\mu^i_{(2)}$: $(0, dx^2/ds_{(2)}, 0)$ are two unit vectors directed along the X^1- and X^2-coordinate lines, respectively, then

$$\cos\theta_{12} = g_{ij}\lambda^i_{(1)}\mu^j_{(2)} = \frac{g_{12}\,dx^1\,dx^2}{ds_{(1)}\,ds_{(2)}} = \frac{g_{12}}{\sqrt{g_{11}g_{22}}}.$$

Since g_{11}, g_{22}, g_{33} never vanish (see equation 44.9), we deduce from (44.10) a

THEOREM. *A necessary and sufficient condition that a given curvilinear coordinate system X be orthogonal is that $g_{ij} = 0$, for $i \neq j$, at every point of the region R.*

From the definition of the element of volume dV in curvilinear coordinates,

$$dV = \pm \left|\frac{\partial y^i}{\partial x^j}\right| dx^1\,dx^2\,dx^3,$$

where $\pm \left|\frac{\partial y^i}{\partial x^j}\right|$ is the absolute value of the Jacobian J of the transformation connecting the cartesian variables y^i with the curvilinear x^i, we can readily deduce that

$$dV = dy^1\,dy^2\,dy^3 = \sqrt{g}\,dx^1\,dx^2\,dx^3. \tag{44.11}$$

For,

$$J^2 = \left|\frac{\partial y^i}{\partial x^j}\right| \cdot \left|\frac{\partial y^i}{\partial x^j}\right| = \left|\frac{\partial y^\alpha}{\partial x^i}\frac{\partial y^\alpha}{\partial x^j}\right| = |g_{ij}| = g,$$

where we made use of the definition for g_{ij} (see equation 44.3), and of the rule for multiplication of determinants. The determinant g is a relative scalar of weight 2 (cf. Sec. 28) since $\sqrt{g}$ is a scalar density.

From developments of this section we see that the metric properties of E_3, referred to a curvilinear coordinate system X, are completely determined by the tensor g_{ij}. Accordingly, this tensor is called the *metric tensor*, and the quadratic form $ds^2 = g_{ij}\, dx^i\, dx^j$ is termed the *fundamental quadratic form.*

45. Reciprocal Base Systems. Covariant and Contravariant Vectors

In this section we interpret the main results of Sec. 44 in the language and notation of the elementary vector analysis introduced in Chapter 1. Let a cartesian system of axes (Fig. 18) be determined by a set of orthonormal base vectors $\mathbf{b}_1, \mathbf{b}_2, \mathbf{b}_3$; then the position vector $\mathbf{r}$ of any point $P(y^1, y^2, y^3)$ can be represented in the form

$$\mathbf{r} = \mathbf{b}_i y^i, \qquad (i = 1, 2, 3). \tag{45.1}$$

Since the base vectors $\mathbf{b}_i$ are independent of the position of the point $P(y^1, y^2, y^3)$, we deduce from (45.1) that

$$d\mathbf{r} = \mathbf{b}_i\, dy^i. \tag{45.2}$$

By definition the square of the element of arc between the points (y^1, y^2, y^3) and $(y^1 + dy^1, y^2 + dy^2, y^3 + dy^3)$ is given by the formula

$$ds^2 = d\mathbf{r} \cdot d\mathbf{r}. \tag{45.3}$$

The substitution from (45.2) in (45.3) gives

$$\begin{aligned} ds^2 &= \mathbf{b}_i \cdot \mathbf{b}_j\, dy^i\, dy^j \\ &= \delta_{ij}\, dy^i\, dy^j \\ &= dy^i\, dy^i, \end{aligned}$$

a familiar expression for the square of the element of arc in orthogonal cartesian coordinates.

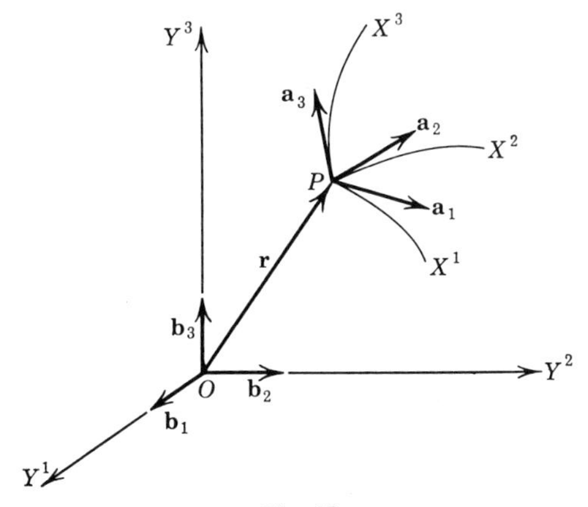

Fig. 18

Let a set of equations of transformation

$$x^i = x^i(y^1, y^2, y^3), \qquad (i = 1, 2, 3),$$

define a curvilinear coordinate system X. The position vector $\mathbf{r}$ can now be regarded as a function of coordinates x^i, and we write

$$d\mathbf{r} = \frac{\partial \mathbf{r}}{\partial x^i}\, dx^i, \tag{45.4}$$

and

$$\begin{aligned} ds^2 = d\mathbf{r} \cdot d\mathbf{r} &= \frac{\partial \mathbf{r}}{\partial x^i} \cdot \frac{\partial \mathbf{r}}{\partial x^j}\, dx^i\, dx^j \\ &= g_{ij}\, dx^i\, dx^j \end{aligned}$$

where

$$g_{ij} = \frac{\partial \mathbf{r}}{\partial x^i} \cdot \frac{\partial \mathbf{r}}{\partial x^j}. \tag{45.5}$$

The geometrical meaning of the vector $\partial \mathbf{r}/\partial x^i$ is simple; it is a base vector directed tangentially to the X^i-coordinate curve. We set

$$\frac{\partial \mathbf{r}}{\partial x^i} = \mathbf{a}_i, \tag{45.6}$$

and rewrite (45.4) and (45.5) as

$$d\mathbf{r} = \mathbf{a}_i\, dx^i \tag{45.7}$$

and

$$g_{ij} = \mathbf{a}_i \cdot \mathbf{a}_j.$$

We observe that the base vectors $\mathbf{a}_i$ are no longer independent of the coordinates (x^1, x^2, x^3).

The use of covariant notation for the base vectors $\mathbf{a}_i$ and $\mathbf{b}_i$ can be justified by observing from (45.2) and (45.7) that

$$\begin{aligned} \mathbf{a}_j\, dx^j &= \mathbf{b}_i\, dy^i \\ &= \mathbf{b}_i \frac{\partial y^i}{\partial x^j}\, dx^j. \end{aligned}$$

We see that the base vectors $\mathbf{a}_j$ transform according to the law for the transformation of components of covariant vectors,

$$\mathbf{a}_j = \frac{\partial y^i}{\partial x^j}\, \mathbf{b}_i,$$

since the dx^j's are arbitrary.

The components of base vectors $\mathbf{a}_i$, when referred to the X-coordinate system, are

$$\mathbf{a}_1\colon\ (a_1, 0, 0), \qquad \mathbf{a}_2\colon\ (0, a_2, 0), \qquad \mathbf{a}_3\colon\ (0, 0, a_3),$$

and we note that they are not necessarily unit vectors, since, in general (see equation 45.5),

$$g_{11} = \mathbf{a}_1 \cdot \mathbf{a}_1 \neq 1 \qquad g_{22} = \mathbf{a}_2 \cdot \mathbf{a}_2 \neq 1, \qquad g_{33} = \mathbf{a}_3 \cdot \mathbf{a}_3 \neq 1.$$

If the curvilinear coordinate system X is orthogonal, then

$$g_{ij} = \mathbf{a}_i \cdot \mathbf{a}_j = |\mathbf{a}_i|\,|\mathbf{a}_j| \cos \theta_{ij} = 0, \qquad \text{if } i \neq j.$$

This is the result stated in the theorem of Sec. 44.

We note that any vector $\mathbf{A}$ can be written in the form $\mathbf{A} = k\,d\mathbf{r}$, where k is a suitable scalar. Since $d\mathbf{r} = (\partial\mathbf{r}/\partial x^i)\,dx^i$, we have

$$\begin{aligned} \mathbf{A} &= \frac{\partial \mathbf{r}}{\partial x^i}(k\,dx^i) \\ &= \mathbf{a}_i A^i, \end{aligned}$$

where $A^i \equiv k\,dx^i$. The numbers A^i are the contravariant components of the vector $\mathbf{A}$, and the vectors $A^1\mathbf{a}_1$, $A^2\mathbf{a}_2$, $A^3\mathbf{a}_3$ form the edges of the parallelepiped whose diagonal is $\mathbf{A}$. Since the $\mathbf{a}_i$ are *not* unit vectors in general, we see that the lengths of edges of this parallelepiped, or the *physical components* of $\mathbf{A}$, are determined by the formulas

$$A^1\sqrt{g_{11}}, \qquad A^2\sqrt{g_{22}}, \qquad A^3\sqrt{g_{33}},$$

since $g_{11} = \mathbf{a}_1 \cdot \mathbf{a}_1$, $g_{22} = \mathbf{a}_2 \cdot \mathbf{a}_2$, $g_{33} = \mathbf{a}_3 \cdot \mathbf{a}_3$.

Let us introduce next three noncoplanar vectors

$$\text{(45.8)} \qquad \mathbf{a}^1 = \frac{\mathbf{a}_2 \times \mathbf{a}_3}{[\mathbf{a}_1\mathbf{a}_2\mathbf{a}_3]}, \qquad \mathbf{a}^2 = \frac{\mathbf{a}_3 \times \mathbf{a}_1}{[\mathbf{a}_1\mathbf{a}_2\mathbf{a}_3]}, \qquad \mathbf{a}^3 = \frac{\mathbf{a}_1 \times \mathbf{a}_2}{[\mathbf{a}_1\mathbf{a}_2\mathbf{a}_3]},$$

where $\mathbf{a}_2 \times \mathbf{a}_3$, etc., denote the vector product[3] of $\mathbf{a}_2$ and $\mathbf{a}_3$, and $[\mathbf{a}_1\mathbf{a}_2\mathbf{a}_3]$ is the *triple scalar product* $\mathbf{a}_1 \cdot \mathbf{a}_2 \times \mathbf{a}_3$.

It is obvious from the definitions 45.8 that $\mathbf{a}^i \cdot \mathbf{a}_j = \delta_j{}^i$, and it is easily verified that $[\mathbf{a}_1\mathbf{a}_2\mathbf{a}_3] = \sqrt{g}$, where $g = |g_{ij}|$, and that the triple scalar products $[\mathbf{a}^1\mathbf{a}^2\mathbf{a}^3]$ and $[\mathbf{a}_1\mathbf{a}_2\mathbf{a}_3]$ are reciprocally related, so that $[\mathbf{a}^1\mathbf{a}^2\mathbf{a}^3] = 1/\sqrt{g}$. Moreover,

$$\text{(45.9)} \qquad \mathbf{a}_1 = \frac{\mathbf{a}^2 \times \mathbf{a}^3}{[\mathbf{a}^1\mathbf{a}^2\mathbf{a}^3]}, \qquad \mathbf{a}_2 = \frac{\mathbf{a}^3 \times \mathbf{a}^1}{[\mathbf{a}^1\mathbf{a}^2\mathbf{a}^3]}, \qquad \mathbf{a}_3 = \frac{\mathbf{a}^1 \times \mathbf{a}^2}{[\mathbf{a}^1\mathbf{a}^2\mathbf{a}^3]},$$

[3] We recall that $\mathbf{a}_1 \times \mathbf{a}_2$ is a vector of length $a_1a_2\,|\sin(\mathbf{a}_1, \mathbf{a}_2)|$, and so oriented that $\mathbf{a}_1, \mathbf{a}_2$ and $\mathbf{a}_1 \times \mathbf{a}_2$ form a right-handed system. The triple scalar product $[\mathbf{a}_1\mathbf{a}_2\mathbf{a}_3]$, on the other hand, is numerically equal to the volume of the parallelepiped constructed on the vectors $\mathbf{a}_1$, $\mathbf{a}_2$, $\mathbf{a}_3$. If $\{\mathbf{a}_i\}$ is a set of base vectors in E_n, the reciprocal basis $\{\mathbf{a}^i\}$ is determined by $\mathbf{a}_i \cdot \mathbf{a}^j = \delta_i{}^j$.

as can be readily checked with the aid of (45.8). In view of this it is natural to call the system of vectors $\mathbf{a}^1$, $\mathbf{a}^2$, $\mathbf{a}^3$ the *reciprocal base system.*

We observe that *if the vectors* $\mathbf{a}_1$, $\mathbf{a}_2$, $\mathbf{a}_3$ *are unit vectors associated with an orthogonal cartesian system of coordinates, then the reciprocal system of vectors defines the same system of coordinates.*

Using the reciprocal base system, we can write the differential of a vector $\mathbf{r}$ in the form $d\mathbf{r} = \mathbf{a}^i \, dx_i$, where the dx_i are the appropriate components of $d\mathbf{r}$. Then

$$\begin{aligned} ds^2 = d\mathbf{r} \cdot d\mathbf{r} &= (\mathbf{a}^i \, dx_i) \cdot (\mathbf{a}^j \, dx_j) \\ &= \mathbf{a}^i \cdot \mathbf{a}^j \, dx_i \, dx_j \\ &= g^{ij} \, dx_i \, dx_j, \end{aligned}$$

where

$$g^{ij} \equiv \mathbf{a}^i \cdot \mathbf{a}^j = g^{ji}. \tag{45.10}$$

It is not difficult to check that the coefficients g^{ij}, defined by the formula 45.10, coincide with the quantities g^{ij} defined earlier. Thus, making use of formulas 45.8 and 45.9, we can readily show that $g_{i\alpha}g^{j\alpha} = \delta_i^{\,j}$, and the solution of this system of equations for the $g^{j\alpha}$ gives $g^{j\alpha} = G^{j\alpha}/g$, where $G^{j\alpha}$ is the cofactor of the element $g_{j\alpha}$ in the determinant $|g_{ij}|$. Thus the definition of g^{ij} given in Sec. 44 follows as a theorem from the definition 45.10.

The system of base vectors determined by (45.8) can be used to represent an arbitrary vector $\mathbf{A}$ in the form $\mathbf{A} = \mathbf{a}^i A_i$, where the A_i are the covariant components of $\mathbf{A}$. If we form the scalar product of the vector $A_i\mathbf{a}^i$ with the base vector $\mathbf{a}_j$, and note that the latter is directed along the X^j-coordinate line, we get $A_i\mathbf{a}^i \cdot \mathbf{a}_j = A_i \, \delta_j^{\,i} = A_j$. *Thus* $A_j/\sqrt{g_{jj}}$ (no sum on j) *is the length of the orthogonal projection of the vector* $\mathbf{A}$ *on the tangent to the* X^j*-coordinate curve at the point* P (*Fig.* 19), *whereas* $A^j\sqrt{g_{jj}}$ *is the length of the edge of the parallelepiped whose diagonal is the vector* $\mathbf{A}$.

Since

$$\mathbf{A} = \mathbf{a}_i A^i = \mathbf{a}^i A_i,$$

we have

$$\mathbf{a}_i \cdot \mathbf{a}_j A^i = \mathbf{a}^i \cdot \mathbf{a}_j A_i,$$

or

$$g_{ij} A^i = \delta_i^j A_i = A_j.$$

We see that *the vector obtained by lowering the index in* A^i *is precisely the covariant vector* A_i. The two sets of quantities A^i and A_i are thus seen to represent the same vector $\mathbf{A}$ referred to two different base systems. As has already been noted, the distinction between the covariant and

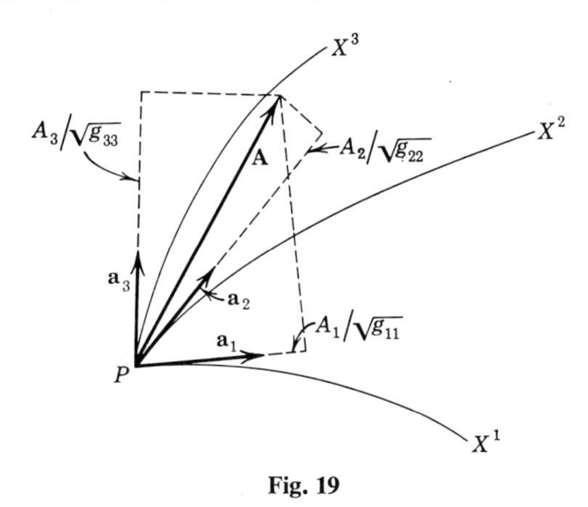

Fig. 19

contravariant components of $\mathbf{A}$ disappears whenever the base vectors $\mathbf{a}_i$ are orthonormal.

Similarly, if we consider the coefficients A_{ijk} in a multilinear form $A_{ijk}\mathbf{a}^i\mathbf{a}^j\mathbf{a}^k$ and require that $A_{ijk}\mathbf{a}^i\mathbf{a}^j\mathbf{a}^k = A^{ijk}\mathbf{a}_i\mathbf{a}_j\mathbf{a}_k$, then the set of quantities $\{A^{ijk}\}$ represents the same tensor A when referred to the basis $\{\mathbf{a}_i\}$. All associated tensors (see Sec. 30) represent the same tensor A in suitable base systems.

46. On the Meaning of Covariant Derivatives

Let $\mathbf{A}$ be a vector localized at some point $P(y^1, y^2, y^3)$ of E_3 referred to an orthogonal cartesian frame Y. If at every point of some region R about P we have a uniquely defined vector $\mathbf{A}$, we refer to the totality of vectors $\mathbf{A}$ in R as a *vector field*. We suppose that the components of $\mathbf{A}$ are continuously differentiable functions of y^i in R, and, if we introduce a curvilinear system of coordinates X by means of the transformation

$$T: \quad x^i = x^i(y^1, y^2, y^3),$$

the corresponding components $A^i(x)$ will be continuously differentiable functions of the point (x^1, x^2, x^3) determined by the position vector $\mathbf{r}(x^1, x^2, x^3)$. In the notation of Sec. 45, the base vectors in the X-reference frame are $\mathbf{a}_i = \partial\mathbf{r}/\partial x^i$, so that $\mathbf{A}$ has the representation

$$\mathbf{A} = A^i\mathbf{a}_i. \tag{46.1}$$

We will be concerned with the calculation of the vector change $\Delta\mathbf{A}$ in $\mathbf{A}$ as the point $P(x^1, x^2, x^3)$ assumes a different position

$$P'(x^1 + \Delta x^1, x^2 + \Delta x^2, x^3 + \Delta x^3).$$

From (46.1) we have

$$\begin{aligned}\Delta\mathbf{A} &= (A^i + \Delta A^i)(\mathbf{a}_i + \Delta\mathbf{a}_i) - A^i\mathbf{a}_i \\ &= \Delta A^i\mathbf{a}_i + A^i\,\Delta\mathbf{a}_i + (\Delta A^i)(\Delta\mathbf{a}_i).\end{aligned}$$

As in ordinary calculus we denote the principal part of the change by $d\mathbf{A}$, and write

$$d\mathbf{A} = \mathbf{a}_i\,dA^i + A^i\,d\mathbf{a}_i. \tag{46.2}$$

This formula states that the differential change in $\mathbf{A}$ arises from two sources:

(*a*) Change in the components A^i as the values (x^1, x^2, x^3) are changed.
(*b*) Change in the base vectors $\mathbf{a}_i$ as the position of the point (x^1, x^2, x^3) is altered.

The partial derivative of $\mathbf{A}$ with respect to x^j is defined as the limit of the quotient,

$$\lim_{\Delta x^j\to 0}\frac{\Delta\mathbf{A}}{\Delta x^j} = \frac{\partial\mathbf{A}}{\partial x^j},$$

and it follows from the expression for the increment $\Delta\mathbf{A}$ that

$$\frac{\partial\mathbf{A}}{\partial x^j} = \frac{\partial A^i}{\partial x^j}\mathbf{a}_i + \frac{\partial\mathbf{a}_i}{\partial x^j}A^i. \tag{46.3}$$

We show next that the vector defined by formula 46.3 is identical with the covariant derivative of the vector A^i. First we establish the identity

$$\frac{\partial\mathbf{a}_i}{\partial x^j} = \begin{Bmatrix}\alpha\\ ij\end{Bmatrix}\mathbf{a}_\alpha. \tag{46.4}$$

We recall that $g_{ij} = \mathbf{a}_i\cdot\mathbf{a}_j$. Hence

$$\frac{\partial g_{ij}}{\partial x^k} = \frac{\partial\mathbf{a}_i}{\partial x^k}\cdot\mathbf{a}_j + \frac{\partial\mathbf{a}_j}{\partial x^k}\cdot\mathbf{a}_i.$$

Permuting the indices in this formula, we get

$$\frac{\partial g_{ik}}{\partial x^j} = \frac{\partial\mathbf{a}_i}{\partial x^j}\cdot\mathbf{a}_k + \frac{\partial\mathbf{a}_k}{\partial x^j}\cdot\mathbf{a}_i,$$

$$\frac{\partial g_{jk}}{\partial x^i} = \frac{\partial\mathbf{a}_j}{\partial x^i}\cdot\mathbf{a}_k + \frac{\partial\mathbf{a}_k}{\partial x^i}\cdot\mathbf{a}_j.$$

If we assume that T is of class C^2, then[4]

$$\frac{\partial \mathbf{a}_i}{\partial x^j} = \frac{\partial \mathbf{a}_j}{\partial x^i}.$$

We form

$$[ij, k] = \frac{1}{2}\left(\frac{\partial g_{ik}}{\partial x^j} + \frac{\partial g_{jk}}{\partial x^i} - \frac{\partial g_{ij}}{\partial x^k}\right),$$

and obtain

$$\frac{\partial \mathbf{a}_i}{\partial x^j} \cdot \mathbf{a}_k = [ij, k]. \tag{46.5}$$

It follows from (46.5) that

$$\frac{\partial \mathbf{a}_i}{\partial x^j} = [ij, k]\mathbf{a}^k.$$

Hence

$$\begin{aligned} \frac{\partial \mathbf{a}_i}{\partial x^j} \cdot \mathbf{a}^\alpha &= [ij, k]\mathbf{a}^k \cdot \mathbf{a}^\alpha \\ &= [ij, k]g^{k\alpha} \\ &= \begin{Bmatrix} \alpha \\ ij \end{Bmatrix}, \end{aligned}$$

from which it follows that

$$\frac{\partial \mathbf{a}_i}{\partial x^j} = \begin{Bmatrix} \alpha \\ ij \end{Bmatrix}\mathbf{a}_\alpha.$$

This establishes the identity 46.4.

Inserting this result in (46.3), we get

$$\begin{aligned} \frac{\partial \mathbf{A}}{\partial x^j} &= \frac{\partial A^i}{\partial x^j}\mathbf{a}_i + \begin{Bmatrix} \alpha \\ ij \end{Bmatrix} A^i \mathbf{a}_\alpha \\ &= \left[\frac{\partial A^\alpha}{\partial x^j} + \begin{Bmatrix} \alpha \\ ij \end{Bmatrix} A^i\right]\mathbf{a}_\alpha, \end{aligned}$$

and the expression in the bracket is precisely $A^\alpha_{,j}$. Thus

$$\frac{\partial \mathbf{A}}{\partial x^j} = A^\alpha_{,j}\mathbf{a}_\alpha. \tag{46.6}$$

It follows from (46.6) that the covariant derivative $A^\alpha_{,j}$ of the vector A^α is a vector whose components are precisely the components of $\partial \mathbf{A}/\partial x^j$ referred to the base system $\mathbf{a}_i$.

[4] For, $\mathbf{a}_i = \dfrac{\partial \mathbf{r}}{\partial x^i}$ and $\dfrac{\partial \mathbf{a}_i}{\partial x^j} = \dfrac{\partial}{\partial x^j}\left(\dfrac{\partial \mathbf{r}}{\partial x^i}\right) = \dfrac{\partial}{\partial x^i}\left(\dfrac{\partial \mathbf{r}}{\partial x^j}\right) = \dfrac{\partial \mathbf{a}_j}{\partial x^i}$.

We can also show that if $\mathbf{A}$ is represented in the form

$$\mathbf{A} = A_\alpha \mathbf{a}^\alpha, \tag{46.7}$$

then

$$\frac{\partial \mathbf{A}}{\partial x^j} = A_{\alpha,j}\mathbf{a}^\alpha. \tag{46.8}$$

From $\mathbf{a}^i \cdot \mathbf{a}_j = \delta^i_j$ we have

$$\frac{\partial \mathbf{a}^i}{\partial x^k} \cdot \mathbf{a}_j + \mathbf{a}^i \cdot \frac{\partial \mathbf{a}_j}{\partial x^k} = 0.$$

Therefore

$$\begin{aligned}\frac{\partial \mathbf{a}^i}{\partial x^k} \cdot \mathbf{a}_j &= -\mathbf{a}^i \cdot \frac{\partial \mathbf{a}_j}{\partial x^k} \\ &= -\mathbf{a}^i \cdot \mathbf{a}_\alpha \begin{Bmatrix} \alpha \\ jk \end{Bmatrix}\end{aligned}$$

by (46.4). Since $\mathbf{a}^i \cdot \mathbf{a}_\alpha = \delta^i_\alpha$, the foregoing result is epuivalent to

$$\frac{\partial \mathbf{a}^i}{\partial x^k} \cdot \mathbf{a}_j = -\begin{Bmatrix} i \\ jk \end{Bmatrix}.$$

Hence

$$\frac{\partial \mathbf{a}^i}{\partial x^k} = -\begin{Bmatrix} i \\ jk \end{Bmatrix}\mathbf{a}^j. \tag{46.9}$$

The differentiation of (46.7) with respect to x^k and the substitution from (46.9) lead at once to (46.8).

We observe that, if the Christoffel symbols vanish identically in R, the reference frame associated with these symbols is cartesian (see Theorem I, Sec. 39), and, in this case, the base vectors $\mathbf{a}_i$ are independent of the coordinates x^i. The formula 46.3 then states that $\dfrac{\partial \mathbf{A}}{\partial x^j} = \dfrac{\partial A^i}{\partial x^j}\mathbf{a}_i$, and hence $A^i_{,j} = \dfrac{\partial A^i}{\partial x^j}$.

47. Intrinsic Differentiation

Let a vector field $\mathbf{A}(x)$ be defined in some region of E_3, and let

$$C\colon \quad x^i = x^i(t), \qquad t_1 \le t \le t_2,$$

be a curve in that region. The vectors $\mathbf{A}(x)$, defined over the one-dimensional manifold C, depend on the parameter t, and if $\mathbf{A}(x)$ is a differentiable vector and the $x^i(t)$ belong to the class C^1, then

$$\frac{d\mathbf{A}}{dt} = \frac{\partial \mathbf{A}}{\partial x^j} \cdot \frac{dx^j}{dt}.$$

By virtue of (46.6) this can be written

$$\frac{d\mathbf{A}}{dt} = A^{\alpha}_{,j} \frac{dx^j}{dt} \mathbf{a}_{\alpha}$$

$$= \left[\frac{dA^{\alpha}}{dt} + \begin{Bmatrix} \alpha \\ ij \end{Bmatrix} A^i \frac{dx^j}{dt} \right] \mathbf{a}_{\alpha}.$$

The vector $\delta A^{\alpha}/\delta t$, defined by the formula

$$\text{(47.1)} \qquad \frac{\delta A^{\alpha}}{\delta t} \equiv \frac{dA^{\alpha}}{dt} + \begin{Bmatrix} \alpha \\ ij \end{Bmatrix} A^i \frac{dx^j}{dt}, \qquad (\alpha = 1, 2, 3),$$

is called the *absolute* or *intrinsic* derivative of A^{α} with respect to the parameter t.

Following McConnell[5] we will make free use of intrinsic differentiation in the treatment of geometry of curves and surfaces.

If the vector field A^{α} is defined in the neighborhood of C, as well as on C, we can write

$$\frac{\delta A^{\alpha}}{\delta t} = A^{\alpha}_{,\beta} \frac{dx^{\beta}}{dt},$$

and it follows that the familiar rules for differentiation of sums, products, etc., remain valid for the process of intrinsic differentiation. If A is a scalar, then, obviously, $\delta A/\delta t = dA/dt$.

The extension of the process of intrinsic differentiation to tensors of rank greater than one is immediate. Thus we write

$$\frac{\delta A^i_{jk}}{\delta t} \equiv \frac{dA^i_{jk}}{dt} + \begin{Bmatrix} i \\ \alpha\beta \end{Bmatrix} A^{\alpha}_{jk} \frac{dx^{\beta}}{dt} - \begin{Bmatrix} \alpha \\ j\beta \end{Bmatrix} A^i_{\alpha k} \frac{dx^{\beta}}{dt} - \begin{Bmatrix} \alpha \\ k\beta \end{Bmatrix} A^i_{j\alpha} \frac{dx^{\beta}}{dt}.$$

We observe that, since $\dfrac{\delta g_{ij}}{\delta t} = 0$, the fundamental tensors g_{ij} and g^{ij} can be taken outside the sign of intrinsic differentiation.

Problems

1. Prove that $\dfrac{d}{dt}(g_{ij}A^iA^j) = 2g_{ij}A^i \dfrac{\delta A^j}{\delta t}$.

2. Show that $A_{i,j} - A_{j,i} = \dfrac{\partial A_i}{\partial x^j} - \dfrac{\partial A_j}{\partial x^i}$.

3. Show that $\dfrac{d}{dt}(g_{ij}A^iB^j) = g_{ij} \dfrac{\delta A^i}{\delta t} B^j + g_{ij}A^i \dfrac{\delta B^j}{\delta t}$.

4. If $A_i = g_{ij}A^j$, show that $A_{i,k} = g_{i\alpha}A^{\alpha}_{,k}$.

[5] Compare A. J. McConnell, *Absolute Differential Calculus*, pp. 156–162.

5. Show that $\dfrac{\partial}{\partial x^k}(g_{ij}A^iB^j) = A_{i,k}B^i + A^iB_{i,k}$.

6. Prove that if A is the magnitude of A^i, then $A_{,j} = A_{i,j}A^i/A$.

7. If y^i are rectangular cartesian coordinates, show that in E_3

$$[\alpha\beta, \gamma] = \frac{\partial^2 y^i}{\partial x^\alpha \partial x^\beta}\frac{\partial y^i}{\partial x^\gamma} \quad \text{and} \quad \begin{Bmatrix} \gamma \\ \alpha\beta \end{Bmatrix} = \frac{\partial^2 y^i}{\partial x^\alpha \partial x^\beta}\frac{\partial x^\gamma}{\partial y^i}.$$

These formulas are often found to be more convenient for the computation of Christoffel's symbols than the defining formulas 31.1 and 31.2.

48. Parallel Vector Fields

Consider a curve (Fig. 20),

$$C: \quad x^i = x^i(t), \qquad t_1 \le t \le t_2, \qquad (i = 1, 2, 3),$$

drawn in some region of E_3, and a vector **A** localized at some point P of C. We suppose that the functions $x^i(t)$ are of class C^1. If we construct at every point of C a vector equal to **A** in magnitude and parallel to it in direction, we obtain what is known as a *parallel field* of vectors along the curve C. We will deduce a set of necessary and sufficient conditions for a vector field to be parallel.

If **A** is a parallel field along C, then the vectors **A** do not change along the curve and we can write $d\mathbf{A}/dt = 0$. It follows, upon noting (47.1), that the components A^i of **A** satisfy a set of simultaneous differential equations $\delta A^i/\delta t = 0$, or, when written out in full,

$$\frac{dA^i}{dt} + \begin{Bmatrix} i \\ \alpha\beta \end{Bmatrix} A^\alpha \frac{dx^\beta}{dt} = 0. \tag{48.1}$$

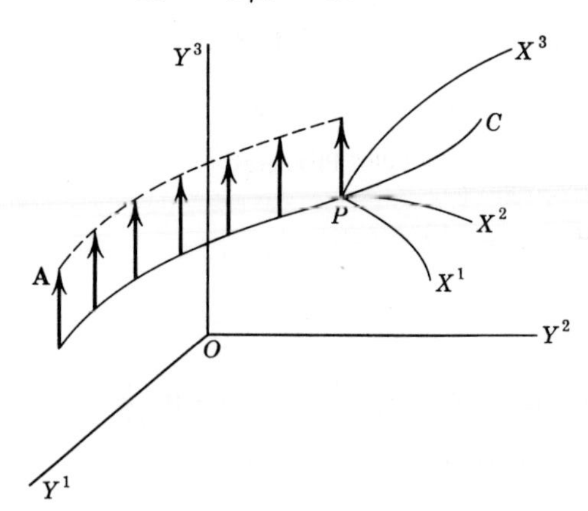

Fig. 20

We can show, conversely, that every solution of the system 48.1 yields a parallel vector field along C. Indeed, from the theory of differential equations it is known that this system of three first-order differential equations has a unique solution when the values of the components A^i are specified at a given point of C. But it was shown previously that the vector field formed by constructing a family of vectors of fixed lengths, parallel to a given vector, satisfies the system. Hence every solution of equation 48.1 satisfying the initial conditions must form a parallel field along C.

Let $A^i(t)$ and $B^i(t)$ be any two solutions of the system 48.1. We verify that the lengths of vectors A^i and B^i indeed do not change as we move along the curve. Moreover, the angle θ between the vectors A^i and B^i remains fixed as the parameter t is allowed to change. To prove this we note that (Sec. 44) $\mathbf{A} \cdot \mathbf{B} = AB \cos \theta = g_{ij}A^iB^j$, and, if $g_{ij}A^iB^j$ is to remain constant along C, then $\dfrac{d}{dt}(g_{ij}A^iB^j) = 0$. But $g_{ij}A^iB^j$ is an invariant, and, since the g_{ij} behave like constants in the process of covariant differentiation, we can write

$$\begin{aligned} \frac{d}{dt}(g_{ij}A^iB^j) &= \frac{\delta}{\delta t}(g_{ij}A^iB^j) \\ &= g_{ij}\frac{\delta A^i}{\delta t}B^j + g_{ij}A^i\frac{\delta B^j}{\delta t}. \end{aligned}$$

Since, by hypothesis, the fields A^i and B^i satisfy (48.1), $\delta A^i/\delta t = 0$ and $\delta B^i/\delta t = 0$, and we conclude that $g_{ij}A^iB^j$ is constant along C. It follows directly from this result that, if $A^i = B^i$, then $g_{ij}A^iA^j = A^2$ is constant along C, and this implies that $\theta =$ constant.

The notion of a parallel vector field along a curve can be extended to define parallel vector fields over three-dimensional Euclidean manifolds. Thus consider any point $P(x)$ and a vector $\mathbf{A}$ localized at P. If we construct at every point of the manifold a vector equal to $\mathbf{A}$ in magnitude and parallel to it in direction, there will result a parallel vector field in the space of three dimensions. If a curve C is drawn passing through P, the vectors A^i of the field lying on C will form a parallel field along C, and will thus satisfy (48.1). However, since vectors A^i are defined at every point (x^i) of the manifold, we can write

$$\frac{dA^i}{dt} = \frac{\partial A^i}{\partial x^k}\frac{dx^k}{dt},$$

so that equations 48.1 assume the form

$$\left(\frac{\partial A^i}{\partial x^k} + \begin{Bmatrix} i \\ \alpha k \end{Bmatrix} A^\alpha\right)\frac{dx^k}{dt} = 0.$$

This must be true for all curves passing through P, that is, for all values of dx^k/dt. Accordingly, the parallel vector field in E_3 satisfies the system of equations

$$\frac{\partial A^i}{\partial x^k} + \left\{ {i \atop \alpha k} \right\} A^\alpha = 0, \qquad \text{or} \qquad A^i_{,k} = 0.$$

The converse follows, as previously, from the existence and uniqueness of solutions of such systems of differential equations.

The condition for a parallel displacement of a covariant vector A_i is

$$\frac{\partial A_i}{\partial x^k} - \left\{ {\alpha \atop ik} \right\} A_\alpha = 0, \qquad \text{or} \qquad A_{i,k} = 0.$$

This follows from the observation that $A_{i,k} = g_{i\alpha} A^\alpha_{,k}$ whenever $A_i = g_{ij} A^j$.

49. Geometry of Space Curves

Let the parametric equations of the curve C in E_3 be

$$C: \quad x^i = x^i(t), \qquad t_1 \leq t \leq t_2, \qquad (i = 1, 2, 3).$$

The square of the length of an element of C is given by

$$ds^2 = g_{ij}\, dx^i\, dx^j, \tag{49.1}$$

and the length of arc s of C is defined by the integral

$$s = \int_{t_1}^{t_2} \sqrt{g_{ij} \frac{dx^i}{dt} \frac{dx^j}{dt}}\, dt. \tag{49.2}$$

From (49.1) we see that

$$1 = g_{ij} \frac{dx^i}{ds} \frac{dx^j}{ds}, \tag{49.3}$$

and, if we set $dx^i/ds = \lambda^i$, equation 49.3 can be written as

$$g_{ij} \lambda^i \lambda^j = 1. \tag{49.4}$$

Thus the vector $\boldsymbol{\lambda}$, with components λ^i, is a unit vector. Moreover, $\boldsymbol{\lambda}$ is tangent to C, since its components $\bar{\lambda}^i$, when the curve C is referred to a rectangular cartesian reference frame Y, become $\bar{\lambda}^i = dy^i/ds$. These are precisely the direction cosines of the tangent vector to the curve C. We shall assume throughout this discussion that the curve C is of class C^2, so that it has a continuously turning tangent at all points of C.

Consider a pair of unit vectors $\boldsymbol{\lambda}$ and $\boldsymbol{\mu}$ (with components λ^i and μ^i, respectively) at any point P of C (Fig. 21). We suppose that $\boldsymbol{\lambda}$ is tangent

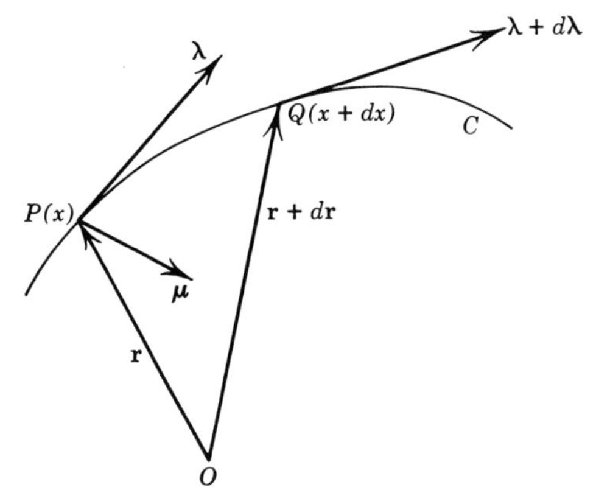

Fig. 21

to C at P. The cosine of the angle θ between $\boldsymbol{\lambda}$ and $\boldsymbol{\mu}$ is given by the formula

$$\cos\theta = g_{ij}\lambda^i\mu^j, \tag{49.5}$$

and, if $\boldsymbol{\lambda}$ and $\boldsymbol{\mu}$ are orthogonal, (49.5) requires that

$$g_{ij}\lambda^i\mu^j = 0. \tag{49.6}$$

Any vector $\boldsymbol{\mu}$ satisfying equation 49.6 is said to be *normal* to C at P.

If we take the intrinsic derivative with respect to the arc parameter s of the quadratic relation 49.4, and recall that the g_{ij}'s behave in covariant differentiation like constants, we obtain

$$g_{ij}\frac{\delta\lambda^i}{\delta s}\lambda^j + g_{ij}\frac{\delta\lambda^j}{\delta s}\lambda^i = 0.$$

Since g_{ij} is symmetric, the foregoing result can be written in the form $g_{ij}\lambda^i\dfrac{\delta\lambda^j}{\delta s} = 0$. We see that the vector $\dfrac{\delta\lambda^j}{\delta s}$ either vanishes or is normal to C, and if it does not vanish we denote the unit vector codirectional with $\dfrac{\delta\lambda^j}{\delta s}$ by μ^j, and write

$$\mu^j = \frac{1}{\varkappa}\frac{\delta\lambda^j}{\delta s}, \tag{49.7}$$

where $\varkappa > 0$ is so chosen as to make μ^i a unit vector.

The vector μ^j, determined by the formula 49.7, is called the *principal normal vector* to the curve C at the point P, and $\varkappa$ is the *curvature* of C at the point in question.

The plane determined by the tangent vector $\boldsymbol{\lambda}$ and the principal normal vector $\boldsymbol{\mu}$ is called the *osculating plane* to the curve C at P.

Since $\boldsymbol{\mu}$ is a unit vector,

$$g_{ij}\mu^i\mu^j = 1, \tag{49.8}$$

and we can treat this quadratic relation just as we did $g_{ij}\lambda^i\lambda^j = 1$ and deduce the orthogonality of vectors $\dfrac{\delta\mu^j}{\delta s}$ and μ^j; that is, $g_{ij}\mu^i\dfrac{\delta\mu^j}{\delta s} = 0$. Moreover, differentiating intrinsically the orthogonality relation 49.6, we get

$$g_{ij}\frac{\delta\lambda^i}{\delta s}\mu^j + g_{ij}\lambda^i\frac{\delta\mu^j}{\delta s} = 0,$$

or

$$\begin{aligned} g_{ij}\lambda^i\frac{\delta\mu^j}{\delta s} &= -g_{ij}\frac{\delta\lambda^i}{\delta s}\mu^j \\ &= -\varkappa g_{ij}\mu^i\mu^j \\ &= -\varkappa, \end{aligned}$$

where we used equation 49.7 and the quadratic relation 49.8. Thus

$$g_{ij}\lambda^i\frac{\delta\mu^j}{\delta s} = -\varkappa, \tag{49.9}$$

and, since $g_{ij}\lambda^i\lambda^j = 1$, we can write (49.9) in the form

$$g_{ij}\lambda^i\left(\frac{\delta\mu^j}{\delta s} + \varkappa\lambda^j\right) = 0.$$

which shows that the vector $\dfrac{\delta\mu^j}{\delta s} + \varkappa\lambda^j$ is orthogonal to λ^i. This result shows that if we define a *unit vector* $\boldsymbol{\nu}$, with components ν^j, by the formula

$$\nu^j = \frac{1}{\tau}\left(\frac{\delta\mu^j}{\delta s} + \varkappa\lambda^j\right), \tag{49.10}$$

the vector $\boldsymbol{\nu}$ will be orthogonal to both $\boldsymbol{\lambda}$ and $\boldsymbol{\mu}$. We agree to choose the sign of τ in such a way that

$$\sqrt{g}\, e_{ijk}\lambda^i\mu^j\nu^k = 1, \tag{49.11}$$

so that the triad of unit vectors $\boldsymbol{\lambda}$, $\boldsymbol{\mu}$, $\boldsymbol{\nu}$ forms, at each point P of C, a right-handed system of axes.[6]

[6] We deduce from (41.2), and from the definition of the triple scalar product (Sec. 45), that

$$e_{ijk}\lambda^i\mu^j\nu^k = \begin{vmatrix} \lambda^1 & \mu^1 & \nu^1 \\ \lambda^2 & \mu^2 & \nu^2 \\ \lambda^3 & \mu^3 & \nu^3 \end{vmatrix} = \frac{1}{\sqrt{g}}\boldsymbol{\lambda}\cdot\boldsymbol{\mu}\times\boldsymbol{\nu}.$$

Since e_{ijk} is a relative tensor of weight -1 (Sec. 41), and $g = \left| \dfrac{\partial y^i}{\partial x^j} \right|^2$, it follows that $\epsilon_{ijk} \equiv \sqrt{g}\, e_{ijk}$ is an absolute tensor, and hence the left-hand member of (49.11) is an invariant. An algorithm of division suggests that ν^k in (49.11) is determined by the formula

$$\nu^k = \epsilon^{ijk}\lambda_i\mu_j, \tag{49.12}$$

where λ_i and μ_i are the associated vectors $g_{i\alpha}\lambda^\alpha$ and $g_{i\alpha}\mu^\alpha$, and

$$\epsilon^{ijk} \equiv \frac{1}{\sqrt{g}} e^{ijk}$$

is an absolute tensor. The validity of this expression follows from an observation that (49.12) satisfies the conditions of orthogonality $g_{ij}\lambda^i\nu^j = 0$, $g_{ij}\mu^i\nu^j = 0$, and the equation 49.11 determining the orientation of the unit vector $\boldsymbol{\nu}$ relative to $\boldsymbol{\lambda}$ and $\boldsymbol{\mu}$. The number τ appearing in equation 49.10 is called the *torsion* of C at P, and the vector $\boldsymbol{\nu}$ is the *binormal.*

In order to reconcile these definitions with the usual definitions of the principal normal and curvature given in elementary vector analysis, we recall the formula 46.6, $\partial\mathbf{A}/\partial x^i = A^\alpha_{,i}\mathbf{a}_\alpha$, and note that if the vector field $\mathbf{A}$ is defined along C, we can write

$$\frac{\partial \mathbf{A}}{\partial x^i}\frac{dx^i}{ds} = A^\alpha_{,i}\frac{dx^i}{ds}\mathbf{a}_\alpha.$$

Using the definition of intrinsic derivative, $\dfrac{\delta A^\alpha}{\delta s} = A^\alpha_{,i}\dfrac{dx^i}{ds}$, we can write the preceding result as

$$\frac{d\mathbf{A}}{ds} = \frac{\delta A^\alpha}{\delta s}\mathbf{a}_\alpha. \tag{49.13}$$

Let $\mathbf{r}$ be the position vector of the point P on C; then the tangent vector $\boldsymbol{\lambda}$ is determined by

$$\frac{d\mathbf{r}}{ds} = \lambda^i\mathbf{a}_i = \boldsymbol{\lambda},$$

and (49.13) gives for the *curvature vector*

$$\frac{d^2\mathbf{r}}{ds^2} = \frac{d\boldsymbol{\lambda}}{ds} = \frac{\delta\lambda^\alpha}{\delta s}\mathbf{a}_\alpha \equiv \mathbf{c}, \tag{49.14}$$

where $\mathbf{c}$ is a vector perpendicular[7] to $\boldsymbol{\lambda}$.

[7] Since $\boldsymbol{\lambda}\cdot\boldsymbol{\lambda} = 1$, $\boldsymbol{\lambda}\cdot d\boldsymbol{\lambda}/ds = 0$.

With each point P of C we can associate a constant $\varkappa$, such that $\mathbf{c}/\varkappa = \boldsymbol{\mu}$ is a unit vector. We can now rewrite (49.14) in the form

$$\begin{aligned}\boldsymbol{\mu} &= \frac{1}{\varkappa}\frac{d\boldsymbol{\lambda}}{ds} \\ &= \frac{1}{\varkappa}\frac{\delta\lambda^\alpha}{\delta s}\mathbf{a}_\alpha \\ &= \mu^\alpha\mathbf{a}_\alpha,\end{aligned}$$

where, in the last step, we have made use of the formula 49.7.

50. Serret-Frenet Formulas

This section contains a set of three remarkable formulas, generally known as Frenet's formulas, which characterize, in the small, all essential geometric properties of space curves. Two of these formulas have already been derived in Sec. 49. They are

$$\frac{\delta\lambda^i}{\delta s} = \varkappa\mu^i, \qquad \varkappa > 0, \tag{50.1}$$

and

$$\frac{\delta\mu^i}{\delta s} = \tau\nu^i - \varkappa\lambda^i. \tag{50.2}$$

The first of these gives the rate of turning of the tangent vector $\boldsymbol{\lambda}$ as the point moves along the curve, and the second that of the principal normal $\boldsymbol{\mu}$. The third formula,

$$\frac{\delta\nu^i}{\delta s} = -\tau\mu^i, \tag{50.3}$$

to be derived next, specifies the rate of turning of the binormal as the point P moves along the curve.

If we differentiate equation

$$\nu^k = \epsilon^{ijk}\lambda_i\mu_j \tag*{[49.12]}$$

intrinsically, we get

$$\frac{\delta\nu^k}{\delta s} = \epsilon^{ijk}\frac{\delta\lambda_i}{\delta s}\mu_j + \epsilon^{ijk}\lambda_i\frac{\delta\mu_j}{\delta s}, \tag{50.4}$$

since the covariant derivatives of ϵ^{ijk} are zero.[8] Lowering the indices in (50.1) and (50.2) we get $\dfrac{\delta\lambda_i}{\delta s} = \varkappa\mu_i$ and $\dfrac{\delta\mu_i}{\delta s} = \tau\nu_i - \varkappa\lambda_i$; and inserting

[8] For the ϵ^{ijk}'s are constants in a cartesian system, hence $\epsilon^{ijk}_{,l} = 0$, and this is a tensor equation!

these values in (50.4) yields

$$\begin{aligned}\frac{\delta \nu^k}{\delta s} &= \epsilon^{ijk}\varkappa\mu_i\mu_j + \epsilon^{ijk}\lambda_i(\tau\nu_j - \varkappa\lambda_j)\\ &= \tau\epsilon^{ijk}\lambda_i\nu_j\\ &= -\tau\mu^k,\end{aligned}$$

since $\epsilon^{ijk}\lambda_i\lambda_j = \epsilon^{ijk}\mu_i\mu_j = 0$, because the ϵ^{ijk} are skew-symmetric, and $\mu^k = \epsilon^{ikj}\lambda_i\nu_j$. This establishes (50.3).

Formulas 50.1, 50.2, and 50.3, when written out explicitly in terms of Christoffel's symbols, assume the forms

$$\frac{d\lambda^i}{ds} + \begin{Bmatrix} i \\ jk \end{Bmatrix} \lambda^j \frac{dx^k}{ds} = \varkappa\mu^i, \qquad \text{or} \qquad \frac{d^2x^i}{ds^2} + \begin{Bmatrix} i \\ jk \end{Bmatrix} \frac{dx^j}{ds}\frac{dx^k}{ds} = \varkappa\mu^i,$$

$$(50.5) \qquad \frac{d\mu^i}{ds} + \begin{Bmatrix} i \\ jk \end{Bmatrix} \mu^j \frac{dx^k}{ds} = -(\varkappa\lambda^i - \tau\nu^i),$$

$$\frac{d\nu^i}{ds} + \begin{Bmatrix} i \\ jk \end{Bmatrix} \nu^j \frac{dx^k}{ds} = -\tau\mu^i.$$

Except for position of the curve C in space, the system 50.5 determines the curve uniquely when continuous functions $\varkappa(s)$ and $\tau(s)$ are specified along C.

We conclude this section by considering an example illustrating the use of Frenet's formulas. Consider a curve, defined in cylindrical coordinates by equations

$$x^1 = a,\quad x^2 = \theta(s),\quad x^3 = 0.$$

This curve is a circle of radius a. The square of the element of arc in cylindrical coordinates is

$$ds^2 = (dx^1)^2 + (x^1)^2(dx^2)^2 + (dx^3)^2,$$

so that $g_{11} = 1$, $g_{22} = (x^1)^2$, $g_{33} = 1$, $g_{ij} = 0$, if $i \neq j$, and it is easy to verify that the nonvanishing Christoffel symbols are

$$\begin{Bmatrix} 1 \\ 22 \end{Bmatrix} = -x^1, \qquad \begin{Bmatrix} 2 \\ 12 \end{Bmatrix} = \begin{Bmatrix} 2 \\ 21 \end{Bmatrix} = \frac{1}{x^1}.$$

The components of the tangent vector $\boldsymbol{\lambda}$ to the circle C are $\lambda^i = dx^i/ds$, so that $\lambda^1 = 0$, $\lambda^2 = d\theta/ds$, $\lambda^3 = 0$. Since $\boldsymbol{\lambda}$ is a unit vector, $g_{ij}\lambda^i\lambda^j = 1$ at all points of C, and this requires that

$$(x^1)^2\left(\frac{d\theta}{ds}\right)^2 = a^2\left(\frac{d\theta}{ds}\right)^2 = 1.$$

Therefore $(d\theta/ds)^2 = 1/a^2$, and the first formula in (50.5) yields

$$\begin{cases} \varkappa\mu^1 = \dfrac{d\lambda^1}{ds} + \begin{Bmatrix} 1 \\ jk \end{Bmatrix} \lambda^j \dfrac{dx^k}{ds} = \begin{Bmatrix} 1 \\ 22 \end{Bmatrix} \lambda^2 \dfrac{dx^2}{ds} = -\dfrac{1}{a}, \\ \varkappa\mu^2 = \dfrac{d\lambda^2}{ds} + \begin{Bmatrix} 2 \\ jk \end{Bmatrix} \lambda^j \dfrac{dx^k}{ds} = \begin{Bmatrix} 2 \\ 21 \end{Bmatrix} \lambda^2 \dfrac{dx^1}{ds} = 0, \\ \varkappa\mu^3 = \dfrac{d\lambda^3}{ds} + \begin{Bmatrix} 3 \\ jk \end{Bmatrix} \lambda^j \dfrac{dx^k}{ds} = 0. \end{cases}$$

Since $\boldsymbol{\mu}$ is a unit vector, $g_{ij}\mu^i\mu^j = 1$, and it follows that $\varkappa = 1/a$ and $\mu^1 = -1$, $\mu^2 = 0$, $\mu^3 = 0$.

An entirely analogous calculation shows that $\tau = 0$ and $\nu^1 = 0$, $\nu^2 = 0$, $\nu^3 = 1$.

Problems

1. Find the curvature and torsion at any point of the circular helix C whose equations in cylindrical coordinates are

$$C: \quad x^1 = a, \qquad x^2 = \theta, \qquad x^3 = k\theta.$$

Show that the tangent vector $\boldsymbol{\lambda}$ at every point of C makes a constant angle with the direction of the X^3-axis. Consider C also in the form $y^1 = a\cos\theta$, $y^2 = a\sin\theta$, $y^3 = k\theta$, where the coordinates y^i are rectangular cartesian.

2. Show that

$$\frac{\delta^2\lambda^i}{\delta s^2} = \frac{d\varkappa}{ds}\mu^i + \varkappa(\tau\nu^i - \varkappa\lambda^i)$$

$$\frac{\delta^2\mu^i}{\delta s^2} = \frac{d\tau}{ds}\nu^i - (\varkappa^2 + \tau^2)\mu^i - \frac{d\varkappa}{ds}\lambda^i,$$

$$\frac{\delta^2\nu^i}{\delta s^2} = \tau(\varkappa\lambda^i - \tau\nu^i) - \frac{d\tau}{ds}\mu^i.$$

3. Using results of Problem 1, show that the ratio of the curvature $\varkappa$ to the torsion τ is a constant. Show from Frenet's formulas that whenever $\tau/\varkappa =$ constant, and the coordinates are cartesian, $\nu^i = c\lambda^i + b^i$, where c and b^i are constants. From this result it follows that $\lambda^i b^i =$ constant, so that the curve makes a constant angle with the lines whose direction ratios are b^i. In other words, the curve is a cylindrical helix. This theorem is due to Bertrand.

4. When C is specified in the form

$$C: \quad y^i = y^i(s),$$

where the y^i are orthogonal cartesian coordinates and s is the arc parameter, show that

$$\varkappa^2 = [(y^1)'']^2 + [(y^2)'']^2 + [(y^3)'']^2$$

and

$$\tau\varkappa^2 = \begin{vmatrix} (y^1)' & (y^2)' & (y^3)' \\ (y^1)'' & (y^2)'' & (y^3)'' \\ (y^1)''' & (y^2)''' & (y^3)''' \end{vmatrix}.$$

5. Write equations of Problem 2 in cartesian coordinates y^i and show that when $\tau = 0$ and $\varkappa =$ constant along C, the equations of C are

$$y^i = A^i \cos \varkappa s + B^i \sin \varkappa s + C^i,$$

where $A^iA^i = B^iB^i = 1/\varkappa^2$, $A^iB^i = 0$. Thus C is a circle.

6. Let C be a cylindrical helix determined by

$$C: \begin{cases} y^1 = \phi(\sigma), \\ y^2 = \psi(\sigma), \\ y^3 = k\sigma, \quad k = \text{constant}, \end{cases}$$

where σ is the arc parameter of the directrix curve C' in the y^1y^2-plane, so that $(d\sigma)^2 = (dy^1)^2 + (dy^2)^2$. Note that $(ds)^2 = (1 + k^2)(d\sigma)^2$ and show that

$$\tau = \frac{\begin{vmatrix} \phi' & \psi' & k \\ \phi'' & \psi'' & 0 \\ \phi''' & \psi''' & 0 \end{vmatrix}}{\varkappa^2} \cdot \frac{1}{(1 + k^2)^3},$$

$$\varkappa = \frac{\phi'\psi'' - \psi'\phi''}{1 + k^2},$$

and verify that $\tau/\varkappa = k$.

51. Equations of a Straight Line

Let A^i be a vector field defined along a curve C in E_3, where C is given parametrically as

$$C: \quad x^i = x^i(s), \qquad s_1 \le s \le s_2, \qquad (i = 1, 2, 3),$$

s being the arc parameter. If the vector field A^i is parallel, then it follows from Sec. 48 that $\delta A^i/\delta s = 0$, or

$$\frac{dA^i}{ds} + \begin{Bmatrix} i \\ \alpha\beta \end{Bmatrix} A^\alpha \frac{dx^\beta}{ds} = 0. \tag{51.1}$$

We shall make use of equation 51.1 to obtain the equations of a straight line in general curvilinear coordinates. The characteristic property of straight lines is that the tangent vector $\boldsymbol{\lambda}$ to a straight line is directed along the straight line, so that the totality of tangent vectors $\boldsymbol{\lambda}$ forms a parallel vector field. Thus the field of tangent vector $\lambda^i = dx^i/ds$ must satisfy (51.1), and we have

$$\frac{\delta\lambda^i}{\delta s} = \frac{d^2x^i}{ds^2} + \begin{Bmatrix} i \\ \alpha\beta \end{Bmatrix} \frac{dx^\alpha}{ds}\frac{dx^\beta}{ds} = 0.$$

The equation

$$\frac{d^2x^i}{ds^2} + \begin{Bmatrix} i \\ \alpha\beta \end{Bmatrix} \frac{dx^\alpha}{ds}\frac{dx^\beta}{ds} = 0 \tag{51.2}$$

is the equation sought. In cartesian coordinates the Christoffel symbols vanish and we obtain the familiar form of differential equations of straight lines. From the geometric interpretation of the curvature $\varkappa$ as a measure of the rate of turning of the tangent line to a curve, we are led to define the curvature of a straight line to be zero. This definition is consistent with the first of Frenet's formulas 50.1.

52. Curvilinear Coordinates on a Surface

In the remainder of this chapter we will study the properties of surfaces imbedded in a three-dimensional Euclidean space. It will be shown that certain of these properties can be phrased independently of the space in which the surface is immersed and that they are concerned solely with the structure of the differential quadratic form for the element of arc of a curve drawn on the surface. All such properties of surfaces are termed the *intrinsic* properties, and the geometry based on the study of this differential quadratic form is called the *intrinsic geometry* of the surface.

We find it convenient to refer the space in which the surface is imbedded to a set of orthogonal cartesian axes Y, and regard the locus of points satisfying the equation

$$F(y^1, y^2, y^3) = 0 \tag{52.1}$$

as an analytical definition of a surface S. We suppose that only two of the variables y^i in (52.1) are independent and that the specification of, say, y^1 and y^2 in some region of the Y^1Y^2-plane determines uniquely a real number y^3 such that the left-hand member in (52.1) reduces to zero. If we suppose that $F(y^1, y^2, y^3)$, regarded as a function of three independent variables, is of class C^1 in some region R about the point $P_0(y_0^{\,1}, y_0^{\,2}, y_0^{\,3})$ with $\left.\dfrac{\partial F}{\partial y^3}\right|_{P_0} \neq 0$ and $F(y_0^{\,1}, y_0^{\,2}, y_0^{\,3}) = 0$, then the fundamental theorem on implicit functions guarantees the existence of a unique solution $y^3 = f(y^1, y^2)$, such that $y_0^{\,3} = f(y_0^{\,1}, y_0^{\,2})$.

The definition of the surface by means of a single equation 52.1 is less convenient than the one introduced by Gauss, who defined the surface as a locus of points satisfying three equations of the type

$$y^i = y^i(u^1, u^2), \tag{52.2}$$

where $u_1^{\,1} \leq u^1 \leq u_2^{\,1}$ and $u_1^{\,2} \leq u^2 \leq u_2^{\,2}$, and the y^i are real functions of class C^1 in the region of definition of the independent parameters u^1, u^2. In order to reconcile these two different definitions we shall require that the functions $y^i(u^1, u^2)$ be such that the Jacobian matrix

$$\begin{bmatrix} \dfrac{\partial y^1}{\partial u^1} & \dfrac{\partial y^2}{\partial u^1} & \dfrac{\partial y^3}{\partial u^1} \\[2ex] \dfrac{\partial y^1}{\partial u^2} & \dfrac{\partial y^2}{\partial u^2} & \dfrac{\partial y^3}{\partial u^2} \end{bmatrix} \tag{52.3}$$

be of rank two, so that not all the determinants of the second order selected from this matrix vanish identically in the region of definition of parameters u^i. This requirement ensures that it is possible to solve two equations in (52.2) for u^1 and u^2 in terms of some pair of variables y^i, and the substitution of these solutions in the remaining equation leads to an equation of the form $y^3 = y^3(y^1, y^2)$. It should be remarked that, if any two determinants formed from the matrix 52.3 vanish identically, then the third one also vanishes, provided that the surface S is not a plane parallel to one of the coordinate planes.

Since u^1 and u^2 are independent variables, the locus defined by equations 52.2 is two-dimensional, and these equations give the coordinates y^i of a point on the surface when u^1 and u^2 are assigned particular values. This point of view leads one to consider the surface as a two-dimensional manifold S imbedded in a three-dimensional enveloping space E_3. We can also study surfaces without reference to the surrounding space, and consider parameters u^1 and u^2 as coordinates of points in the surface. A familiar example of this is the use of the latitude and longitude as coordinates of points on the surface of the earth.

If we assign to u^1 in (52.2) some fixed value $u^1 = c$ (Fig. 22) we obtain as a locus the one-dimensional manifold

$$y^i = y^i(c, u^2), \qquad (i = 1, 2, 3),$$

which is a curve lying on the surface S defined by equations 52.2. We shall call this curve the u^2-curve. Similarly, setting $u^2 =$ constant in (52.2) defines the u^1-curve, along which only u^1 varies. By assigning to u^1 and u^2 a succession of fixed values, we obtain a net of curves, on the surface,

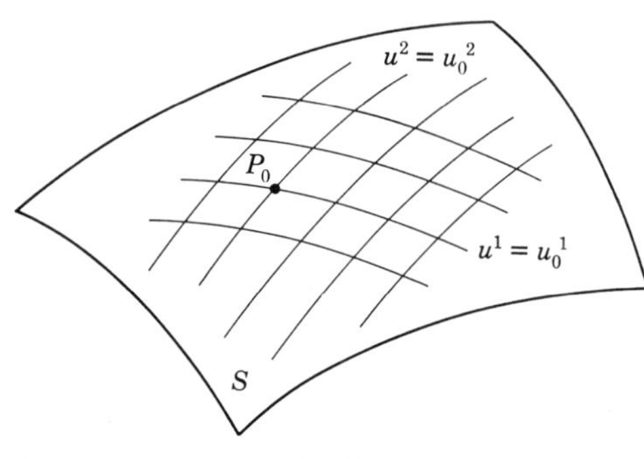

Fig. 22

which are termed the *coordinate curves*. The intersection of a pair of coordinate curves obtained by setting $u^1 = u_0^{\ 1}$, $u^2 = u_0^{\ 2}$ determines a point P_0. The variables u^1, u^2 determining the point P on S are called the *curvilinear*, or *Gaussian*, *coordinates on the surface*.

Obviously the parametric representation of a surface in the form 52.2 is not unique, and there are infinitely many curvilinear coordinate systems which can be used to locate points on a given surface S. Thus, if one introduces a transformation

$$
(52.4) \qquad \begin{cases} u^1 = u^1(\bar{u}^1, \bar{u}^2), \\ u^2 = u^2(\bar{u}^1, \bar{u}^2), \end{cases}
$$

where the $u^\alpha(\bar{u}^1, \bar{u}^2)$ are of class C^1 and are such that the Jacobian $J = \dfrac{\partial(u^1, u^2)}{\partial(\bar{u}^1, \bar{u}^2)}$ does not vanish in some region of the variables $\bar{u}^i$, then one can insert the values from (52.4) in (52.2) and obtain a different set of parametric equations,

$$
(52.5) \qquad y^i = f^i(\bar{u}^1, \bar{u}^2), \qquad (i = 1, 2, 3),
$$

defining the same surface S. Equations 52.4 can be looked upon as representing a *transformation of coordinates in the surface* precisely in the same way as equations $x^i = x^i(\bar{x}^1, \bar{x}^2, \bar{x}^3)$, $(i = 1, 2, 3)$ were viewed as defining a transformation of coordinates in E_3.

53. Intrinsic Geometry. First Fundamental Quadratic Form. Metric Tensor

We remarked in Sec. 52 that the properties of surfaces that can be described without reference to the space in which the surface is imbedded are termed *intrinsic* properties. A study of intrinsic properties is made to depend on a certain quadratic differential form describing the metric character of the surface. We proceed to derive this quadratic form.

It will be convenient to adopt certain conventions concerning the meaning of the indices to be used in this and remaining sections of this chapter. We will be dealing with two distinct sets of variables: those referring to the space E_3 in which the surface is imbedded (these are three in number), and with two curvilinear coordinates u^1 and u^2 referring to the two-dimensional manifold S. In order not to confuse these sets of variables we shall use Latin letters for the indices referring to the space variables and Greek letters for the surface variables. Thus Latin indices will assume values 1, 2, 3, and Greek indices will have the range of values 1, 2. A transformation T of space coordinates from one system X to another system $\bar{X}$ will be written as

$$T:\quad x^i = x^i(\bar{x}^1, \bar{x}^2, \bar{x}^3);$$

a transformation of Gaussian surface coordinates, such as described by equations 52.4, will be denoted by

$$u^\alpha = u^\alpha(\bar{u}^1, \bar{u}^2).$$

A repeated Greek index in any term denotes the summation from 1 to 2; a repeated Latin index represents the sum from 1 to 3. Unless a statement to the contrary is made, we shall suppose that all functions appearing in the discussion of the remainder of this chapter are of class C^2 in the regions of their definition.

Consider a surface S defined by

$$y^i = y^i(u^1, u^2), \tag{53.1}$$

where the y^i are the orthogonal cartesian coordinates covering the space E_3 in which the surface S is imbedded, and a curve C on S defined by

$$u^\alpha = u^\alpha(t), \qquad t_1 \leq t \leq t_2, \tag{53.2}$$

where the u^α's are the Gaussian coordinates covering S. Viewed from the surrounding space, the curve defined by (53.2) is a curve in a three-dimensional Euclidean space and its element of arc is given by the formula

$$ds^2 = dy^i\, dy^i. \tag{53.3}$$

From (53.1) we have

$$dy^i = \frac{\partial y^i}{\partial u^\alpha}\, du^\alpha; \tag{53.4}$$

where, as is clear from (53.2),

$$du^\alpha = \frac{du^\alpha}{dt}\, dt.$$

Substituting from (53.4) in (53.3), we get

$$ds^2 = \frac{\partial y^i}{\partial u^\alpha}\frac{\partial y^i}{\partial u^\beta}\, du^\alpha\, du^\beta$$

$$= a_{\alpha\beta}\, du^\alpha\, du^\beta,$$

where

$$a_{\alpha\beta} \equiv \frac{\partial y^i}{\partial u^\alpha}\frac{\partial y^i}{\partial u^\beta}. \tag{53.5}$$

The expression for ds^2, namely,

$$ds^2 = a_{\alpha\beta}\, du^\alpha\, du^\beta, \tag{53.6}$$

is the square of the *linear element of C lying on the surface S*, and the right-hand member of (53.6) is called the *first fundamental quadratic form of the surface*. The length of arc of the curve defined by (53.2) is given by the formula

$$s = \int_{t_1}^{t_2} \sqrt{a_{\alpha\beta}\dot{u}^\alpha \dot{u}^\beta}\, dt,$$

where $\dot{u}^\alpha = du^\alpha/dt$. Since in a nontrivial case $ds^2 > 0$, it follows at once from (53.6) upon setting $u^2 =$ constant and $u^1 =$ constant in turn, that $ds^2_{(1)} = a_{11}(du^1)^2$ and $ds^2_{(2)} = a_{22}(du^2)^2$. Thus a_{11} and a_{22} are positive functions of u^1 and u^2.

Consider a transformation of surface coordinates

$$u^\alpha = u^\alpha(\bar{u}^1, \bar{u}^2), \tag{53.7}$$

with a nonvanishing Jacobian $J = \left|\frac{\partial u^\alpha}{\partial \bar{u}^\beta}\right|$. It follows from (53.7) that

$$du^\alpha = \frac{\partial u^\alpha}{\partial \bar{u}^\gamma}\, d\bar{u}^\gamma,$$

and hence (53.6) yields

$$ds^2 = a_{\alpha\beta}\frac{\partial u^\alpha}{\partial \bar{u}^\gamma}\frac{\partial u^\beta}{\partial \bar{u}^\delta}\, d\bar{u}^\gamma\, d\bar{u}^\delta.$$

If we set

$$\bar{a}_{\gamma\delta} = a_{\alpha\beta}\frac{\partial u^\alpha}{\partial \bar{u}^\gamma}\frac{\partial u^\beta}{\partial \bar{u}^\delta},$$

we see that the set of quantities $a_{\alpha\beta}$ represents a symmetric covariant tensor of rank two with respect to the admissible transformations 53.7 of surface coordinates. The fact that the $a_{\alpha\beta}$ are components of a tensor is also evident from (53.6), since ds^2 is an invariant and the quantities $a_{\alpha\beta}$ are symmetric. The tensor $a_{\alpha\beta}$ is called the *covariant metric tensor* of the surface.

Since the form 53.6 is positive definite, the determinant

$$a = \begin{vmatrix} a_{11} & a_{12} \\ a_{21} & a_{22} \end{vmatrix} > 0,$$

and we can define the reciprocal tensor $a^{\alpha\beta}$ (see Sec. 30) by the formula $a^{\alpha\beta}a_{\beta\gamma} = \delta^\alpha_\gamma$. Thus we have

$$a^{11} = \frac{a_{22}}{a}, \qquad a^{12} = a^{21} = \frac{-a_{12}}{a}, \qquad a^{22} = \frac{a_{11}}{a}.$$

The contravariant tensor $a^{\alpha\beta}$ is called the *contravariant metric tensor.*

We can repeat, almost verbatim, the contents of Sec. 44 concerning metric properties of our two-dimensional space S. Thus the direction of a linear element in the surface can be specified either by the *direction cosines* dy^i/ds, $(i = 1, 2, 3)$, or by the *direction parameters*

$$\lambda^\alpha = \frac{du^\alpha}{ds}. \tag{53.8}$$

For,

$$\frac{dy^i}{ds} = \frac{\partial y^i}{\partial u^\alpha}\frac{du^\alpha}{ds}$$

and the du^α/ds are uniquely determined when the direction cosines dy^i/ds are specified, and conversely. We define the length of the *surface vector* A^α, that is, the vector determined by $A^1(u^1, u^2)$ and $A^2(u^1, u^2)$, by the formula[9]

$$A = \sqrt{a_{\alpha\beta}A^\alpha A^\beta}.$$

It follows from (53.6) that

$$\begin{aligned} 1 &= a_{\alpha\beta}\frac{du^\alpha}{ds}\frac{du^\beta}{ds} \\ &= a_{\alpha\beta}\lambda^\alpha\lambda^\beta, \end{aligned}$$

so that the direction parameters λ^α are components of a unit vector. The covariant vector

$$\lambda_\beta \equiv a_{\alpha\beta}\lambda^\alpha \tag{53.9}$$

is sometimes called the *direction moment*, and it is clear from (53.9) that

$$a^{\gamma\beta}\lambda_\beta = a^{\gamma\beta}a_{\alpha\beta}\lambda^\alpha = \delta_\alpha{}^\gamma\lambda^\alpha = \lambda^\gamma,$$

and that

$$\lambda^\alpha\lambda_\alpha = a_{\alpha\beta}\lambda^\beta\lambda^\alpha. \tag{53.10}$$

[9] The components $\bar{A}^i$ of the vector A^α, as viewed from the enveloping space E_3, are given by $\bar{A}^i = \dfrac{\partial y^i}{\partial u^\alpha}A^\alpha$, and it is clear that $\bar{A}^i\bar{A}^i = \dfrac{\partial y^i}{\partial u^\alpha}\dfrac{\partial y^i}{\partial u^\beta}A^\alpha A^\beta = a_{\alpha\beta}A^\alpha A^\beta$.

54. Angle between Two Intersecting Curves in a Surface. Element of Surface Area

The equations of a curve C drawn on the surface S can be written in the form

$$C: \quad u^\alpha = u^\alpha(t), \qquad t_1 \le t \le t_2.$$

Since the $u^\alpha(t)$ are assumed to be of class C^2, the curve C has a continuously turning tangent. Let C_1 and C_2 be two such curves intersecting at the point P of S (Fig. 23). We take the equations of S, referred to orthogonal cartesian axes Y, in the form

$$y^i = y^i(u^1, u^2), \tag{54.1}$$

and denote the direction cosines of the tangent lines to C_1 and C_2 at P by ξ^i and η^i, respectively. The cosine of the angle θ between C_1 and C_2, calculated by a geometer in the enveloping space E_3, is

$$\cos\theta = \xi^i \eta^i. \tag{54.2}$$

However,

$$\xi^i = \frac{\partial y^i}{\partial u^\alpha}\frac{d_1 u^\alpha}{ds_1} = \frac{d_1 y^i}{ds_1},$$

$$\eta^i = \frac{\partial y^i}{\partial u^\beta}\frac{d_2 u^\beta}{ds_2} = \frac{d_2 y^i}{ds_2},$$

where the subscripts 1 and 2 refer to the elements of arc of C_1 and C_2, respectively. Using the definition 53.8, we can write the unit vectors in the directions of the tangents to C_1 and C_2 as

$$\lambda^\alpha = \frac{d_1 u^\alpha}{ds_1}, \qquad \mu^\alpha = \frac{d_2 u^\alpha}{ds_2},$$

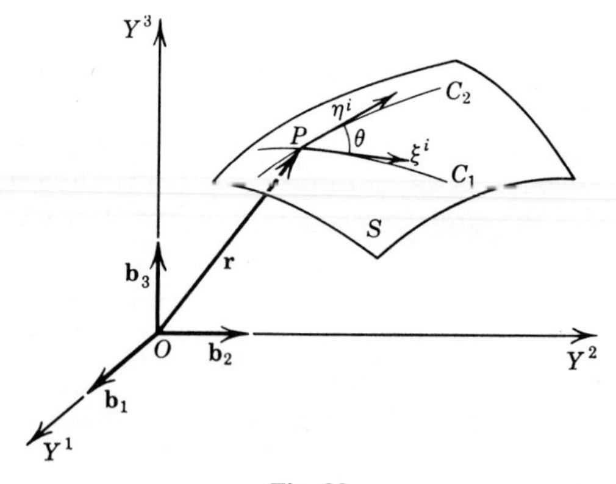

Fig. 23

and

$$\xi^i = \frac{\partial y^i}{\partial u^\alpha} \lambda^\alpha, \qquad \eta^i = \frac{\partial y^i}{\partial u^\beta} \mu^\beta. \tag{54.3}$$

Inserting in (54.2) the expressions from (54.3), we get

$$\cos\theta = \frac{\partial y^i}{\partial u^\alpha}\frac{\partial y^i}{\partial u^\beta}\lambda^\alpha\mu^\beta,$$

and since

$$a_{\alpha\beta} = \frac{\partial y^i}{\partial u^\alpha}\frac{\partial y^i}{\partial u^\beta},$$

the foregoing expression can be written

$$\cos\theta = a_{\alpha\beta}\lambda^\alpha\mu^\beta. \tag{54.4}$$

If the curves C_1 and C_2 are orthogonal,

$$a_{\alpha\beta}\lambda^\alpha\mu^\beta = 0. \tag{54.5}$$

In particular, if the surface vectors λ^α and μ^β are taken along the coordinate curves ($\lambda^1 = 1/\sqrt{a_{11}}$, $\lambda^2 = 0$, $\mu^1 = 0$, $\mu^2 = 1/\sqrt{a_{22}}$), then it follows from (54.5) that the *coordinate curves will form an orthogonal net if, and only if, $a_{12} = 0$ at every point of the surface.*

We can give a pictorial interpretation of these results in the manner of Sec. 45. Thus, if $\mathbf{r}$ denotes the position vector of any point P on the surface S, and the $\mathbf{b}_i$ are the unit vectors directed along the orthogonal coordinate axes Y, then equations 53.1 of the surface S can be written in vector form (see Fig. 24) as

$$\mathbf{r}(u^1, u^2) = \mathbf{b}_i y^i(u^1, u^2).$$

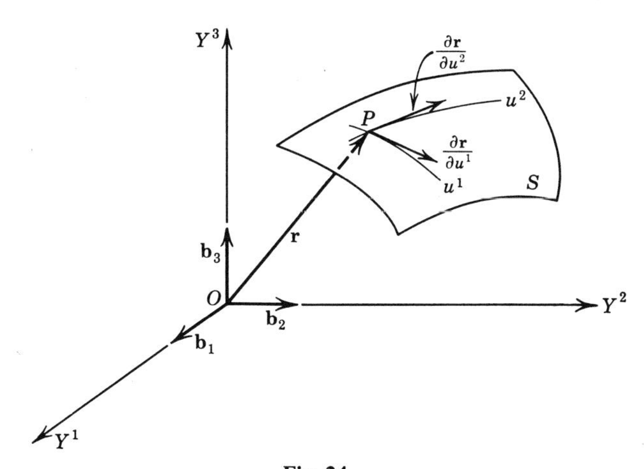

Fig. 24

It follows from this representation of S that

$$ds^2 = d\mathbf{r} \cdot d\mathbf{r} = \frac{\partial \mathbf{r}}{\partial u^\alpha} \cdot \frac{\partial \mathbf{r}}{\partial u^\beta} du^\alpha du^\beta$$
$$= a_{\alpha\beta} du^\alpha du^\beta,$$

where

$$a_{\alpha\beta} = \frac{\partial \mathbf{r}}{\partial u^\alpha} \cdot \frac{\partial \mathbf{r}}{\partial u^\beta}. \tag{54.6}$$

Setting $\partial \mathbf{r}/\partial u^\alpha = \mathbf{a}_\alpha$, where $\mathbf{a}_1$ and $\mathbf{a}_2$ are obviously tangent vectors to the coordinate curves, we see that

$$a_{11} = \mathbf{a}_1 \cdot \mathbf{a}_1, \qquad a_{12} = \mathbf{a}_1 \cdot \mathbf{a}_2, \qquad a_{22} = \mathbf{a}_2 \cdot \mathbf{a}_2.$$

In the notation of (54.3) the space components of $\mathbf{a}_1$ and $\mathbf{a}_2$ are ξ^i and η^i, respectively.

We can define an element of area $d\sigma$ of the surface S by the formula

$$d\sigma = |\mathbf{a}_1 \times \mathbf{a}_2| \, du^1 \, du^2,$$

and it is readily verified that the right-hand member of this expression can be written

$$d\sigma = \sqrt{a_{11}a_{22} - a_{12}^2} \, du^1 \, du^2$$
$$= \sqrt{a} \, du^1 \, du^2. \tag{54.7}$$

This formula has precisely the same structure as the expression 44.11 for the volume element.

It follows from Sec. 40 that the skew-symmetric e-systems, in a two-dimensional manifold, can be defined by the formulas

$$e_{11} = e_{22} = e^{11} = e^{22} = 0, \qquad e^{12} = -e^{21} = e_{12} = -e_{21} = 1,$$

and, since these systems are relative tensors (see Sec. 41), the expressions

$$\epsilon_{\alpha\beta} = \sqrt{a} \, e_{\alpha\beta} \qquad \text{and} \qquad \epsilon^{\alpha\beta} = \frac{1}{\sqrt{a}} e^{\alpha\beta}$$

are absolute tensors. Using the ϵ-symbols, we can write the sine of the angle θ between two unit vectors λ^α, μ^α in the form

$$\epsilon_{\alpha\beta}\lambda^\alpha\mu^\beta = \sin\theta,$$

which is numerically equal to the area of the parallelogram constructed on the unit vectors λ^α and μ^α. It follows from this result that *a necessary and sufficient condition for the orthogonality of two surface unit-vectors* λ^α *and* μ^α *is* $|\epsilon_{\alpha\beta}\lambda^\alpha\mu^\beta| = 1$.

Problems

1. Show that the cosine of the angle θ between the coordinate curves u^1 and u^2 on S is $\cos\theta = a_{12}/\sqrt{a_{11}a_{22}}$.

2. Find the element of area of the surface of the sphere of radius r if the equations of the surface are given in the form:

$$y^1 = r\sin u^1 \cos u^2, \quad y^2 = r\sin u^1 \sin u^2, \quad y^3 = r\cos u^1,$$

where the y^i are orthogonal cartesian coordinates. (Note that in this case $a_{11} = r^2$, $a_{12} = 0$, $a_{22} = r^2\sin^2 u^1$.)

55. Fundamental Concepts of Calculus of Variations

The most celebrated problem of intrinsic geometry of surfaces is concerned with the determination of curves of shortest length joining two specified points on the surface. This is the problem of geodesics. This problem has such profound implications on the formulation of the fundamental principles of optics, dynamics, and mechanics of deformable media that it is desirable to treat it in greater generality than one would if concerned solely with the geometry of surfaces imbedded in E_3. To do this we shall draw on certain concepts in the calculus of variations. Since we will be concerned with the study of extremal properties of integrals, we shall recall some salient facts about the problem of relative maxima and minima of functions of several independent variables.

Let $f(x^1, x^2, \ldots, x^n)$ be a continuous function of n independent variables x^i defined in a bounded, closed region R. We are interested in determining a point $P(x)$ of R at which f attains an extreme value in comparison with the values of f in a certain neighborhood of the point $P(x)$. There is no doubt about the existence of maximum or minimum of f since it is known that *every function continuous in a bounded closed region attains its maximum and minimum values either in the interior or on the boundary of the region.*[10] Moreover, if

$$f(x^1, \ldots, x^n)$$

is a differentiable function, then at interior points of the region, where the function attains its extreme values, $\partial f/\partial x^i = 0$, $(i = 1, 2, \ldots, n)$. The vanishing of the derivatives of $f(x^1, x^2, \ldots, x^n)$ obviously is not a sufficient condition for an extremum. We will call the points of the region R at which $\partial f/\partial x^i$ vanish simultaneously the *stationary points* of $f(x^1, \ldots, x^n)$. The determination of stationary points is studied in advanced calculus, and we assume that this subject is quite familiar to the reader.

Calculus of variations is also concerned with the determination of extreme or stationary values of certain expressions, but there is an

[10] This theorem is due to Weierstrass.

important distinction in that in calculus of variations we deal with extremes of functionals rather than functions of a finite number of variables. By a functional we understand a function depending on the changes of one or several functions, which assume the roles of the arguments. As an example of a functional consider the formula

$$s = \int_{x_0}^{x_1} \sqrt{1 + (dy/dx)^2}\, dx,$$

defining the length of a plane curve $y = y(x)$ joining the points whose abscissas are x_0 and x_1. Here the value of s depends on the behavior of the functional argument $y(x)$, and the class of functions $y(x)$ on which the functional s depends is in some measure arbitrary. Thus, one might consider the problem of determining the extremes of s when $y(x)$ is an arbitrary continuous function with a piecewise continuous first derivative.

In the study of extremes of continuous functions $f(x^1, \ldots, x^n)$, of a finite number of independent variables x^i, we must specify the region R within which f is defined, whereas in the study of extremes of functionals we must characterize the class of *admissible functional arguments*. For example, we may demand that the functional arguments possess certain properties of continuity, or behave in some specified fashion at the end points of the interval, and so on. We will be concerned with relative extremes of functionals, that is, extremes relative to a certain "neighborhood" of functional arguments for which the functional takes on an extreme value, just as we were with relative maxima and minima of functions. In order to make the notion of the *neighborhood of a function* precise, we introduce a

DEFINITION. *A function* $g(x^1, x^2, \ldots, x^n)$ *belongs to the h-neighborhood of the function* $f(x^1, \ldots, x^n)$, *provided that* $|f - g| < h$, $h > 0$, *for all values of the independent variables* $x^1, x^2, \ldots, x^n$ *in the interior of* R.

With the aid of this definition, we can formulate the fundamental problem of the calculus of variations as follows: *Find, within the class of admissible functional arguments, those functions f that yield extreme values for the functional under consideration in comparison with the values given the functional by functions belonging to some h neighborhood of f.*

A word concerning the difficulties inherent in this problem is in order. We have already remarked that in the theory of maxima and minima of continuous functions of several independent variables the existence of extreme values is guaranteed by the theorem of Weierstrass. In the problem of calculus of variations, on the other hand, it may happen that the problem is formulated without internal inconsistencies, and yet it has no solution because of the limitations imposed on the class of admissible functional arguments. For example, let it be required to join two given points on the X-axis by the shortest curve with continuous curvature so that the curve is orthogonal to the X-axis at the end points. This problem has no solution because the length of every admissible curve is always greater than the length of the straight line joining the given points. We can always find a curve of admissible type whose length differs from the length of the straight line by as little as desired so that there exists a lower bound of the functional, but this lower bound is not the minimum attained for any curve of the class of curves under consideration. It follows from this example that in each variational problem we are confronted with the question of the existence of a solution of the problem.

In order to deduce the differential equations furnishing a set of necessary conditions for an extremum of a functional, we need the following **fundamental lemma of calculus of variations.**

If the integral $\int_{t_1}^{t_2} \xi(t)M(t)\,dt$, *where* $M(t)$ *is a continuous function of* t *in the interval* $t_1 \leq t \leq t_2$, *vanishes for every choice of the function* $\xi(t)$ *of class* C^n *in* $t_1 \leq t \leq t_2$, *and which is such that* $\xi(t_1) = \xi(t_2) = 0$, *then* $M(t)$ *is identically zero in the interval* $t_1 \leq t \leq t_2$.

We shall prove the lemma by assuming that $M(t) \not\equiv 0$ and reaching a contradiction. Assume $M(t) \neq 0$ at some point t' of $t_1 < t < t_2$, and suppose that $M(t') > 0$. Since $M(t)$ is continuous, there exists a number $\delta > 0$ such that $M(t) > 0$ in the interval $(t' - \delta, t' + \delta)$. Define a function $\xi(t)$ as follows:

$$\begin{aligned} &\xi(t) \equiv 0, \quad \text{in } t_1 \leq t \leq \tau_1, \quad \text{where } \tau_1 = t' - \delta, \\ &\xi(t) \equiv 0, \quad \text{in } \tau_2 \leq t \leq t_2, \quad \text{where } \tau_2 = t' + \delta, \\ &\xi(t) \equiv (t - \tau_1)^{2n+2}(t - \tau_2)^{2n+2}, \qquad \text{in } \tau_1 \leq t \leq \tau_2. \end{aligned}$$

The function $\xi(t)$ is surely of class C^n in (t_1, t_2) and $\xi(t_1) = \xi(t_2) = 0$. For this function, however,

$$\int_{t_1}^{t_2} \xi(t)M(t)\,dt = \int_{\tau_1}^{\tau_2} \xi(t)M(t)\,dt > 0,$$

since the integrand is always positive in $\tau_1 < t < \tau_2$. Thus we reach a contradiction, and hence our assumption that $M(t) \not\equiv 0$ is not tenable.

56. Euler's Equation in the Simplest Case

The simplest problem of the calculus of variations is concerned with the determination of extremals of a functional

$$J(x) = \int_{t_1}^{t_2} F(t, x, \dot{x})\,dt, \tag{56.1}$$

where $F(t, x, \dot{x})$ is a prescribed real function of its real arguments t, x, and $\dot{x} \equiv dx/dt$. We shall suppose that $F(t, x, \dot{x})$ is of class C^2, in some region R of the plane (x, t), for all values of $\dot{x}$.[11] Concerning the class of admissible functions $x(t)$, we suppose that the values $x(t_1)$ and $x(t_2)$ are prescribed in advance and that $x(t)$ is also of class C^2 in $t_1 \leq t \leq t_2$.

Our problem is to find a function

$$x = f(t), \qquad t_1 \leq t \leq t_2,$$

called an *extremal* for the integral 56.1, such that $J(x)$ for $x = f(t)$ assumes an extreme value in comparison with the values given to J by the admissible functions in a sufficiently small h-neighborhood of the function $x = f(t)$. In other words, admissible functions $x(t)$ are such that $|x(t) - f(t)| < h$ for $t_1 \leq t \leq t_2$. We shall deduce next a necessary condition for an extremum of J. Consider a function $\xi(t)$ of class C^2, such that $\xi(t_1) = \xi(t_2) = 0$, and form a set of functions

$$\bar{x}(t) = x(t) + \epsilon\xi(t) \equiv x + \delta x,$$

where ϵ is an arbitrary numerical parameter near zero. The functions $\bar{x}(t)$ clearly assume the same values at the end points of the interval (t_1, t_2) as $x(t)$. We shall call the $\bar{x}(t)$ the *varied functions*, and the quantity $\epsilon\xi(t) \equiv \delta x$ the *variation of the function* $x = f(t)$. For sufficiently small values of ϵ all varied functions $\bar{x}(t)$ will be contained in the h-neighborhood of the extremal $x = f(t)$. Consequently, the integral 56.1,

$$J(\bar{x}) = J(x + \epsilon\xi) \equiv \Phi(\epsilon),$$

considered as a function of ϵ, will have an extreme value for $\epsilon = 0$. A necessary condition that this be so is $\Phi'(0) = 0$.

Because of the restrictions imposed on functions under consideration, the integral

$$\Phi(\epsilon) = \int_{t_1}^{t_2} F(t, x + \epsilon\xi, \dot{x} + \epsilon\dot{\xi})\, dt$$

can be differentiated under the integral sign, and we obtain as a necessary condition for an extremum the equation

$$\Phi'(0) = \int_{t_1}^{t_2} (F_x\xi + F_{\dot{x}}\dot{\xi})\, dt = 0, \tag{56.2}$$

[11] These restrictions are more severe than necessary, but we have in mind certain geometrical problems in which the continuity of second derivatives is a desirable property.

which must be true for every $\xi(t)$ satisfying the conditions laid down in the definition of $\xi(t)$. Integrating (56.2) by parts, we get

$$\int_{t_1}^{t_2} F_x\xi(t)\,dt + F_{\dot{x}}\xi(t)\bigg|_{t_1}^{t_2} - \int_{t_1}^{t_2} \xi(t)\frac{dF_{\dot{x}}}{dt}\,dt = 0, \tag{56.3}$$

and, since $\xi(t_1) = \xi(t_2) = 0$, the foregoing equation simplifies to

$$\int_{t_1}^{t_2} \xi(t)\left[F_x - \frac{dF_{\dot{x}}}{dt}\right] dt = 0. \tag{56.4}$$

Since $\xi(t)$ satisfies the restrictions imposed on $\xi(t)$ in the lemma of Sec. 55, we deduce from (56.4) that a necessary condition for an extremum of (56.1) is that $x(t)$ satisfy the differential equation

$$F_x - \frac{dF_{\dot{x}}}{dt} = 0. \tag{56.5}$$

Expanding (56.5) we obtain

$$F_{\dot{x}\dot{x}}\frac{d^2x}{dt^2} + F_{\dot{x}x}\frac{dx}{dt} + F_{\dot{x}t} - F_x = 0, \tag{56.6}$$

where the subscripts denote the derivatives of $F(t, x, \dot{x})$ with t, x, and $\dot{x}$ regarded as the independent variables. In order to determine $x(t)$ we must solve this ordinary differential equation subject to the end conditions $x(t_1) = x_1$ and $x(t_2) = x_2$. Equations 56.5 and 56.6 were first deduced by Euler and are called *Euler's equations.*

The expression (see equation 56.2)

$$\epsilon\Phi'(0) = \epsilon\int_{t_1}^{t_2} [\xi(t)F_x + \dot{\xi}(t)F_{\dot{x}}]\,dt,$$

which is akin to the differential of the function $\Phi(\epsilon)$ evaluated at $\epsilon = 0$, is called the *first variation of the integral J*, and is denoted by the symbol δJ. Thus

$$\delta J \equiv \epsilon\Phi'(0).$$

Taking into account the left-hand member of equation 56.3 and the definition of δJ, we can write

$$\delta J = [F_{\dot{x}}\,\delta x]_{t=t_1}^{t=t_2} + \int_{t_1}^{t_2}\left(F_x - \frac{d}{dt}F_{\dot{x}}\right)\delta x\,dt, \tag{56.7}$$

where $\delta x \equiv \epsilon\xi(t)$. Since the right-hand member of (56.7) vanishes when $x(t)$ is an extremal, we can state a

THEOREM. *A necessary condition for an extremum of the functional $J(x)$ is the vanishing of its first variation.*

57. Euler's Equations for a Functional of Several Arguments

Consider next the case of a functional J depending on several functional arguments x^i, $(i = 1, 2, \ldots, n)$, where[12]

$$J = \int_{t_1}^{t_2} F(t, x^1, x^2, \ldots, x^n, \dot{x}^1, \dot{x}^2, \ldots, \dot{x}^n)\, dt. \tag{57.1}$$

As in Sec. 56 we assume that F is a real function of class C^2 in the $2n + 1$-dimensional space of the real variables $t, x^1, \ldots, x^n, \dot{x}^1, \ldots, \dot{x}^n$.

We suppose that there exists a set of functions

$$x^i = f^i(t), \qquad t_1 \leq t \leq t_2, \qquad (i = 1, 2, \ldots, n), \tag{57.2}$$

whose values at the end points of the interval are known, and which are such that (57.1) assumes an extreme value in comparison with the values given to J by a class of admissible functions belonging to the h-neighborhood of (57.2). We introduce n arbitrary functions $\xi^i = \xi^i(t)$, $t_1 \leq t \leq t_2$, of class C^2 which vanish for $t = t_1$ and $t = t_2$, and construct a family of admissible functions

$$\bar{x}^i = x^i(t) + \epsilon\xi^i(t), \tag{57.3}$$

where the parameter ϵ is so chosen as to make the varied paths 57.3 lie in the h neighborhood of the curve 57.2.

As in Sec. 56 we form the function

(57.4)

$$\Phi(\epsilon) = \int_{t_1}^{t_2} F(t, x^1 + \epsilon\xi^1, \ldots, x^n + \epsilon\xi^n, \dot{x}^1 + \epsilon\dot{\xi}^1, \ldots, \dot{x}^n + \epsilon\dot{\xi}^n)\, dt,$$

which, by hypothesis, has an extremum for $\epsilon = 0$; hence,

$$\left.\frac{d\Phi}{d\epsilon}\right|_{\epsilon=0} = 0. \tag{57.5}$$

It follows that

$$\delta J = \epsilon \int_{t_1}^{t_2} [(F_{x^1}\xi^1 + F_{\dot{x}^1}\dot{\xi}^1) + \cdots + (F_{x^n}\xi^n + F_{\dot{x}^n}\dot{\xi}^n)]\, dt = 0, \tag{57.6}$$

and the integration by parts gives

$$\delta J = \epsilon\left[F_{\dot{x}^1}\xi^1 \Big|_{t_1}^{t_2} + \cdots + F_{\dot{x}^n}\xi^n \Big|_{t_1}^{t_2} + \int_{t_1}^{t_2} \xi^1 \left(F_{x^1} - \frac{d}{dt} F_{\dot{x}^1}\right) dt + \cdots + \int_{t_1}^{t_2} \xi^n \left(F_{x^n} - \frac{d}{dt} F_{\dot{x}^n}\right) dt \right] = 0.$$

[12] To ensure the independence of the integral 57.1 of special modes of parametrization, we suppose that $F(t,x,\dot{x})$ is positively homogeneous of degree one in the $\dot{x}^i$ (See Sec. 43).

Since the ξ^i are arbitrary and vanish at the end points of the interval, we conclude from the fundamental lemma that

$$F_{x^i} - \frac{d}{dt} F_{\dot{x}^i} = 0, \qquad (i = 1, 2, \ldots, n), \tag{57.7}$$

or

$$F_{x^i} - \dot{x}^j F_{\dot{x}^i x^j} - \ddot{x}^j F_{\dot{x}^i \dot{x}^j} - F_{\dot{x}^i t} = 0.$$

This set of n ordinary differential equations of second order is called the *Euler equations* for the variational problem associated with the functional 57.1. Thus, to obtain the set of functions 57.2, we must determine the solution of the system 57.7 satisfying the end conditions

$$x_1{}^i = f^i(t_1), \qquad x_2{}^i = f^i(t_2), \qquad (i = 1, 2, \ldots, n). \tag{57.8}$$

The problem discussed in this section appears to be entirely analogous to the simpler one treated in Sec. 56, but there is a distinction in that the vanishing of the first variation of (57.1) is a necessary condition not only for an extremum but also for a mixed maximum and minimum, the so-called minimax. An integral $J(x^1, \ldots, x^n)$ may attain a maximum when the function $x^1(t)$ is varied and a minimum in the course of the variation of $x^2(t)$. The saddle point of a hyperbolic paraboloid, studied in the elementary theory of maxima and minima, is a simple illustration of this circumstance. We will call the solutions of Euler's equations 57.7 satisfying the end conditions 57.8 the *extremals of the functional J*. This term will be used regardless of the nature of the stationary value assumed by the functional J, be it a maximum, minimum, or neither.

In our derivation of Euler's equations 57.7, we assumed that the variables x^i are independent. When the x^i are constrained by a set of $k < n$ functional relations of the form

$$\phi_j(t, x^1, x^2, \ldots, x^n) = 0, \qquad (j = 1, 2, \ldots, k),$$

the set of appropriate Euler's equations can be deduced by considering the free extremum of a certain new functional introduced by Lagrange.

To clarify the essential differences in the problems of free and constrained extrema, consider the functional

$$J = \int_{t_1}^{t_2} F(t, x^1, x^2, \dot{x}^1, \dot{x}^2)\, dt, \tag{57.9}$$

in which the variables are constrained by the relation

$$\phi(t, x^1, x^2) = 0. \tag{57.10}$$

We suppose that the extremal,

$$x^i = x^i(t), \qquad t_1 \le t \le t_2, \qquad (i = 1, 2),$$

satisfies the end condition of the type 57.8. When the constraining condition 57.10 is written in the form

$$\phi(x, y, z) = 0 \tag{57.11}$$

by setting $x = t$, $y = x^1$, $z = x^2$, equation 57.11 can be thought to represent a surface referred to a set of cartesian xyz-axes. The extremal must lie on this surface and we suppose that (57.11) can be solved for z in the neighborhood of the extremal to yield a differentiable function

$$z = f(x, y). \tag{57.12}$$

The substitution from (57.12) in (57.9) then yields an integral of the form

$$J = \int_{x_1}^{x_2} \mathscr{F}(x, y, y')\, dx, \tag{57.13}$$

in which the variables x and y are independent, and we can obtain the Euler equation by minimizing (57.13) on a set of admissible paths that satisfy the end conditions $y(x_1) = y_1$, $y(x_2) = y_2$. This is the problem of the free extremum already considered in Sec. 56.

However, such reduction of the problem of constrained extremum to the problem of a free extremum of the functional 57.13 is usually inconvenient because an explicit solution 57.12 of equation 57.11 may prove unwieldy. In this event we can follow a procedure similar to that of the Lagrange multiplier method of obtaining the relative extreme values of functions of several variables constrained by relations of the type 57.11.

We suppose that $d\phi/dz \neq 0$, so that it is theoretically possible to obtain the solution of (57.11). If this is so, (57.9) can be rewritten in the form

$$J = \int_{x_1}^{x_2} F(x, y, y', f, f_x + f_y y')\, dx, \tag{57.14}$$

since $z' = f_x + f_y y'$.

The integrand in (57.14) is a function of x, y, and y', which we denote by

$$\mathscr{F}(x, y, y') \equiv F(x, y, y', f, f_x + f_y y'). \tag{57.15}$$

Thus the Euler equation associated with the integral 57.14 is

$$\frac{\partial \mathscr{F}}{\partial y} - \frac{d}{dx}\frac{\partial \mathscr{F}}{\partial y'} = 0. \tag{57.16}$$

On noting (57.15), we see that

$$\frac{\partial \mathscr{F}}{\partial y} = F_y + F_z f_y + F_{z'}(f_{xy} + f_{yy} y'),$$

$$\frac{\partial \mathscr{F}}{\partial y'} = F_{y'} + F_{z'} f_y,$$

so that (57.16) yields

$$F_y + F_z f_y + F_{z'}(f_{xy} + f_{yy}y') - \frac{dF_{y'}}{dx} - f_y \frac{dF_{z'}}{dx} - F_{z'}(f_{xy} + f_{yy}y') = 0,$$

or

$$F_y + f_y\left(F_z - \frac{dF_{z'}}{dx}\right) - \frac{dF_{y'}}{dx} = 0.$$

Thus

$$f_y = -\frac{\dfrac{dF_{y'}}{dx} - F_y}{\dfrac{dF_{z'}}{dx} - F_z}, \tag{57.17}$$

since f_y is assumed to be defined along the extremal.

On the other hand, the differentiation of (57.11) yields

$$\phi_y + \phi_z f_y = 0,$$

so that

$$f_y = -\frac{\phi_y}{\phi_z}. \tag{57.18}$$

The expressions for f_y, given by (57.18) and (57.17) *along the extremal*, represent the same function of x; hence

$$\frac{\dfrac{dF_{y'}}{dx} - F_y}{\phi_y} = \frac{\dfrac{dF_{z'}}{dx} - F_z}{\phi_z} \equiv \lambda(x), \tag{57.19}$$

where $\lambda(x)$ denotes the common value of the ratios.[13]

It follows from (57.19) that the necessary conditions for the extremum of the integral 57.9 are

$$\begin{aligned} \frac{dF_{y'}}{dx} - [F_y + \lambda(x)\phi_y] &= 0, \\ \frac{dF_{z'}}{dx} - [F_z + \lambda(x)\phi_z] &= 0. \end{aligned} \tag{57.20}$$

When we revert to the original notation by setting $x = t, y = x^1$, $z = x^2$, we obtain a pair of equations

$$\frac{dF_{\dot{x}^i}}{dt} - F_{x^i} - \lambda(t)\frac{\partial \phi}{\partial x^i} = 0, \qquad (i = 1, 2), \tag{57.21}$$

[13] We note that if both $\phi_y = 0$ and $\phi_z = 0$, equation 57.11 does not define a surface.

which have the structure of Euler's equations 57.7 for the variational problem associated with the *free extremum* of the integral

$$\int_{t_1}^{t_2} [F(t, x, \dot{x}) - \lambda(t)\phi(t, x)]\, dt.$$

Similar considerations apply to the problem of minimizing the integral 57.1 in which the n variables x^i are constrained by a set of $k < n$ relations

$$\phi_j(t, x^1, \ldots, x^n) = 0, \qquad (j = 1, \ldots, k). \tag{57.22}$$

If the matrix $\partial\phi_j/\partial x^i$ is of rank k, the system of differential equations for the extremal is

$$\frac{dG_{\dot{x}^i}}{dt} - G_{x^i} = 0, \tag{57.23}$$

where

$$G = F + \lambda_j(t)\phi_j(t, x), \quad (j = 1, \ldots, k).$$

We note in conclusion that the constraining relations 57.22 do not involve the derivatives $\dot{x}^i$. Such constraints are called *holonomic* to distinguish them from constraints of the form

$$\phi_j(t, x, \dot{x}) = 0 \tag{57.24}$$

which are nonholonomic. *Nonholonomic* constraints arise in the study of dissipative dynamical systems, and we shall encounter them in Chapter 4. It is clear from the foregoing discussion that equations corresponding to (57.23), when nonholonomic constraints are present, must involve not only the multipliers $\lambda_i(t)$ but also the derivatives $\dot{\lambda}_i(t)$.

Problems

1. Consider the variational problems in (56.1) with Euler's equation 56.6. Note that if $F(t, x, \dot{x})$ does not contain x explicitly then (56.5) yields at once the first integral in the form $F_{\dot{x}} = \text{constant}$. Show that when $F(t, x, \dot{x})$ does not contain t explicitly then the first integral of (56.6) is $F - \dot{x}F_{\dot{x}} = \text{constant}$. *Hint*. Let $F = F(x, \dot{x})$, compute $\frac{d}{dt}(F - \dot{x}F_{\dot{x}})$ and make use of (56.6).

2. Note the hint in Problem 1 and show that the Euler equation 56.5 can be written in the form

$$\frac{d}{dt}(F - \dot{x}F_{\dot{x}}) = F_t.$$

3. The Euler equation 57.16 is an identity whenever $\mathscr{F}(x, y, y') = M(x, y) + N(x, y)y'$, with $M_y = N_x$. In this case (57.13) becomes

$$J = \int_{x_1}^{x_2} M\, dx + N\, dy$$

and this integral is independent of the path. Thus, every curve joining the given endpoints is an extremal for (57.13).

58. Geodesics in R_n

We are now in a position to discuss the problem of finding curves of minimum length joining a pair of given points on the surface. We will carry out our calculation for the case of the n-dimensional Riemannian manifolds, since our results will be of interest not only in connection with the geometry of surfaces but also in the study of dynamical trajectories in Chapter 4.

Let metric properties of the n-dimensional manifold R_n be determined by

$$ds^2 = g_{ij}\,dx^i\,dx^j, \qquad (i, j = 1, \ldots, n), \tag{58.1}$$

where $g_{ij} = g_{ji}$ are specified functions of the variables x^i. We suppose that the form 58.1 is positive definite and the functions g_{ij} are of class C^2. The length of a curve C, represented in R_n by equations

$$C\colon\ x^i = x^i(t), \qquad t_1 \le t \le t_2,$$

is given by

$$s = \int_{t_1}^{t_2} \sqrt{g_{\alpha\beta}\dot{x}^\alpha\dot{x}^\beta}\,dt, \qquad (\alpha, \beta = 1, \ldots, n). \tag{58.2}$$

The extremals of the functional 58.2 will be termed *geodesics in* R_n. The function F of Sec. 57 in this case is

$$F = \sqrt{g_{\alpha\beta}\dot{x}^\alpha\dot{x}^\beta}, \tag{58.3}$$

and, to form Euler's equations 57.7, we need to compute F_{x^i} and $F_{\dot{x}^i}$. This computation is straightforward. We deduce from (58.3) that

$$F_{x^j} = \frac{1}{2}(g_{\alpha\beta}\dot{x}^\alpha\dot{x}^\beta)^{-1/2}\frac{\partial g_{\alpha\beta}}{\partial x^j}\dot{x}^\alpha\dot{x}^\beta,$$

and

$$F_{\dot{x}^j} = (g_{\alpha\beta}\dot{x}^\alpha\dot{x}^\beta)^{-1/2}g_{\alpha j}\dot{x}^\alpha.$$

Substituting these expressions in Euler's equations yields

$$\frac{d}{dt}\left[\frac{g_{\alpha j}\dot{x}^\alpha}{\sqrt{g_{\alpha\beta}\dot{x}^\alpha\dot{x}^\beta}}\right] - \frac{\dfrac{\partial g_{\alpha\beta}}{\partial x^j}\dot{x}^\alpha\dot{x}^\beta}{2\sqrt{g_{\alpha\beta}\dot{x}^\alpha\dot{x}^\beta}} = 0. \tag{58.4}$$

Since $ds/dt = \sqrt{g_{\alpha\beta}\dot{x}^\alpha\dot{x}^\beta}$, equation 58.4 can be written in the form

$$\frac{d}{dt}\left(\frac{g_{\alpha j}\dot{x}^\alpha}{ds/dt}\right) - \frac{\dfrac{\partial g_{\alpha\beta}}{\partial x^j}\dot{x}^\alpha\dot{x}^\beta}{2\,ds/dt} = 0,$$

and, carrying out the indicated differentiation, we obtain

$$g_{\alpha j}\ddot{x}^\alpha + \frac{\partial g_{\alpha j}}{\partial x^\beta}\dot{x}^\alpha\dot{x}^\beta - \frac{1}{2}\frac{\partial g_{\alpha\beta}}{\partial x^j}\dot{x}^\alpha\dot{x}^\beta = \frac{g_{\alpha j}\dot{x}^\alpha\, d^2s/dt^2}{ds/dt}.$$

Since the second term in this equation can be written as a sum of two terms, we have

$$g_{\alpha j}\ddot{x}^\alpha + \frac{1}{2}\left(\frac{\partial g_{\alpha j}}{\partial x^\beta} + \frac{\partial g_{\beta j}}{\partial x^\alpha} - \frac{\partial g_{\alpha\beta}}{\partial x^j}\right)\dot{x}^\alpha\dot{x}^\beta = \frac{g_{\alpha j}\dot{x}^\alpha\, d^2s/dt^2}{ds/dt}.$$

But $[\alpha\beta, j] \equiv \frac{1}{2}\left(\frac{\partial g_{\alpha j}}{\partial x^\beta} + \frac{\partial g_{\beta j}}{\partial x^\alpha} - \frac{\partial g_{\alpha\beta}}{\partial x^j}\right)$, so that the foregoing equation assumes the form

$$g_{\alpha j}\ddot{x}^\alpha + [\alpha\beta, j]\dot{x}^\alpha\dot{x}^\beta = g_{\alpha j}\dot{x}^\alpha \frac{d^2s/dt^2}{ds/dt}. \tag{58.5}$$

These are the desired equations of geodesics. If we choose the parameter t to be the arc length s of the curve, that is, if we set

$$\frac{ds}{dt} = \sqrt{g_{\alpha\beta}\dot{x}^\alpha\dot{x}^\beta} = 1,$$

the system 58.5 simplifies to read

$$g_{\alpha j}\ddot{x}^\alpha + [\alpha\beta, j]\dot{x}^\alpha\dot{x}^\beta = 0. \tag{58.6}$$

In equation 58.6, dots denote the differentiation with respect to the arc parameter s.

If we multiply equation 58.6 by the tensor g^{ij} and sum, we obtain a simple form of the equations of geodesics in R_n.

$$\ddot{x}^i + \left\{ {i \atop \alpha\beta} \right\}\dot{x}^\alpha\dot{x}^\beta = 0, \qquad (i = 1, 2, \ldots, n), \quad (\alpha, \beta = 1, \ldots, n). \tag{58.7}$$

We observe that the form of these equations is identical with equations 51.2 defining the straight line in E_3. Since (58.7) is an ordinary second-order differential equation it possesses a unique solution when the values $x^i(s)$ and the first derivatives dx^i/ds are prescribed arbitrarily at a given point $x^i(s_0)$.

If we regard a given surface S as a Riemannian two-dimensional manifold R_2, covered by Gaussian coordinates u^α, then (58.7) assumes the form

$$\frac{d^2u^\gamma}{ds^2} + \left\{ {\gamma \atop \alpha\beta} \right\}\frac{du^\alpha}{ds}\frac{du^\beta}{ds} = 0, \qquad (\alpha, \beta, \gamma = 1, 2). \tag{58.8}$$

Hence at each point of S there exists a unique geodesic with an arbitrarily prescribed direction $\lambda^\alpha = du^\alpha/ds$. It is not difficult to prove that, if there

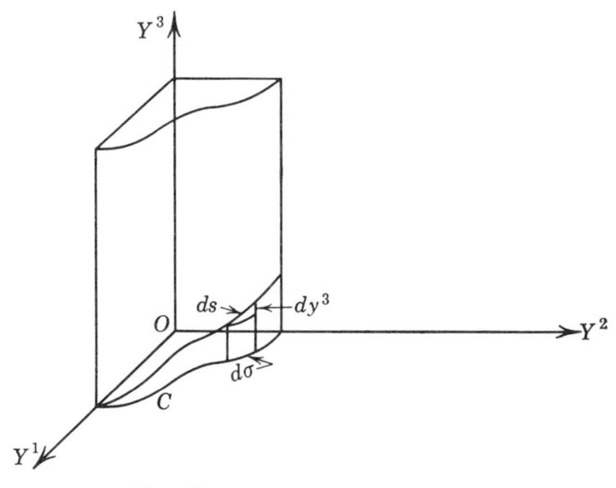

Fig. 25

exists a unique solution $u^\alpha(s)$, passing through two given points on S, then the curve $u^\alpha(s)$ is the curve of shortest length joining these points.[14]

If the manifold R_n is Euclidean, a coordinate system exists in which the Christoffel symbols vanish. In this case equations 58.7 become $d^2x^i/ds^2 = 0$. The general solution of this equation is $x^i = A^i s + B^i$. *Thus the geodesics in* E_n *are straight lines.*

As another illustration, consider the problem of determining geodesics on an arbitrary cylinder immersed in E_3. We choose the Y^3-axis parallel to the generators of the cylinder and let the trace of the cylinder on the Y^1Y^2-plane be given by equations

$$C: \quad \begin{cases} y^1 = \phi(\sigma), \\ y^2 = \psi(\sigma), \end{cases}$$

where σ is the arc length of C. (See Fig. 25.) Since

$$(d\sigma)^2 = (dy^1)^2 + (dy^2)^2,$$

an element of arc ds of the geodesic is given by

$$(ds)^2 = (d\sigma)^2 + (dy^3)^2,$$

so that $a_{11} = a_{22} = 1$, $a_{12} = a_{21} = 0$. Hence equations 58.8 reduce to

$$\frac{d^2\sigma}{ds^2} = 0, \qquad \text{for } \gamma = 1,$$

and

$$\frac{d^2y^3}{ds^2} = 0, \qquad \text{for } \gamma = 2.$$

[14] See, for example, L. P. Eisenhart, *Differential Geometry* (1940), p. 175.

We thus obtain

$$\begin{cases} \sigma = As + B, \\ y^3 = A_1 s + B_1. \end{cases}$$

If $A \neq 0$, we can write these equations in the form

$$y^3 = C_1\sigma + C_2,$$

where C_1 and C_2 are arbitrary constants.

The equations of the geodesics are therefore

$$y^1 = \phi(\sigma),$$
$$y^2 = \psi(\sigma),$$
$$y^3 = C_1\sigma + C_2,$$

and hence the curve is a helix, whose pitch is determined by C_1. The constant C_2 determines the origin for the arc parameter σ.

We can show in a similar way that the geodesics on the surface of a sphere are arcs of great circles. (See Problem 1.) However, as an illustration of the use of equations 57.23 we consider the functional

$$\text{(58.9)} \qquad J = \int_{t_1}^{t_2} \sqrt{\dot{x}^i\dot{x}^i}\, dt, \qquad (i = 1, 2, 3),$$

representing the length of arc of a curve C in cartesian coordinates. If C is to lie on a sphere of radius a, the constraining relation 57.10 has the form

$$\text{(58.10)} \qquad \phi \equiv x^i x^i - a^2 = 0,$$

when the center of the sphere is at the origin. The function $G = F + \lambda\phi$, introduced in Sec. 57, is

$$\text{(58.11)} \qquad G = \sqrt{\dot{x}^i\dot{x}^i} - \lambda(t)(x^i x^i - a^2)$$

and equations 57.23 take the form

$$\text{(58.12)} \qquad \frac{d}{dt}\frac{dx^i}{ds} + 2\lambda(t)x^i = 0, \qquad (i = 1, 2, 3),$$

where $ds = \sqrt{\dot{x}^i\dot{x}^i}\, dt$.

On eliminating $\lambda(t)$ from the first two equations in the set 58.12, we get

$$x^2 d\left(\frac{dx^1}{ds}\right) - x^1 d\left(\frac{dx^2}{ds}\right) = 0,$$

or

$$d\left(x^2\frac{dx^1}{ds} - x^1\frac{dx^2}{ds}\right) = 0.$$

Therefore

$$x^2 \frac{dx^1}{ds} - x^1 \frac{dx^2}{ds} = C_1, \tag{58.13}$$

where C_1 is a constant of integration.

Similarly, the elimination of $\lambda(t)$ from the last two equations in (58.12) yields on integration

$$x^3 \frac{dx^1}{ds} - x^1 \frac{dx^3}{ds} = C_2, \tag{58.14}$$

where C_2 is a constant.

From (58.13) and (58.14) we find

$$C_1 \frac{x^3\,dx^1 - x^1\,dx^3}{(x^1)^2} = C_2 \frac{x^2\,dx^1 - x^1\,dx^2}{(x^1)^2}$$

or

$$C_1\, d\left(\frac{x^3}{x^1}\right) = C_2\, d\left(\frac{x^2}{x^1}\right) \tag{58.15}$$

and the integration of equation 58.15 gives

$$C_3 x^1 - C_2 x^2 + C_1 x^3 = 0, \tag{58.16}$$

which represents a plane through the origin. Equation 58.16 together with the constraining relation 58.10 shows that the solution of the system 58.12 is an arc of a great circle.

Problems

1. In an orthogonal cartesian frame Y, the sphere of radius a is determined by equations

$$\begin{cases} y^1 = a \cos u^1 \cos u^2, \\ y^2 = a \cos u^1 \sin u^2, \\ y^3 = a \sin u^1. \end{cases}$$

In this case $ds^2 = a^2\,(du^1)^2 + a^2\,(\cos u^1)^2\,(du^2)^2$ and

$$s = a \int_{u_0{}^1}^{u_1{}^1} \sqrt{1 + \cos^2 u^1 (\dot{u}^2)^2}\, du^1,$$

where $\dot{u}^2 = du^2/du^1$. Show that the geodesics are great circles.

2. Find the geodesics on the surface

$$y^1 = u^1 \cos u^2, \qquad y^2 = u^1 \sin u^2, \qquad y^3 = 0,$$

imbedded in E_3. The coordinates y^i are orthogonal cartesian.

3. Show that, if we set $Q = a_{\alpha\beta}\dot{u}^\alpha\dot{u}^\beta$ where $\dot{u}^\alpha = du^\alpha/ds$, the equations of the geodesics 58.8 in R_2 can be written

$$\frac{d}{ds}\left(\frac{\partial Q}{\partial \dot{u}^\gamma}\right) - \frac{\partial Q}{\partial u^\gamma} = 0.$$

Hence the solutions of these equations for $\ddot{u}^\gamma$ should yield $-\begin{Bmatrix}\gamma\\ \alpha\beta\end{Bmatrix}\dot{u}^\alpha\dot{u}^\beta$, as can be seen from (58.8). This suggests a different means for computing the symbols $\begin{Bmatrix}\gamma\\ \alpha\beta\end{Bmatrix}$ in any particular coordinate system. Use this method to calculate the Christoffel symbols for the coordinate system in Problem 1 by determining the coefficients of $\dot{u}^\alpha\dot{u}^\beta$ in the solutions for the second derivatives of u^γ with respect to s.

59. Geodesic Coordinates

We have seen (Sec. 39) that, if a Riemannian space R_n is Euclidean, a coordinate system exists in which the components g_{ij} of the metric tensor are constants throughout the space. This implies that in such a coordinate system $\partial g_{ij}/\partial x^k \equiv 0$. The vanishing of these partial derivatives is equivalent to the vanishing of all Christoffel symbols, since[15] $[ij, k] = \frac{1}{2}\left(\frac{\partial g_{ik}}{\partial x^j} + \frac{\partial g_{jk}}{\partial x^i} - \frac{\partial g_{ij}}{\partial x^k}\right)$. If R_n is not Euclidean, then the Christoffel symbols do not vanish at all points of R_n, but it is possible to find a coordinate system, in fact infinitely many, in which they vanish at any given point P of R_n. Such coordinates are called *geodesic for that particular point*, or *locally cartesian* at P.

Thus consider some surface net with coordinates u^α and consider the point $P(u_0^1, u_0^2)$ on S. If v^α are the coordinates of some other net on S, then

$$u^\alpha = u^\alpha(v^1, v^2), \qquad (\alpha = 1, 2). \tag{59.1}$$

The second derivative formula 32.5 yields the relation

$$\frac{\partial^2 u^\alpha}{\partial v^\lambda\, \partial v^\mu} + {}_u\begin{Bmatrix}\alpha\\ \beta\gamma\end{Bmatrix}\frac{\partial u^\beta}{\partial v^\lambda}\frac{\partial u^\gamma}{\partial v^\mu} = {}_v\begin{Bmatrix}\nu\\ \lambda\mu\end{Bmatrix}\frac{\partial u^\alpha}{\partial v^\nu}. \tag{59.2}$$

However, if there exists a transformation of coordinates 59.1 such that the Christoffel symbols ${}_v\begin{Bmatrix}\nu\\ \lambda\mu\end{Bmatrix}$ vanish at P, then for *that particular point*

$$\frac{\partial^2 u^\alpha}{\partial v^\lambda\, \partial v^\mu} + {}_u\begin{Bmatrix}\alpha\\ \beta\gamma\end{Bmatrix}\frac{\partial u^\beta}{\partial v^\lambda}\frac{\partial u^\gamma}{\partial v^\mu} = 0. \tag{59.3}$$

[15] See also Theorem I, Sec. 39.

We exhibit next a solution of this equation yielding a particular transformation 59.1 to a coordinate system v^α in which the Christoffel symbols vanish at P. It is the second-degree polynomial

$$u^\alpha = u_P{}^\alpha + v^\alpha - \frac{1}{2}\left\{\begin{matrix}\alpha\\ \lambda\mu\end{matrix}\right\}_P v^\lambda v^\mu, \tag{59.4}$$

where the $u_P{}^\alpha$ is the value of u^α at P and the $\left\{\begin{matrix}\alpha\\ \lambda\mu\end{matrix}\right\}_P$ are the values of the Christoffel symbols at P. To verify that (59.4) satisfies (59.3), we compute

$$\frac{\partial u^\alpha}{\partial v^\mu} = \delta^\alpha_\mu - \left\{\begin{matrix}\alpha\\ \lambda\mu\end{matrix}\right\}_P v^\lambda \tag{59.5}$$

and

$$\frac{\partial^2 u^\alpha}{\partial v^\lambda\,\partial v^\mu} = -\left\{\begin{matrix}\alpha\\ \lambda\mu\end{matrix}\right\}_P. \tag{59.6}$$

From (59.4) we see that the point P, in new coordinates, is given by $v^\alpha = 0$, and hence at the point P, equation 59.5 yields $\left.\dfrac{\partial u^\alpha}{\partial v^\mu}\right|_P = \delta^\alpha_\mu$. Inserting values from this equation and (59.6) in (59.3), we see that it is satisfied at P. Hence the new variables indeed are geodesic coordinates at P.

We conclude this section by a remark that there is an extension of this result by Fermi, who proved that in every Riemannian manifold R_n there exists a coordinate system such that the coordinates are geodesic at all points of an arbitrarily prescribed analytic curve.[16]

60. Parallel Vector Fields in a Surface

The concept of parallel vector fields along a curve imbedded in E_3 (Sec. 48) was generalized by Levi-Civita to curves imbedded in n-dimensional Riemannian manifolds. As an illustration of the usefulness of the concept, consider a surface S immersed in E_3 and a curve C on S. We take equations of C in the form

$$C:\quad u^\alpha = u^\alpha(t), \qquad t_1 \le t \le t_2,$$

and suppose that the metric properties of S are governed by the tensor $a_{\alpha\beta}$. If A^α is a *surface vector field* defined along C, we can calculate the surface intrinsic derivative

$$\frac{\delta A^\alpha}{\delta t} \equiv \frac{dA^\alpha}{dt} + \left\{\begin{matrix}\alpha\\ \beta\gamma\end{matrix}\right\} A^\beta \frac{du^\gamma}{dt}. \tag{60.1}$$

[16] A derivation of explicit equations of transformation for this case, which include (59.4) as a special case, was given by Levi-Civita in a paper entitled "Sur l'écart géodésique," *Mathematische Annalen*, **97** (1926–27), pp. 291–320.

This is identical in form with the left-hand member of equation 48.1 defining the parallel vector field along a space curve. Accordingly, we take the differential equation

$$\frac{dA^\alpha}{dt} + \begin{Bmatrix} \alpha \\ \beta\gamma \end{Bmatrix} A^\beta \frac{du^\gamma}{dt} = 0, \tag{60.2}$$

which determines a unique vector field when the components of the vector are specified at an arbitrary point of C as the definition of the parallel vector field along a curve C on the surface S. If the parameter t is chosen as the arc length s, equation 60.2 reads

$$\frac{dA^\alpha}{ds} + \begin{Bmatrix} \alpha \\ \beta\gamma \end{Bmatrix} A^\beta \frac{du^\gamma}{ds} = 0, \tag{60.3}$$

and if A^α is taken to be the unit tangent vector to C so that

$$A^\alpha = \frac{du^\alpha}{ds} \equiv \lambda^\alpha$$

with $a_{\alpha\beta}\lambda^\alpha\lambda^\beta = 1$, then (60.3) yields

$$\frac{d^2u^\alpha}{ds^2} + \begin{Bmatrix} \alpha \\ \beta\gamma \end{Bmatrix} \frac{du^\beta}{ds}\frac{du^\gamma}{ds} = 0. \tag{60.4}$$

This equation is recognized as the equation of a geodesic on S, and hence one can enunciate a

THEOREM. *The vector obtained by the parallel propagation of the tangent vector to a geodesic always remains tangent to the geodesic.*

From uniqueness of solution of (60.4) it follows that the property of tangency of a parallel vector field to a surface curve is both a necessary and sufficient condition for a geodesic.

In the Euclidean plane geodesics are straight lines, and the parallel vector field formed by the tangents to a straight line traces out the same straight line. On the surface of the sphere the geodesic is an arc of a great circle joining two given points on the sphere, and the corresponding vector field is the field of tangents to the geodesic. From the last example it is clear that parallelism with respect to a surface curve differs from the parallelism with respect to a space curve imbedded in E_3, since vectors obtained by a parallel propagation, along the surface curve C, need not be parallel in the Euclidean sense. However, it is easy to prove that the lengths of vectors forming a parallel field with respect to C remain constant. Indeed, word-for-word repetition of the proof given in Sec. 48 leads to the conclusion that the angle between two vectors propagated in parallel fashion remains unchanged, and it follows, as it did in Sec. 48, that the vectors forming a parallel field are constant in magnitude. A corollary of this result is that the vector field obtained by a parallel propagation of a surface vector along a geodesic makes equal angles with the geodesic.

It should be noted that the concept of parallelism in Riemannian manifolds is defined relative to a given curve. A surface vector A^α, specified at a point P of S, when propagated in parallel manner along a given curve C to a point Q, need not coincide with the vector obtained by the parallel propagation along a different path joining P and Q. Moreover, if a closed curve C, enclosing a simply connected region of S, is drawn, and a parallel vector field is constructed starting with some point P on C, then the vector obtained by traversing the closed path need not coincide with the initial vector. The angle between the initial and final vector measures another intrinsic property of S, known as the *Gaussian curvature* of S. This property is introduced in a somewhat different way[17] in Sec. 62.

61. Isometric Surfaces

The properties of surfaces with which we have been concerned so far hinged entirely on the study of the first fundamental quadratic form

$$ds^2 = a_{\alpha\beta}\, du^\alpha\, du^\beta. \tag{61.1}$$

These properties constitute a body of what is known as the *intrinsic geometry of surfaces*. They take no account of the distinguishing characteristics of surfaces as they might appear to an observer located in the surrounding space. Two surfaces, a cylinder and a cone, for example, appear to be entirely different when viewed from the enveloping space, and yet their intrinsic geometries are completely indistinguishable since metric properties of cylinders and cones can be described by the identical expressions for the square of the element of arc. If a coordinate system exists on each of the two surfaces such that the linear elements on them are characterized by the same metric coefficients $a_{\alpha\beta}$, the surfaces are called *isometric*. Obviously the surfaces of the cylinder and cone are isometric with the Euclidean plane, since these surfaces can be rolled out, or developed, on the plane without changing the lengths of arc elements, and hence without altering the measurements of angles and areas.

In the following section we introduce an important scalar invariant, known as the Gaussian curvature, which will enable us to determine the circumstances under which a given surface is developable, that is, isometric with the Euclidean plane.

[17] Compare L. P. Eisenhart, *Introduction to Differential Geometry*, Princeton University Press, p. 200.

As an illustration of an isometric nondevelopable surface consider the catenoid

$$
\begin{aligned}
S_1\colon\quad y^1 &= v^1 \cos v^2,\\
y^2 &= v^1 \sin v^2,\\
y^3 &= a \cosh^{-1} \frac{v^1}{a},
\end{aligned}
$$

obtained by revolving the catenary $y^2 = \cosh (y^3/a)$ about the Y^3-axis. We will show that the surface S_1 is isometric with the surface of the helicoid defined by

$$
\begin{aligned}
S_2\colon\quad y^1 &= u^1 \cos u^2,\\
y^2 &= u^1 \sin u^2,\\
y^3 &= au^2.
\end{aligned}
$$

The first fundamental form $ds^2 = dy^i\, dy^i$ for S_1 is easily found to be

$$ds^2 = a_{\alpha\beta}\, dv^\alpha\, dv^\beta = \frac{(v^1)^2}{(v^1)^2 - a^2} (dv^1)^2 + (v^1)^2 (dv^2)^2, \tag{61.2}$$

so that

$$a_{11} = \frac{(v^1)^2}{(v^1)^2 - a^2}, \qquad a_{12} = 0, \qquad a_{22} = (v^1)^2 \equiv a^2 + [(v^1)^2 - a^2].$$

For the surface S_2, we find

$$ds^2 = a_{\alpha\beta}\, du^\alpha\, du^\beta = (du^1)^2 + [a^2 + (u^1)^2] (du^2)^2, \tag{61.3}$$

so that

$$a_{11} = 1, \qquad a_{12} = 0, \qquad a_{22} = a^2 + (u^1)^2.$$

Now, if we set in (61.2)

$$(v^1)^2 - a^2 = (u^1)^2, \qquad v^2 = u^2,$$

we obtain

$$ds^2 = (du^1)^2 + [(u^1)^2 + a^2](du^2)^2.$$

Since this is identical with (61.3), the surfaces S_1 and S_2 are isometric. It follows from discussion in the next section that these surfaces are not developable.

62. The Riemann-Christoffel Tensor and the Gaussian Curvature

The formulas of Sec. 37 describing the properties of Riemann-Christoffel tensors in n-dimensional manifolds simplify considerably when n is set equal to 2. Thus, if we are given the first fundamental form,

$$ds^2 = a_{\alpha\beta}\, du^\alpha\, du^\beta, \tag{62.1}$$

of the surface S, we can form the Christoffel symbols with respect to this surface, and the corresponding Riemann tensor

$$(62.2) \qquad R_{\alpha\beta\gamma\delta} = \begin{vmatrix} \dfrac{\partial}{\partial u^\gamma} & \dfrac{\partial}{\partial u^\delta} \\ [\beta\gamma, \alpha] & [\beta\delta, \alpha] \end{vmatrix} + \begin{vmatrix} \begin{Bmatrix} \lambda \\ \beta\gamma \end{Bmatrix} & \begin{Bmatrix} \lambda \\ \beta\delta \end{Bmatrix} \\ [\alpha\gamma, \lambda] & [\alpha\delta, \lambda] \end{vmatrix}.$$

We recall that this tensor is skew-symmetric in the first two and last two indices, so that, for the surface S,

$$(62.3) \quad R_{\alpha\alpha\beta\gamma} = R_{\alpha\beta\gamma\gamma} = 0, \qquad R_{1212} = R_{2121} = -R_{2112} = -R_{1221}.$$

Hence every nonvanishing component of the Riemann tensor is equal to R_{1212} or to its negative.

We define the quantity K by the formula

$$(62.4) \qquad K = \frac{R_{1212}}{a},$$

where $a = |a_{\alpha\beta}|$, and call it the *Gaussian curvature* or the *total curvature* of the surface S. Since only metric coefficients $a_{\alpha\beta}$ are involved in this definition, the properties described by K are intrinsic properties of the surface S.

If we introduce the two-dimensional ϵ-tensors,

$$\epsilon_{\alpha\beta} = \sqrt{a}\, e_{\alpha\beta} \quad \text{and} \quad \epsilon^{\alpha\beta} = \frac{e^{\alpha\beta}}{\sqrt{a}},$$

where the $e_{\alpha\beta}$'s are the alternating e-systems (see Sec. 40), and note relations 62.3, we can write equation 62.4 as

$$(62.5) \qquad R_{\alpha\beta\gamma\delta} = K\epsilon_{\alpha\beta}\epsilon_{\gamma\delta}.$$

Since $\epsilon^{\alpha\beta}\epsilon_{\alpha\beta} = 2$, we can solve (62.5) for K and obtain

$$(62.6) \qquad K = \tfrac{1}{4}R_{\alpha\beta\gamma\delta}\epsilon^{\alpha\beta}\epsilon^{\gamma\delta}.$$

These equations show that the Gaussian curvature is an invariant.

Now, when a surface S is isometric with the Euclidean plane, there exists on S a coordinate system with respect to which $a_{11} = a_{22} = 1$, $a_{12} = 0$. It is obvious that in this case $R_{\alpha\beta\gamma\delta} = 0$ in this particular coordinate system, and since $R_{\alpha\beta\gamma\delta}$ is a tensor, it must vanish in every coordinate system.

Conversely, if the Riemann tensor vanishes at all points of the surface, Theorem II of Sec. 39 guarantees that there exist coordinate systems on the surface such that $a_{11} = a_{22} = 1$, $a_{12} = 0$.

Thus we have a

THEOREM. *A necessary and sufficient condition that a surface S be isometric with the Euclidean plane is that the Riemann tensor (or the Gaussian curvature of S) be identically zero.*

Consider next an invariant

$$R = a^{\mu\nu} R_{\mu\nu}, \tag{62.7}$$

where

$$R_{\mu\nu} = R^{\alpha}_{\mu\nu\alpha} = a^{\lambda\alpha} R_{\lambda\mu\nu\alpha} \tag{62.8}$$

is the Ricci tensor introduced in Sec. 38.[18]

If we multiply (62.8) by $a^{\mu\nu}$ and sum, we get

$$R = a^{\mu\nu} R_{\mu\nu} = a^{\lambda\alpha} a^{\mu\nu} R_{\lambda\mu\nu\alpha}, \tag{62.9}$$

and recalling (62.3) we see that (62.9) is equivalent to

$$R = -2R_{1212}(a^{11}a^{22} - a^{12}a^{12}). \tag{62.10}$$

Since

$$a^{11} = \frac{a_{22}}{a}, \qquad a^{22} = \frac{a_{11}}{a}, \qquad a^{12} = \frac{-a_{12}}{a},$$

we have

$$R = -2\frac{R_{1212}}{a}. \tag{62.11}$$

Comparing (62.11) with (62.4) we see that

$$R = -2K.$$

The invariant R is sometimes called the *Einstein curvature* of S.

We shall give a more revealing geometrical interpretation of the Gaussian curvature in Sec. 72, where the surface S is viewed from the enveloping space.

Problems

1. Use formulas 62.2 and 62.4 to show that, if the system of coordinates is orthogonal, then

$$K = -\frac{1}{2\sqrt{a}}\left[\frac{\partial}{\partial u^1}\left(\frac{1}{\sqrt{a}}\frac{\partial a_{22}}{\partial u^1}\right) + \frac{\partial}{\partial u^2}\left(\frac{1}{\sqrt{a}}\frac{\partial a_{11}}{\partial u^2}\right)\right].$$

2. Calculate the total curvature of the manifold whose quadratic form is

$$ds^2 = a^2 \sin^2 u^1 (du^2)^2 + a^2(du^1)^2.$$

[18] We recall that $R^{\alpha}_{\mu\nu\beta} = a^{\lambda\alpha} R_{\lambda\mu\nu\beta}$.

3. Determine whether the surface of a helicoid given by

$$\begin{cases} y^1 = u^1 \cos u^2, \\ y^2 = u^1 \sin u^2, \\ y^3 = au^2, \end{cases}$$

is developable.

4. Show that, for a surface of revolution defined by

$$\begin{cases} y^1 = u^1 \cos u^2, \\ y^2 = u^1 \sin u^2, \\ y^3 = f(u^1), \end{cases}$$

$$K = \frac{f'f''}{u^1[1 + (f')^2]^2}, \text{ when } f \text{ is of class } C^2.$$

5. Show that the surface defined by

$$\begin{cases} y^1 = f_1(u^1), \\ y^2 = f_2(u^1), \\ y^3 = u^2, \end{cases}$$

where f_1 and f_2 are differentiable functions, is developable.

6. Show that the formula for the Gaussian curvature K can be written in the form

$$K = \frac{1}{2\sqrt{a}} \left\{ \frac{\partial}{\partial u^1} \left[\frac{a_{12}}{a_{11}\sqrt{a}} \frac{\partial a_{11}}{\partial u^2} - \frac{1}{\sqrt{a}} \frac{\partial a_{22}}{\partial u^1} \right] + \frac{\partial}{\partial u^2} \left[\frac{2}{\sqrt{a}} \frac{\partial a_{12}}{\partial u^1} - \frac{1}{\sqrt{a}} \frac{\partial a_{11}}{\partial u^2} - \frac{a_{12}}{a_{11}\sqrt{a}} \frac{\partial a_{11}}{\partial u^1} \right] \right\}.$$

63. The Geodesic Curvature of Surface Curves

We shall conclude our study of intrinsic geometry of surfaces with a derivation of a formula describing the behavior of the tangent vector to a surface curve. This formula is analogous to Frenet's formula 50.1.

Let C be a surface curve defined parametrically by

$$u^\alpha = u^\alpha(s), \tag{63.1}$$

where s is the arc parameter. Accordingly, at every point of the curve we have the condition

$$a_{\alpha\beta} \frac{du^\alpha}{ds} \frac{du^\beta}{ds} = 1. \tag{63.2}$$

The quantities du^1/ds, du^2/ds obviously determine a tangent vector λ^α to C, and it is clear from (63.2) that

$$\lambda^\alpha = \frac{du^\alpha}{ds} \tag{63.3}$$

is a unit vector. If we differentiate the quadratic relation $a_{\alpha\beta}\lambda^\alpha\lambda^\beta = 1$ intrinsically with respect to s, we obtain $a_{\alpha\beta}\lambda^\alpha(\delta\lambda^\beta/\delta s) = 0$, from which it follows that the surface vector $\delta\lambda^\alpha/\delta s$ is orthogonal to λ^α. Following the line of thought of Sec. 49, we introduce a unit surface vector η^α normal to λ^α, so that

$$\frac{\delta\lambda^\alpha}{\delta s} = \varkappa_g \eta^\alpha, \tag{63.4}$$

where $\varkappa_g$ is a suitable scalar. In order to determine the direction of η^α uniquely we choose η^α in the way analogous to the choice of the triad of vectors in Sec. 49 (equation 49.11), namely, $\epsilon_{\alpha\beta}\lambda^\alpha\eta^\beta = 1$. This choice of the orientation of $\boldsymbol{\lambda}$ and $\boldsymbol{\eta}$ uniquely determines the sign of $\varkappa_g$, and it amounts to saying that the sine of the angle between $\boldsymbol{\lambda}$ and $\boldsymbol{\eta}$ is $+1$. The vector η^α is the unit surface vector orthogonal to the curve C, and the scalar $\varkappa_g$ is called the *geodesic curvature* of C.

We recall that the equation of the geodesic on S (see equation 60.4) can be written as $\delta\lambda^\alpha/\delta s = 0$. Comparing this with (63.4) leads to the conclusion that, if the geodesic curvature $\varkappa_g = 0$, then the curve C is a geodesic, and conversely. Hence the

THEOREM. *A necessary and sufficient condition that a curve on a surface S be a geodesic is that its geodesic curvature be zero.*

As an illustration, we compute the geodesic curvature of the small circle

$$C: \quad u^1 = \text{constant} = u_0^{\ 1} \neq 0, \qquad u^2 = u^2,$$

on the surface of the sphere (Fig. 26)

$$\begin{aligned} S: \quad y^1 &= a \cos u^1 \cos u^2 \\ y^2 &= a \cos u^1 \sin u^2 \\ y^3 &= a \sin u^1. \end{aligned}$$

If the arc-length s of C is measured from the plane $u_2 = 0$, we have $u^2 = s/(a \cos u_0^{\ 1})$, and the equations of C can be written in the form

$$u^1 = u_0^{\ 1}, \qquad u^2 = \frac{s}{a \cos u_0^{\ 1}}. \tag{63.5}$$

From (63.5) we find that the components of the unit tangent vector $\lambda^\alpha = du^\alpha/ds$ along C are

$$\lambda^1 = 0, \qquad \lambda^2 = \frac{1}{a \cos u_0^{\ 1}}, \tag{63.6}$$

so that

$$\frac{\delta\lambda^1}{\delta s} = \frac{d\lambda^1}{ds} + \left\{ {1 \atop \alpha\beta} \right\} \lambda^\alpha \frac{du^\beta}{ds} = 0 + \left\{ {1 \atop 22} \right\} \lambda^2\lambda^2$$

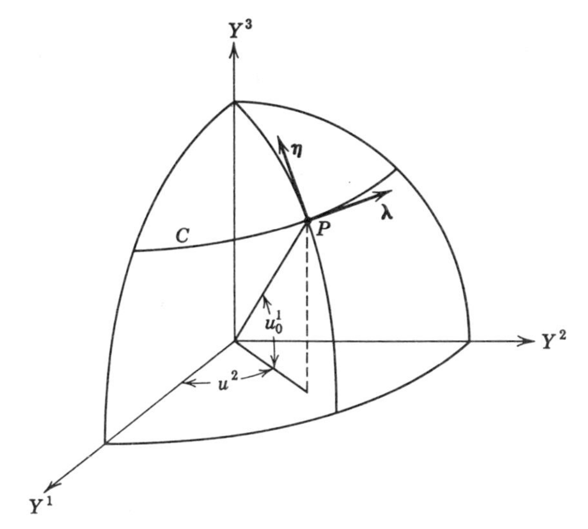

Fig. 26

and

$$\frac{\delta\lambda^2}{\delta s} = \frac{d\lambda^2}{ds} + \begin{Bmatrix} 2 \\ \alpha\beta \end{Bmatrix} \lambda^\alpha \frac{du^\beta}{ds} = 0 + \begin{Bmatrix} 2 \\ 22 \end{Bmatrix} \lambda^2\lambda^2.$$

Since metric coefficients of S are $a_{11} = a^2$, $a_{12} = 0$, $a_{22} = a^2 \cos^2 u^1$, we find, on referring to Problem 3 of Sec. 31, that $\begin{Bmatrix} 1 \\ 22 \end{Bmatrix} = \sin u^1 \cos u^1$, $\begin{Bmatrix} 2 \\ 22 \end{Bmatrix} = 0$. Accordingly, formulas 63.4 yield

$$\frac{\delta\lambda^1}{\delta s} = \varkappa_g \eta^1 = \frac{1}{a^2} \tan u_0{}^1, \quad \frac{\delta\lambda^2}{\delta s} = \varkappa_g \eta^2 = 0. \tag{63.7}$$

Since C is not a geodesic, $\varkappa_g \neq 0$, and we conclude that $\eta^2 = 0$. But η^α is a unit vector so that $a_{\alpha\beta}\eta^\alpha\eta^\beta = 1$, and we find that $\eta^1 = 1/a$. Hence equations 63.7 yield $\varkappa_g = (\tan u_0{}^1)/a$. In Sec. 71 we establish the relation between the geodesic curvature $\varkappa_g$ and the ordinary curvature $\varkappa$ of C.

64. Surfaces in Space

With the exception of occasional references to the surrounding space, our study of geometry of surfaces was carried out from the point of view of a two-dimensional being whose universe is determined by the surface parameters u^1 and u^2. The treatment of surfaces presented in the foregoing was based entirely on the study of the first quadratic differential form. In the discussion of isometric surfaces in Sec. 61, we remarked that a pair of isometric surfaces, a cone and a cylinder for example, which are indistinguishable in intrinsic geometry, appear to be quite distinct to an observer examining them from a reference frame located in the space in which the surfaces are imbedded. An entity that provides a characterization of the shape of the surface as it appears from the enveloping space is the normal line to the surface. The behavior of the normal line as its foot is displaced along the surface depends on the shape of the surface, and it occurred to Gauss to describe certain properties of surfaces with the aid of a quadratic form that depends in a fundamental way on the behavior of the normal line. Before we introduce this new quadratic form, let us recall our point of departure in the study of surfaces in Secs. 52 and 53.

A surface S imbedded in E_3 was defined by three parametric equations

$$y^i = y^i(u^1, u^2), \qquad (i = 1, 2, 3), \tag{64.1}$$

where the y^i are orthogonal cartesian coordinates of the reference frame located in the space surrounding S. An element of arc ds of a curve lying on S is determined by the formula

$$ds^2 = a_{\alpha\beta}\, du^\alpha\, du^\beta, \tag{64.2}$$

where

$$a_{\alpha\beta} = \frac{\partial y^i}{\partial u^\alpha}\frac{\partial y^i}{\partial u^\beta}.$$

The choice of cartesian variables y^i in the space enveloping the surface is clearly not essential, and we could have equally well referred the points of E_3 to a curvilinear coordinate system X related to Y by the transformation $x^i = x^i(y^1, y^2, y^3)$. Now, relative to the frame X, the line element in E_3 is given by

$$ds^2 = g_{ij}\, dx^i\, dx^j, \tag{64.3}$$

where $g_{ij} = \dfrac{\partial y^k}{\partial x^i}\dfrac{\partial y^k}{\partial x^j}$, and the set of equations 64.1 for the surface S can be written as

$$S\colon\quad x^i = x^i(u^1, u^2). \tag{64.4}$$

It follows from this representation of S that

$$dx^i = \frac{\partial x^i}{\partial u^\alpha}\, du^\alpha, \tag{64.5}$$

and hence the expression for the surface element of arc (64.3) assumes the form

$$
\begin{aligned}
ds^2 &= g_{ij}\, dx^i\, dx^j \\
&= g_{ij} \frac{\partial x^i}{\partial u^\alpha} \frac{\partial x^j}{\partial u^\beta}\, du^\alpha\, du^\beta.
\end{aligned}
$$

A comparison of this with equation 64.2 leads to the conclusion that

$$
a_{\alpha\beta} = g_{ij} \frac{\partial x^i}{\partial u^\alpha} \frac{\partial x^j}{\partial u^\beta}, \qquad (i, j, = 1, 2, 3), (\alpha, \beta = 1, 2). \tag{64.6}
$$

We note that the foregoing formulas depend on both the Latin and Greek indices, and we recall that the Latin indices run from 1 to 3 and refer to the surrounding space, whereas the Greek indices assume values 1 and 2 and are associated with the surface S imbedded in E_3. Furthermore, the dx^i and g_{ij}'s are tensors *with respect to the transformations induced on the space variables* x^i, whereas such quantities as du^α and $a_{\alpha\beta}$ are *tensors with respect to the transformation of Gaussian surface coordinates* u^α. Equation 64.6 is a curious one since it contains partial derivatives, $\partial x^i/\partial u^\alpha$, depending on both Latin and Greek indices. Since both $a_{\alpha\beta}$ and g_{ij} in (64.6) are tensors, this formula suggests that $\partial x^i/\partial u^\alpha$ can be regarded either as *a contravariant space vector or as a covariant surface vector*. Let us investigate this set of quantities more closely.

Let us take a small displacement on the surface S, specified by the *surface vector* du^α. The same displacement, as is clear from (64.5), is described by the *space vector* with components

$$
dx^i = \frac{\partial x^i}{\partial u^\alpha}\, du^\alpha. \tag*{[64.5]}
$$

The left-hand member of this expression is independent of the Greek indices, and hence it is invariant relative to a change of the surface coordinates u^α. Since du^α is an arbitrary surface vector, we conclude that

$$
\frac{\partial x^i}{\partial u^\alpha} \tag{64.7}
$$

is a *covariant surface vector*. On the other hand, if we change the space coordinates, the du^α, being a surface vector, is invariant relative to this change, so that (64.7) must be a *contravariant space vector*. Hence we can write (64.7) as

$$
x^i_\alpha \equiv \frac{\partial x^i}{\partial u^\alpha}, \tag{64.8}
$$

where the indices properly describe the tensor character of this set of quantities.

A simple geometrical significance of the set of quantities 64.8 can be deduced from Fig. 27. Let $\mathbf{r}$ be the position vector of an arbitrary point P on S. The point P is determined by a pair of Gaussian coordinates (u^1, u^2), or by a triplet of space coordinates (x^1, x^2, x^3). Accordingly, the vector $\mathbf{r}$ can be viewed as a function of the space variables x^i satisfying equations 64.4. Thus

$$\frac{\partial \mathbf{r}}{\partial u^\alpha} = \frac{\partial \mathbf{r}}{\partial x^i} \frac{\partial x^i}{\partial u^\alpha}. \tag{64.9}$$

But $\partial \mathbf{r}/\partial x^i$ are the base vectors $\mathbf{b}_i$ at P, associated with the curvilinear system X, whereas $\partial \mathbf{r}/\partial u^\alpha$ are the base vectors $\mathbf{a}_\alpha$ at P relative to the Gaussian system U.

Hence equations 64.9 yield

$$\mathbf{a}_\alpha = \mathbf{b}_i \frac{\partial x^i}{\partial u^\alpha}. \tag{64.10}$$

It is clear from this representation that $\mathbf{a}_1 = \dfrac{\partial x^i}{\partial u^1} \mathbf{b}_i$ and $\mathbf{a}_2 = \dfrac{\partial x^i}{\partial u^2} \mathbf{b}_i$, so that $\partial x^i/\partial u^\alpha \equiv x_\alpha^{\ i}$, $(\alpha = 1, 2)$, are the contravariant components of the surface base vectors $\mathbf{a}_\alpha$ referred to the base systems $\mathbf{b}_i$. Thus the sets of quantities

$$x_1^i: \left(\frac{\partial x^1}{\partial u^1}, \frac{\partial x^2}{\partial u^1}, \frac{\partial x^3}{\partial u^1}\right) \quad \text{and} \quad x_2^i: \left(\frac{\partial x^1}{\partial u^2}, \frac{\partial x^2}{\partial u^2}, \frac{\partial x^3}{\partial u^2}\right)$$

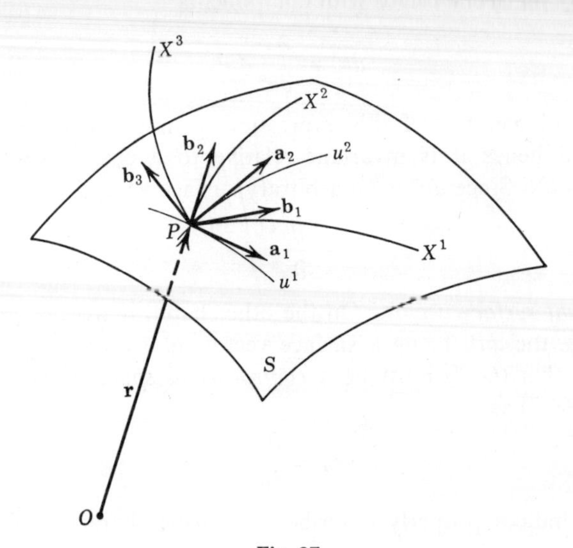

Fig. 27

transform in a contravariant manner relative to the transformation of space coordinates x^i.

We can also show that the three surface vectors

$$x^1_\alpha: \quad \left(\frac{\partial x^1}{\partial u^1}, \frac{\partial x^1}{\partial u^2}\right), \qquad x^2_\alpha: \quad \left(\frac{\partial x^2}{\partial u^1}, \frac{\partial x^2}{\partial u^2}\right), \qquad x^3_\alpha: \quad \left(\frac{\partial x^3}{\partial u^1}, \frac{\partial x^3}{\partial u^2}\right)$$

transform according to the covariant law with respect to the transformation of Gaussian surface coordinates u^α. Indeed, consider a transformation $u^\alpha = u^\alpha(\bar{u}^1, \bar{u}^2)$; then the equations 64.4 of S go over into $x^i = x^i(\bar{u}^1, \bar{u}^2)$, and

$$\frac{\partial x^i}{\partial u^\alpha} = \frac{\partial x^i}{\partial \bar{u}^\beta}\frac{\partial \bar{u}^\beta}{\partial u^\alpha}. \tag{64.11}$$

But $\partial x^i/\partial \bar{u}^\beta \equiv \bar{x}^i_\beta$, and (64.11) yields, for $i = 1, 2, 3$, $x^i_\alpha = (\partial \bar{u}^\beta/\partial u^\alpha)\bar{x}^i_\beta$. This is the covariant law.

Let ds be an element of arc joining a pair of points $P(u^1, u^2)$ and $P(u^1 + du^1, u^2 + du^2)$ on S. The direction of the line element ds is given by the direction parameters $du^\alpha/ds = \lambda^\alpha$. The same direction can be specified by an observer in the enveloping space by means of three parameters $dx^i/ds = \lambda^i$, and it follows from (64.5) that $\lambda^i = x^i_\alpha \lambda^\alpha$. This formula tells us that any surface vector A^α (that is, a vector lying in the tangent plane to S) can be viewed as a space vector with components A^i determined by

$$A^i = x^i_\alpha A^\alpha. \tag{64.12}$$

We shall refer to a vector A^i determined by this formula as a *tangent vector to the surface S*.

65. The Normal Line to the Surface

Let $\mathbf{A}$ and $\mathbf{B}$ be a pair of surface vectors drawn at some point P of S (Fig. 28). According to formula 64.12, they can be represented in the form

$$A^i = x^i_\alpha A^\alpha, \qquad B^i = x^i_\alpha B^\alpha. \tag{65.1}$$

The vector product $\mathbf{A} \times \mathbf{B}$ is the vector normal to the tangent plane determined by the vectors $\mathbf{A}$ and $\mathbf{B}$, and the unit vector $\mathbf{n}$ perpendicular to the tangent plane, so oriented that $\mathbf{A}$, $\mathbf{B}$, and $\mathbf{n}$ form a right-handed system, is

$$\mathbf{n} = \frac{\mathbf{A} \times \mathbf{B}}{|\mathbf{A} \times \mathbf{B}|} = \frac{\mathbf{A} \times \mathbf{B}}{AB\,|\sin\theta|}, \tag{65.2}$$

where θ is the angle between $\mathbf{A}$ and $\mathbf{B}$.

We call the vector $\mathbf{n}$ the *unit normal vector* to the surface S at P. Clearly, $\mathbf{n}$ is a function of coordinates (u^1, u^2), and, as the point $P(u^1, u^2)$ is

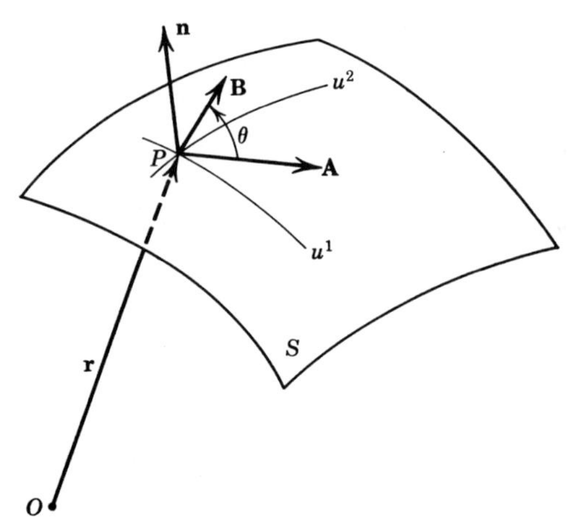

Fig. 28

displaced to a new position $P(u^1 + du^1, u^2 + du^2)$, the vector $\mathbf{n}$ undergoes a change

$$dn = \frac{\partial \mathbf{n}}{\partial u^\alpha} du^\alpha, \tag{65.3}$$

whereas the position vector $\mathbf{r}$ is changed by the amount $d\mathbf{r} = \dfrac{\partial \mathbf{r}}{\partial u^\alpha} du^\alpha$.

Let us form the scalar product

$$d\mathbf{n} \cdot d\mathbf{r} = \frac{\partial \mathbf{n}}{\partial u^\alpha} \cdot \frac{\partial \mathbf{r}}{\partial u^\beta} du^\alpha du^\beta. \tag{65.4}$$

If we define

$$b_{\alpha\beta} = -\frac{1}{2}\left(\frac{\partial \mathbf{n}}{\partial u^\alpha} \cdot \frac{\partial \mathbf{r}}{\partial u^\beta} + \frac{\partial \mathbf{n}}{\partial u^\beta} \cdot \frac{\partial \mathbf{r}}{\partial u^\alpha}\right),$$

so that (65.4) reads

$$d\mathbf{n} \cdot d\mathbf{r} = -b_{\alpha\beta}\, du^\alpha du^\beta, \tag{65.5}$$

the left-hand member of (65.5), being the scalar product of two vectors, is obviously an invariant; moreover, from symmetry with respect to α and β, it is clear that the coefficients of $du^\alpha du^\beta$ in the right-hand member of (65.5) define a covariant tensor of rank two. The quadratic form

$$\mathscr{B} \equiv b_{\alpha\beta}\, du^\alpha du^\beta \tag{65.6}$$

will be shown to play an essential part in the study of surfaces when they are viewed from the surrounding space, just as the first fundamental quadratic form $\mathscr{A} \equiv d\mathbf{r} \cdot d\mathbf{r}$, or

$$\mathscr{A} = a_{\alpha\beta}\, du^\alpha du^\beta$$

did in the study of intrinsic properties of a surface. The differential form 65.6 was introduced by Gauss, and it is called the *second fundamental quadratic form of the surface.*

Since the notation for the unit normal used previously, despite its pictorial suggestiveness, is more cumbersome than the tensor notation, we shall rewrite the defining formula 65.2 in terms of the components x^i_α of the base vectors $\mathbf{a}_\alpha$. We denote the contravariant components of $\mathbf{n}$ by n^i and observe that its covariant components n_i are given by[19]

$$n_i = \frac{\epsilon_{ijk}A^jB^k}{AB\sin\theta} \tag{65.7}$$

and (see Sec. 54)

$$AB\sin\theta = \epsilon_{\alpha\beta}A^\alpha B^\beta. \tag{65.8}$$

Substituting in (65.7) from (65.1) and (65.8), we get

$$(n_i\epsilon_{\alpha\beta} - \epsilon_{ijk}x^j_\alpha x^k_\beta)A^\alpha B^\beta = 0,$$

and, since this relation is valid for all surface vectors, we conclude that

$$n_i\epsilon_{\alpha\beta} = \epsilon_{ijk}x^j_\alpha x^k_\beta. \tag{65.9}$$

Multiplying (65.9) through by $\epsilon^{\alpha\beta}$, and noting that $\epsilon^{\alpha\beta}\epsilon_{\alpha\beta} = 2$, we get the desired result

$$n_i = \tfrac{1}{2}\epsilon^{\alpha\beta}\epsilon_{ijk}x^j_\alpha x^k_\beta. \tag{65.10}$$

It is clear from the structure of this formula that n_i is a space vector which does not depend on the choice of surface coordinates. This fact is also obvious from purely geometric considerations.

66. Tensor Derivatives

In Sec. 67 we shall deduce the second fundamental quadratic form 65.6 analytically by the operation of tensor differentiation of tensor fields which are functions of both surface and space coordinates. The fruitful

[19] Note that the vector product $\mathbf{A} \times \mathbf{B}$ depends on the lengths of the vectors $\mathbf{A}$ and $\mathbf{B}$ and on the angle between them. If we choose an orthogonal cartesian system of axes Y, so that the vectors $\mathbf{A}$ and $\mathbf{B}$ lie in the Y^1Y^2-plane with $\mathbf{A}$ directed along the Y^1-axis, then the cartesian components A^i of $\mathbf{A}$ are $A^1 = A$, $A^2 = 0$, $A^3 = 0$, and the components of $\mathbf{B}$ are $B^1 = B\cos\theta$, $B^2 = B\sin\theta$, $B^3 = 0$. Since in the Y-system $\epsilon_{ijk} = e_{ijk}$,

$$C_i \equiv \epsilon_{ijk}A^jB^k = e_{i12}AB\sin\theta.$$

Hence $C_1 = 0$, $C_2 = 0$, $C_3 = AB\sin\theta$. Thus the C_i define the vector $\mathbf{C} = \mathbf{A} \times \mathbf{B}$ normal to the plane determined by $\mathbf{A}$ and $\mathbf{B}$ whose magnitude is $AB\,|\sin\theta|$. If A^α and B^β are the surface components of $\mathbf{A}$ and $\mathbf{B}$, then $AB\sin\theta = \epsilon_{\alpha\beta}A^\alpha B^\beta$. This result follows immediately from the formula for the sine of the angle between two vectors given in Sec. 54.

concept of tensor differentiation was introduced by A. J. McConnell, whose elegant treatment of surfaces is followed closely in this and several other sections of this chapter.[20]

Let us consider a curve C lying on a given surface S and a vector A^i defined along C. If t is a parameter along C, we can compute the intrinsic derivative $\delta A^i/\delta t$ of A^i, namely,

$$\frac{\delta A^i}{\delta t} = \frac{dA^i}{dt} + {}_g\begin{Bmatrix} i \\ jk \end{Bmatrix} A^j \frac{dx^k}{dt}. \tag{66.1}$$

In formula 66.1 the Christoffel symbols ${}_g\begin{Bmatrix} i \\ jk \end{Bmatrix}$ refer to the space coordinates x^i and are formed from the metric coefficients g_{ij}. This is indicated by the prefix g on the symbol. On the other hand, if we consider a *surface vector* A^α defined along the same curve C, we can form the intrinsic derivative with respect to the surface variables, namely,

$$\frac{\delta A^\alpha}{\delta t} = \frac{dA^\alpha}{dt} + {}_a\begin{Bmatrix} \alpha \\ \beta\gamma \end{Bmatrix} A^\beta \frac{du^\gamma}{dt}. \tag{66.2}$$

In this expression the Christoffel symbols ${}_a\begin{Bmatrix} \alpha \\ \beta\gamma \end{Bmatrix}$ are formed from the metric coefficients $a_{\alpha\beta}$ associated with the Gaussian surface coordinates u^α. A geometric interpretation of these formulas is at hand when the fields A^i and A^α are such that $\delta A^i/\delta t = 0$ and $\delta A^\alpha/\delta t = 0$. In the first equation the vectors A^i form a parallel field with respect to C, *considered as a space curve*, whereas the equation $\delta A^\alpha/\delta t = 0$ defines a parallel field *with respect to C regarded as a surface curve*. The corresponding formulas for the intrinsic derivatives of the covariant vectors A_i and A_α are

$$\frac{\delta A_i}{\delta t} = \frac{dA_i}{dt} - {}_g\begin{Bmatrix} k \\ ij \end{Bmatrix} A_k \frac{dx^j}{dt}, \tag{66.3}$$

and

$$\frac{\delta A_\alpha}{\delta t} = \frac{dA_\alpha}{dt} - {}_a\begin{Bmatrix} \gamma \\ \alpha\beta \end{Bmatrix} A_\gamma \frac{du^\beta}{dt}. \tag{66.4}$$

Consider next a tensor field $T_\alpha{}^i$, which is a contravariant vector with respect to a transformation of space coordinates x^i and a covariant vector relative to a transformation of surface coordinates u^α. An example of a field of this type is a tensor $x_\alpha{}^i = \partial x^i/\partial u^\alpha$ introduced in Sec. 64. If $T_\alpha{}^i$ is defined over a surface curve C, and the parameter along C is t, then $T_\alpha{}^i$ is a function of t. We introduce a parallel vector field A_i along C, regarded

[20] A. J. McConnell, *Absolute Differential Calculus*, Chapters XIV–XVI.

as a space curve, and a parallel vector field B^α along C, viewed as a surface curve, and form an invariant

$$\Phi(t) = T_\alpha^{\ i} A_i B^\alpha.$$

The derivative of $\Phi(t)$ with respect to the parameter t is given by the expression

$$\frac{d\Phi}{dt} = \frac{dT_\alpha^{\ i}}{dt} A_i B^\alpha + T_\alpha^{\ i} \frac{dA_i}{dt} B^\alpha + T_\alpha^{\ i} A_i \frac{dB^\alpha}{dt}, \tag{66.5}$$

which is obviously an invariant relative to both the space and surface coordinates. But, since the fields $A_i(t)$ and $B^\alpha(t)$ are parallel,

$$\frac{dA_i}{dt} = {}_g\begin{Bmatrix} k \\ ij \end{Bmatrix} A_k \frac{dx^j}{dt} \quad \text{and} \quad \frac{dB^\alpha}{dt} = -{}_a\begin{Bmatrix} \alpha \\ \beta\gamma \end{Bmatrix} B^\beta \frac{du^\gamma}{dt},$$

and (66.5) becomes

$$\frac{d\Phi}{dt} = \left[\frac{dT_\alpha^{\ i}}{dt} + {}_g\begin{Bmatrix} i \\ jk \end{Bmatrix} T_\alpha^{\ j} \frac{dx^k}{dt} - {}_a\begin{Bmatrix} \delta \\ \alpha\gamma \end{Bmatrix} T_\delta^{\ i} \frac{du^\gamma}{dt}\right] A_i B^\alpha. \tag{66.6}$$

Since this is invariant for an arbitrary choice of parallel fields A_i and B^α, the quotient law guarantees that the expression in the brackets of (66.6) is a tensor of the same character as $T_\alpha^{\ i}$. We call this tensor, after McConnell, the *intrinsic tensor derivative* of $T_\alpha^{\ i}$ with respect to the parameter t, and write

$$\frac{\delta T_\alpha^{\ i}}{\delta t} \equiv \frac{dT_\alpha^{\ i}}{dt} + {}_g\begin{Bmatrix} i \\ jk \end{Bmatrix} T_\alpha^{\ j} \frac{dx^k}{dt} - {}_a\begin{Bmatrix} \delta \\ \alpha\gamma \end{Bmatrix} T_\delta^{\ i} \frac{du^\gamma}{dt}. \tag{66.7}$$

If the field $T_\alpha^{\ i}$ is defined over the entire surface S, we can argue that, since

$$\frac{\delta T_\alpha^{\ i}}{\delta t} \equiv \left[\frac{\partial T_\alpha^{\ i}}{\partial u^\gamma} + {}_g\begin{Bmatrix} i \\ jk \end{Bmatrix} T_\alpha^{\ j} x_\gamma^{\ k} - {}_a\begin{Bmatrix} \delta \\ \gamma\alpha \end{Bmatrix} T_\delta^{\ i}\right] \frac{du^\gamma}{dt}$$

is a tensor field and du^γ/dt is an arbitrary surface vector (for C is arbitrary), the expression in the bracket is a tensor of the type $T_{\alpha\gamma}^{\ \ i}$. We write

$$T_{\alpha,\gamma}^{i} \equiv \frac{\partial T_\alpha^{\ i}}{\partial u^\gamma} + {}_g\begin{Bmatrix} i \\ jk \end{Bmatrix} T_\alpha^{\ j} x_\gamma^{\ k} - {}_a\begin{Bmatrix} \delta \\ \alpha\gamma \end{Bmatrix} T_\delta^{\ i} \tag{66.8}$$

and call $T_{\alpha,\gamma}^{i}$ the *tensor derivative of* T_α^i *with respect to* u^γ.

The extension of this definition to more complicated tensors is obvious from the structure of the formula 66.8. Thus the tensor derivative of $T_{\alpha\beta}^{\ \ i}$ with respect to u^γ is given by

$$T_{\alpha\beta,\gamma}^{i} = \frac{\partial T_{\alpha\beta}^{i}}{\partial u^\gamma} + {}_g\begin{Bmatrix} i \\ jk \end{Bmatrix} T_{\alpha\beta}^{j} x_\gamma^k - {}_a\begin{Bmatrix} \delta \\ \alpha\gamma \end{Bmatrix} T_{\delta\beta}^{i} - {}_a\begin{Bmatrix} \delta \\ \beta\gamma \end{Bmatrix} T_{\alpha\delta}^{i}. \tag{66.9}$$

If the surface coordinates at any point P of S are geodesic, and the space coordinates are orthogonal cartesian, we see that *at that point* the tensor derivatives reduce to the ordinary derivatives. This leads us to conclude that the operations of tensor differentiation of products and sums follow the usual rules and that the tensor derivatives of g_{ij}, $a_{\alpha\beta}$, ϵ_{ijk}, $\epsilon_{\alpha\beta}$ and their associated tensors vanish. Accordingly, they behave as constants in the tensor differentiation.

67. The Second Fundamental Form of a Surface

The apparatus developed in the preceding section permits us to obtain easily and in the most general form an important set of formulas due to Gauss. We will also deduce with its aid the second fundamental quadratic form of a surface already encountered in Sec. 65.[21]

We begin by calculating the tensor derivative of the tensor $x_\alpha{}^i$, representing the components of the surface base vectors $\mathbf{a}_\alpha$. We have

$$x^i_{\alpha,\beta} = \frac{\partial^2 x^i}{\partial u^\alpha \partial u^\beta} + {}_g\begin{Bmatrix} i \\ jk \end{Bmatrix} x^j_\alpha x^k_\beta - {}_a\begin{Bmatrix} \delta \\ \alpha\beta \end{Bmatrix} x^i_\delta, \tag{67.1}$$

from which we deduce that

$$x^i_{\alpha,\beta} = x^i_{\beta,\alpha}. \tag{67.2}$$

Since the tensor derivative of $a_{\alpha\beta}$ vanishes, we obtain, upon differentiating the relation

$$a_{\alpha\beta} = g_{ij} x^i_\alpha x^j_\beta, \tag{64.6}$$

$$g_{ij} x^i_{\alpha,\gamma} x^j_\beta + g_{ij} x^i_\alpha x^j_{\beta,\gamma} = 0. \tag{67.3}$$

Interchanging α, β, γ cyclically leads to two formulas:

$$g_{ij} x^i_{\beta,\alpha} x^j_\gamma + g_{ij} x^i_\beta x^j_{\gamma,\alpha} = 0, \tag{67.4}$$

$$g_{ij} x^i_{\gamma,\beta} x^j_\alpha + g_{ij} x^i_\gamma x^j_{\alpha,\beta} = 0. \tag{67.5}$$

If we add (67.4) and (67.5), subtract (67.3), and take into account the symmetry relation (67.2), we obtain

$$g_{ij} x^i_{\alpha,\beta} x^j_\gamma = 0. \tag{67.6}$$

This is the orthogonality relation which states that $x^i_{\alpha,\beta}$ is a space vector normal to the surface, and hence it is directed along the unit normal n^i. Consequently, there exists a set of functions $b_{\alpha\beta}$ such that

$$x^i_{\alpha,\beta} = b_{\alpha\beta} n^i. \tag{67.7}$$

[21] Compare A. J. McConnell, *Absolute Differential Calculus* (1931), p. 200.

The quantities $b_{\alpha\beta}$ are the components of a symmetric surface tensor, and the differential quadratic form

$$\mathscr{B} \equiv b_{\alpha\beta}\, du^\alpha\, du^\beta \tag{67.8}$$

is the desired *second fundamental form.*

To demonstrate the equivalence of this definition of the tensor $b_{\alpha\beta}$ with that given in Sec. 65, namely,

$$b_{\alpha\beta} = -\frac{1}{2}\left(\frac{\partial \mathbf{n}}{\partial u^\alpha}\cdot\frac{\partial \mathbf{r}}{\partial u^\beta} + \frac{\partial \mathbf{n}}{\partial u^\beta}\cdot\frac{\partial \mathbf{r}}{\partial u^\alpha}\right),$$

note that the vectors $\mathbf{n}$ and $\mathbf{a}_\alpha = \partial\mathbf{r}/\partial u^\alpha$ are orthogonal and hence

$$\mathbf{n}\cdot\frac{\partial \mathbf{r}}{\partial u^\alpha} = 0 \quad \text{and} \quad \mathbf{n}\cdot\frac{\partial \mathbf{r}}{\partial u^\beta} = 0.$$

Differentiating these two scalar products with respect to u^β and u^α, respectively, and adding, we get

$$\frac{1}{2}\left(\frac{\partial \mathbf{n}}{\partial u^\alpha}\cdot\frac{\partial \mathbf{r}}{\partial u^\beta} + \frac{\partial \mathbf{n}}{\partial u^\beta}\cdot\frac{\partial \mathbf{r}}{\partial u^\alpha}\right) = -\mathbf{n}\cdot\frac{\partial^2 \mathbf{r}}{\partial u^\alpha\,\partial u^\beta}.$$

Hence

$$b_{\alpha\beta} = \mathbf{n}\cdot\frac{\partial^2 \mathbf{r}}{\partial u^\alpha\,\partial u^\beta}. \tag{67.9}$$

However,

$$\frac{\partial \mathbf{r}}{\partial u^\alpha} = \mathbf{a}_\alpha = \mathbf{b}_i x^i_\alpha;$$

therefore

$$\begin{aligned}\frac{\partial^2 \mathbf{r}}{\partial u^\alpha\,\partial u^\beta} &= \mathbf{b}_i\frac{\partial x^i_\alpha}{\partial u^\beta} + \frac{\partial \mathbf{b}_i}{\partial u^\beta}\,x^i_\alpha \\ &= \mathbf{b}_i\frac{\partial x^i_\alpha}{\partial u^\beta} + \frac{\partial \mathbf{b}}{\partial x^j}\,x^i_\alpha x^j_\beta \\ &= \mathbf{b}_i\left(\frac{\partial^2 x^i}{\partial u^\alpha\,\partial u^\beta} + {}_g\left\{\begin{matrix} i \\ jk \end{matrix}\right\} x^j_\alpha x^k_\beta\right),\end{aligned}$$

where, in the last step, we made use of formula 46.4 for the derivative of the base vector $\mathbf{b}_i$.

If we insert in the right-hand member of the foregoing expression from equation 67.1, we get

$$\frac{\partial^2 \mathbf{r}}{\partial u^\alpha\,\partial u^\beta} = \mathbf{b}_i\left(x^i_{\alpha,\beta} + {}_a\left\{\begin{matrix} \delta \\ \alpha\beta \end{matrix}\right\} x^i_\delta\right). \tag{67.10}$$

Multiplying equation 67.10 scalarly by $\mathbf{n}$, and observing that the vectors $\mathbf{b}_i x^i_\delta = \mathbf{a}_\delta$ and $\mathbf{n}$ are orthogonal, we get

$$\begin{aligned} \mathbf{n} \cdot \frac{\partial^2 \mathbf{r}}{\partial u^\alpha \, \partial u^\beta} &= \mathbf{n} \cdot \mathbf{b}_i x^i_{\alpha,\beta} \\ &= x^i_{\alpha,\beta} n_i \\ &= b_{\alpha\beta} \end{aligned}$$

by formula 67.7. This establishes the equivalence of the two definitions of the second fundamental quadratic form.

Equations 67.7 are known as the *formulas of Gauss*. The importance of the form 67.8 in differential geometry stems from the fact that the tensors $a_{\alpha\beta}$ and $b_{\alpha\beta}$, satisfying equations of Gauss and Codazzi (to be derived in Sec. 69), determine the surface to within a rigid body motion in space.

Problems

1. Show that $b_{\alpha\beta} = g_{ij} x^i_{\alpha,\beta} n^j = \frac{1}{2} \epsilon^{\gamma\delta} \epsilon_{ijk} x^i_{\alpha,\beta} x^j_\gamma x^k_\delta$.

2. Show that, in the notation of Sec. 65,

$$b_{\alpha\beta} = -\frac{1}{2}\left(\frac{\partial \mathbf{n}}{\partial u^\alpha} \cdot \frac{\partial \mathbf{r}}{\partial u^\beta} + \frac{\partial \mathbf{n}}{\partial u^\beta} \cdot \frac{\partial \mathbf{r}}{\partial u^\alpha}\right) = -g_{ij} n^i_{,\alpha} x^j_\beta,$$

where $\mathbf{n}$ is the unit normal and $\mathbf{r}$ is the position vector of the point on the surface.

3. When the equation of a surface S, referred to a set of orthogonal cartesian axes, is taken in the form $y^3 = f(y^1, y^2)$, we can write the parametric equations of S in the form $y^1 = u^1, y^2 = u^2, y^3 = f(u^1, u^2)$. If partial derivatives of $f(y^1, y^2)$ are denoted by $f_{y^1} \equiv p, f_{y^2} \equiv q, f_{y^1y^1} \equiv r, f_{y^1y^2} \equiv s, f_{y^2y^2} \equiv t$, then the coefficients $a_{\alpha\beta}$ in $ds^2 = a_{\alpha\beta}\, du^\alpha\, du^\beta$ are

$$a_{11} = 1 + p^2, \qquad a_{12} = pq, \qquad a_{22} = 1 + q^2,$$

whereas the coefficients $b_{\alpha\beta}$ of the second fundamental form are

$$b_{11} = \frac{r}{\sqrt{1 + p^2 + q^2}}, \qquad b_{12} = \frac{s}{\sqrt{1 + p^2 + q^2}}, \qquad b_{22} = \frac{t}{\sqrt{1 + p^2 + q^2}}.$$

Show this and compute $a_{\alpha\beta}$ and $b_{\alpha\beta}$ for the surface of the sphere

$$y^3 = \sqrt{a^2 - (y^1)^2 - (y^2)^2}.$$

4. If equations of S are

$$S\colon \quad y^i = y^i(u^1, u^2),$$

where the y^i are orthogonal cartesian coordinates, and $\mathbf{r}$ is the position vector of the point (y^1, y^2, y^3) on S, then on using subscripts to denote partial derivatives,

$$a_{\alpha\beta} = \mathbf{r}_{u^\alpha} \cdot \mathbf{r}_{u^\beta}, \qquad b_{\alpha\beta} = [\mathbf{r}_{u^\alpha u^\beta} \cdot \mathbf{r}_{u^1} \times \mathbf{r}_{u^2}]/\sqrt{a}.$$

Show this. Use these formula to compute the $a_{\alpha\beta}$ and $b_{\alpha\beta}$ for the surface of revolution $y^1 = u^1 \cos u^2$, $y^2 = u^1 \sin u^2$, $y^3 = f(u^1)$.

68. The Integrability Conditions

In order to get insight into the significance of the tensor $b_{\alpha\beta}$ let us examine more closely the Gauss formulas

$$x^i_{\alpha,\beta} = b_{\alpha\beta} n^i, \tag{68.1}$$

where

$$x^i_{\alpha,\beta} = \frac{\partial^2 x^i}{\partial u^\alpha \, \partial u^\beta} + {}_g\begin{Bmatrix} i \\ jk \end{Bmatrix} x_\alpha{}^j x_\beta{}^k - {}_a\begin{Bmatrix} \delta \\ \alpha\beta \end{Bmatrix} x_\delta{}^i, \tag*{[67.1]}$$

and

$$n_i = \tfrac{1}{2}\epsilon^{\alpha\beta}\epsilon_{ijk} x_\alpha{}^j x_\beta{}^k, \tag*{[65.10]}$$

with $x_\alpha{}^j = \partial x^j/\partial u^\alpha$.

If we insert these expressions in equation 68.1, we obtain a set of second-order partial differential equations, in which the dependent variables x^i are functions of the surface coordinates u^α. The coefficients in these differential equations are functions of metric coefficients g_{ij} of the manifold in which the surface S, defined by

$$x^i = x^i(u^1, u^2), \qquad (i = 1, 2, 3), \tag{68.2}$$

is immersed; they are also functions of $a_{\alpha\beta} = g_{ij}(\partial x^i/\partial u^\alpha)(\partial x^j/\partial u^\beta)$, and $b_{\alpha\beta}$.

If equations 68.2 are given, we can compute $a_{\alpha\beta}$ and $b_{\alpha\beta}$ (see Problem 1, Sec. 67), insert the appropriate expressions in (68.1), and, of course, equations 68.1 will be satisfied identically. On the other hand, if the functions $a_{\alpha\beta}$ and $b_{\alpha\beta}$ are prescribed in advance, equations 68.1 will become *equations of conditions*, and in general they will have no solutions yielding equations 68.2 of the surface S. In order that the tensors $a_{\alpha\beta}$ and $b_{\alpha\beta}$ be related to some surface, it is necessary that the x^i satisfy the integrability conditions,

$$\frac{\partial^2 x_\alpha{}^i}{\partial u^\gamma \, \partial u^\beta} = \frac{\partial^2 x_\alpha{}^i}{\partial u^\beta \, \partial u^\gamma}, \tag{68.3}$$

whenever the functions $x_\alpha{}^i$ are of class C^2. From our discussion of inversion of order of covariant differentiation in Sec. 36, it follows that the condition 68.3 is equivalent to[22] (cf. equation 36.6)

$$x^i_{\alpha,\beta\gamma} - x^i_{\alpha,\gamma\beta} = R^\delta_{.\alpha\beta\gamma} x_\delta{}^i, \tag{68.4}$$

[22] We dispense with the details of computation since they are not essential to the course of argument. See, for example, A. J. McConnell, *Absolute Differential Calculus*, p. 203.

where $R^{\delta}_{.\alpha\beta\gamma}$ is the Riemann tensor of the second kind, formed with the aid of the coefficients $a_{\alpha\beta}$ of the first fundamental quadratic form. Equations 68.4 involve third partial derivatives of the coordinates x^i, and we shall assume from now on that the functions entering in (68.2) are of class C^3.

We shall see that the conditions of integrability 68.4 impose certain restrictions on the possible choices of functions $b_{\alpha\beta}$ and $a_{\alpha\beta}$. These restrictive conditions are known as the *equations of Gauss and Codazzi*. They will be derived in the following section.

69. Formulas of Weingarten and Equations of Gauss and Codazzi[23]

In order to derive the equations of Gauss and Codazzi we need an auxiliary result, due to Weingarten, giving the expressions for the derivatives of the unit normal vector n^i to S. We begin with the relation $g_{ij}n^in^j = 1$, and form its tensor derivative.[24] We have

$$g_{ij}n^i_{,\alpha}n^j + g_{ij}n^in^j_{,\alpha} = 0,$$

or

$$g_{ij}n^in^j_{,\alpha} = 0. \tag{69.1}$$

Equation 69.1 shows that $n^j_{,\alpha}$, considered as a space vector, is orthogonal to the unit normal n^i, and hence it lies in the tangent plane to the surface. Accordingly, it can be represented as a linear form in the base vectors x^i_α,

$$n^i_{,\alpha} = c^\beta_\alpha x^i_\beta\,. \tag{69.2}$$

Since n^i is normal to the surface, we have the orthogonality relation $g_{ij}x^i_\alpha n^j = 0$, whose tensor derivative is

$$g_{ij}x^i_{\alpha,\beta}n^j + g_{ij}x^i_\alpha n^j_{,\beta} = 0. \tag{69.3}$$

But, from (68.1),

$$x^i_{\alpha,\beta} = b_{\alpha\beta}n^i, \tag{69.4}$$

so that the substitution from (69.4) and (69.2) in (69.3) yields

$$b_{\alpha\beta} + g_{ij}x^i_\alpha x^j_\gamma c^\gamma_\beta = 0,$$

and, since $a_{\alpha\beta} = g_{ij}x_\alpha{}^i x_\beta{}^j$, we have

$$b_{\alpha\beta} = -a_{\alpha\gamma}c_\beta{}^\gamma.$$

[23] The treatment given here is patterned after A. J. McConnell, *Absolute Differential Calculus*, pp. 201–205.

[24] We recall that $n,_\alpha{}^i = \dfrac{\partial n^i}{\partial u^\alpha} + {}_g\left\{ \begin{matrix} i \\ jk \end{matrix} \right\} n^j x^k_\alpha$.

Solving this equation for $c_\beta{}^\gamma$, we get

$$c_\beta{}^\gamma = -a^{\alpha\gamma} b_{\alpha\beta},$$

so that equation 69.2 reads

$$n^i_{,\alpha} = -a^{\beta\gamma} b_{\beta\alpha} x_\gamma{}^i. \tag{69.5}$$

These are the *Weingarten formulas* which we will use in deriving the Codazzi equations.

The equations we desire follow from the integrability conditions 68.4, namely,

$$x^i_{\alpha,\beta\gamma} - x^i_{\alpha,\gamma\beta} = R^\delta_{.\alpha\beta\gamma} x_\delta^i. \tag{69.6}$$

We form the tensor derivative of equation 69.4, and use 69.5 to obtain

$$\begin{aligned} x^i_{\alpha,\beta\gamma} &= b_{\alpha\beta,\gamma} n^i + b_{\alpha\beta} n^i_{,\gamma} \\ &= b_{\alpha\beta,\gamma} n^i - b_{\alpha\beta} a^{\delta\lambda} b_{\delta\gamma} x_\lambda{}^i. \end{aligned} \tag{69.7}$$

Substituting from (69.7) in the left-hand member of (69.6), we get

$$x^i_{\alpha,\beta\gamma} - x^i_{\alpha,\gamma\beta} = (b_{\alpha\beta,\gamma} - b_{\alpha\gamma,\beta}) n^i - a^{\delta\lambda}(b_{\alpha\beta} b_{\delta\gamma} - b_{\alpha\gamma} b_{\delta\beta}) x_\lambda{}^i.$$

Hence

$$(b_{\alpha\beta,\gamma} - b_{\alpha\gamma,\beta}) n^i - a^{\delta\lambda}(b_{\alpha\beta} b_{\delta\gamma} - b_{\alpha\gamma} b_{\delta\beta}) x_\lambda{}^i = R^\delta_{.\alpha\beta\gamma} x_\delta^i. \tag{69.8}$$

To obtain the *equations of Codazzi* we multiply (69.8) by n_i, and, since $x^i_\alpha n_i = 0$, we get the desired result

$$b_{\alpha\beta,\gamma} - b_{\alpha\gamma,\beta} = 0. \tag{69.9}$$

To obtain the *equations of Gauss* we multiply (69.8) by $g_{ij} x^j_\rho$ and obtain

$$b_{\rho\beta} b_{\alpha\gamma} - b_{\rho\gamma} b_{\alpha\beta} = R_{\rho\alpha\beta\gamma}. \tag{69.10}$$

Since α, β assume values 1, 2 and $b_{\alpha\beta} = b_{\beta\alpha}$, we see that there are two independent equations of Codazzi and only one independent equation of Gauss.[25] The independent equations of Codazzi are

$$b_{\alpha\alpha,\beta} - b_{\alpha\beta,\alpha} = 0, \qquad (\alpha \neq \beta), \qquad \text{(no sum on } \alpha), \tag{69.11}$$

or, when the covariant derivatives are written out in full with the aid of

$$b_{\alpha\beta,\gamma} = \frac{\partial b_{\alpha\beta}}{\partial u^\gamma} - {}_a\begin{Bmatrix} \delta \\ \alpha\gamma \end{Bmatrix} b_{\delta\beta} - {}_a\begin{Bmatrix} \delta \\ \beta\gamma \end{Bmatrix} b_{\alpha\delta},$$

[25] We recall that

$$R_{\alpha\alpha\beta\gamma} = R_{\alpha\beta\gamma\gamma} = 0, \qquad R_{1212} = R_{2121} = -R_{2112} = -R_{1221}.$$

we get

$$\frac{\partial b_{\alpha\alpha}}{\partial u^\beta} - \frac{\partial b_{\alpha\beta}}{\partial u^\alpha} - b_{\alpha\delta}\begin{Bmatrix}\delta\\ \alpha\beta\end{Bmatrix} + b_{\delta\beta}\begin{Bmatrix}\delta\\ \alpha\alpha\end{Bmatrix} = 0, \qquad \alpha \neq \beta, \quad \text{(no sum on } \alpha). \tag{69.12}$$

The equation of Gauss, on the other hand, is

$$b_{11}b_{22} - b_{12}^2 = R_{1212}. \tag{69.13}$$

This equation relates the coefficients $b_{\alpha\beta}$ and $a_{\alpha\beta}$ in the two fundamental quadratic forms.

The foregoing demonstration shows that if the tensors $a_{\alpha\beta}$ and $b_{\alpha\beta}$ are the fundamental tensors of the surface S: $x^i = x^i(u^1, u^2)$, then equations 69.11 and 69.13 are satisfied. Conversely, it can be shown that if the two sets of functions $a_{\alpha\beta}$ and $b_{\alpha\beta}$ satisfying equations 69.11 and 69.13 are prescribed, and if $a_{\alpha\beta}\, du^\alpha\, du^\beta$ is a positive definite form, then the surface S is determined (locally) to within a rigid body motion in space. The proof[26] of this depends on considering the existence of a solution of a system of differential equations of the type discussed in Sec. 39. We conclude by remarking that if $b_{\alpha\beta} = 0$, then by 65.5 the surface is a plane.

70. The Mean and Total Curvatures of a Surface

If we recall the definition 62.4 of the total curvature K,

$$K = \frac{R_{1212}}{a}, \qquad a = a_{11}a_{22} - a_{12}^2, \tag{62.4}$$

we can write equation 69.13 in the form

$$K = \frac{b_{11}b_{22} - b_{12}^2}{a_{11}a_{22} - a_{12}^2} = \frac{b}{a}. \tag{70.1}$$

Thus the Gaussian curvature is equal to the quotient of the discriminants of the second and first fundamental quadratic forms.

We introduce next another important invariant H, called the *mean curvature of the surface*. This is given by the formula

$$H \equiv \tfrac{1}{2} a^{\alpha\beta} b_{\alpha\beta}, \tag{70.2}$$

and we shall see in Sec. 72 that the invariants K and H are connected in a remarkable way with the ordinary curvatures of certain curves formed by taking normal sections of the surface.

[26] For a detailed discussion, see L. P. Eisenhart, *Introduction to Differential Geometry*, pp. 218–221, where the case of cartesian variables x^i is considered.

71. Curves on a Surface. Theorem of Meusnier

Let equations of a smooth curve C lying on the surface

$$S:\quad x^i = x^i(u^1, u^2) \tag{71.1}$$

be given in the form

$$C:\quad u^\alpha = u^\alpha(s), \tag{71.2}$$

where s is the arc parameter. If the values of $u^\alpha(s)$ are inserted in (71.1), we obtain the space coordinates x^i of C in the form

$$x^i = x^i(s). \tag{71.3}$$

These are the equations of C, regarded as a space curve. The properties of C can then be studied with the aid of the Frenet-Serret formulas 50.1, 50.2, and 50.3 by analyzing the rates of change of the unit tangent vector $\boldsymbol{\lambda}$, the unit principal normal $\boldsymbol{\mu}$, and the unit binormal $\boldsymbol{\nu}$.

On the other hand, if we regard C as a surface curve, defined by (71.2), the components λ^α of the unit tangent vector $\boldsymbol{\lambda}$ are related to the space components λ^i of the same vector by the formulas

$$\lambda^i = \frac{\partial x^i}{\partial u^\alpha}\frac{du^\alpha}{ds} \equiv x_\alpha^{\ i}\lambda^\alpha, \tag{71.4}$$

where

$$\lambda^i = \frac{dx^i}{ds} \quad \text{and} \quad \lambda^\alpha = \frac{du^\alpha}{ds}. \tag{71.5}$$

We also recall equation 63.4,

$$\frac{\delta\lambda^\alpha}{\delta s} = \varkappa_g \eta^\alpha, \tag{71.6}$$

where η^α is the unit normal to C in the tangent plane to the surface, and $\varkappa_g$ is the geodesic curvature of C. (See Fig. 29.)

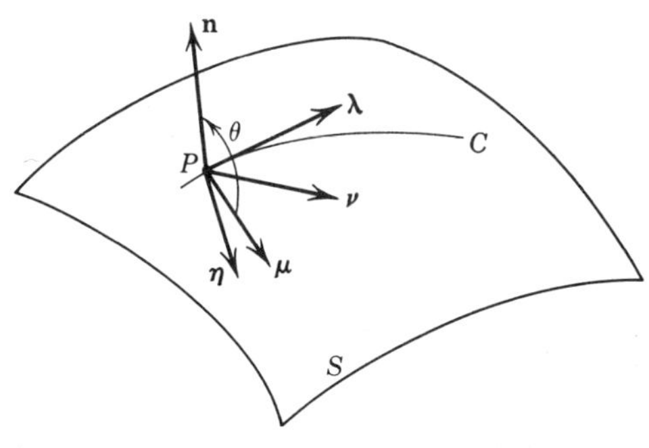

Fig. 29

If we differentiate (71.4) intrinsically with respect to s, we obtain

$$\frac{\delta \lambda^i}{\delta s} = x^i_{\alpha,\beta} \lambda^\alpha \frac{du^\beta}{ds} + x_\alpha{}^i \frac{\delta \lambda^\alpha}{\delta s},$$

which, upon taking into account Frenet's formula 50.1 and equation 71.6, becomes

$$\varkappa \mu^i = x^i_{\alpha,\beta} \lambda^\alpha \lambda^\beta + \varkappa_g x_\alpha{}^i \eta^\alpha.$$

The space components η^i of $\boldsymbol{\eta}$ are $\eta^i = x_\alpha{}^i \eta^\alpha$, and, if we recall Gauss's formula $x^i_{\alpha,\beta} = b_{\alpha\beta} n^i$, the foregoing equation becomes

$$\varkappa \mu^i = b_{\alpha\beta} \lambda^\alpha \lambda^\beta n^i + \varkappa_g \eta^i, \tag{71.7}$$

where n^i is the unit normal to the surface S.

Formula 71.7 states that the principal normal $\boldsymbol{\mu}$ to C lies in the plane of the vectors $\mathbf{n}$ and $\boldsymbol{\eta}$. Since $\mathbf{n}$, $\boldsymbol{\eta}$, and $\boldsymbol{\lambda}$ are orthonormal and $\mathbf{n} \times \boldsymbol{\eta} = \boldsymbol{\lambda}$, we have

$$\epsilon_{ijk} n^j \eta^k = \lambda_i, \tag{71.8}$$

and, since $\boldsymbol{\lambda}$ is orthogonal to the plane of $\mathbf{n}$ and $\boldsymbol{\mu}$, $\boldsymbol{\mu} \times \mathbf{n} = -\sin\theta \boldsymbol{\lambda}$, or

$$\epsilon_{ijk} \mu^j n^k = -\sin\theta \cdot \lambda_i, \tag{71.9}$$

where θ is the angle between $\boldsymbol{\mu}$ and $\mathbf{n}$.

On multiplying (71.9) by $\varkappa$ we get

$$\epsilon_{ijk} \varkappa \mu^j n^k = -\varkappa \sin\theta \cdot \lambda_i,$$

which, on substitution from (71.7), yields

$$\epsilon_{ijk}(b_{\alpha\beta} \lambda^\alpha \lambda^\beta n^j + \varkappa_g \eta^j) n^k = -\varkappa \sin\theta \cdot \lambda_i.$$

But $\epsilon_{ijk} n^j n^k = 0$ and $\epsilon_{ijk} \eta^j n^k = -\lambda_i$ by (71.8), and we conclude that

$$\varkappa_g = \varkappa \sin\theta. \tag{71.10}$$

On the other hand, if we form the scalar product of both members of equation 71.7 with n_i and note that $n_i \mu^i = \cos\theta$, we get

$$\varkappa \cos\theta = b_{\alpha\beta} \lambda^\alpha \lambda^\beta. \tag{71.11}$$

The invariant $b_{\alpha\beta} \lambda^\alpha \lambda^\beta$ in (71.11) has the same value for all curves on S with the same tangent vector $\boldsymbol{\lambda}$ at P. In particular, it has this value for the curve formed by the intersection of the normal plane containing $\mathbf{n}$ and $\boldsymbol{\lambda}$. But for every normal plane section the angle θ is either 0 or π radians, so that for the normal plane section $\varkappa \cos\theta = \varkappa$ or $-\varkappa$; since the right-hand member of (71.11) is an invariant, the value of $\varkappa \cos\theta$ for every curve C tangent to $\boldsymbol{\lambda}$ is equal to the curvature $\varkappa_{(n)}$ of the normal section in the direction $\boldsymbol{\lambda}$. The curvature $\varkappa_{(n)}$ is called the *normal curvature of the surface S* in the direction $\boldsymbol{\lambda}$. We can thus write (71.11) as

$$\varkappa_{(n)} = b_{\alpha\beta}\lambda^\alpha\lambda^\beta, \tag{71.12}$$

where $\varkappa_{(n)} = \varkappa \cos \theta$. Accordingly, Eq. 71.7 can be written as

$$\varkappa\mu^i = \varkappa_{(n)}n^i + \varkappa_g\eta^i.$$

This equation states that $\varkappa_{(n)}$ and $\varkappa_g$ are the components of the curvature vector $\varkappa\mu^i$ in the directions of the vectors n^i and η^i.

The result embodied in formula 71.12 can be stated as

MEUSNIER'S THEOREM. *The radius of curvature $R = 1/\varkappa$ of any curve at a given point on the surface is equal to the product of the radius of curvature $R_{(n)} = 1/\varkappa_{(n)}$ of the corresponding normal section at that point by the cosine of the angle between the normal to the surface and the principal normal to the curve.*

In symbols, we have

$$R = \pm R_{(n)} \cos \theta.$$

If S is a sphere, every normal section is a great circle of the sphere, and if C is any circle drawn on the sphere, then the preceding result becomes obvious from elementary geometric considerations. (See Fig. 30.)

If we recall that $ds^2 = a_{\alpha\beta}\, du^\alpha\, du^\beta$, and $du^\alpha/ds = \lambda^\alpha$, we see that formula 71.2 can be put in the form

$$\varkappa_{(n)} = \frac{b_{\alpha\beta}\, du^\alpha\, du^\beta}{a_{\alpha\beta}\, du^\alpha\, du^\beta} \equiv \frac{\mathscr{B}}{\mathscr{A}}. \tag{71.13}$$

We note that, if the surface is a plane, the normal curvature $\varkappa_{(n)} = 0$ at all points of the plane, and if it is a sphere, $\varkappa_{(n)} = 1/R$, where R is the radius of the sphere. Accordingly, we conclude from (71.13) that for the plane $b_{\alpha\beta} = 0$, and for the sphere $b_{\alpha\beta}\, du^\alpha\, du^\beta = (1/R)a_{\alpha\beta}\, du^\alpha\, du^\beta$ so that $a_{\alpha\beta} = Rb_{\alpha\beta}$ at all points of the sphere.

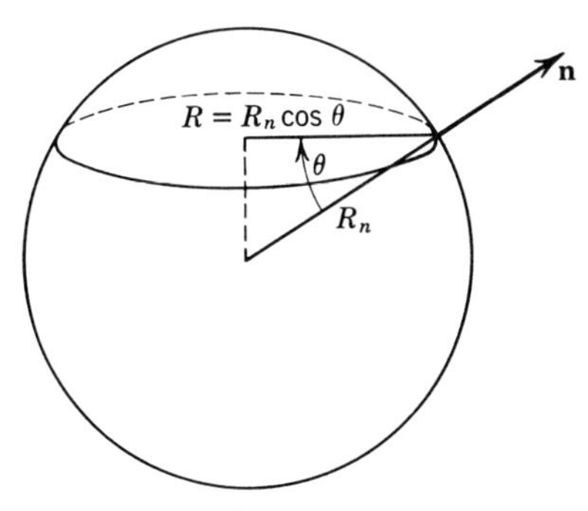

Fig. 30

Problems

1. Prove that the geodesic curvature $\varkappa_g$ and the curvature $\varkappa$ of any surface curve C are connected by the formula $\varkappa_g = \varkappa \sin \theta$, where θ is the angle between the normal to the surface and the principal normal to C.

2. Consider the surface of the right circular cone

$$S\colon \quad y^1 = u^1 \cos u^2, \quad y^2 = u^1 \sin u^2, \quad y^3 = u^1,$$

and the curve

$$C\colon \quad u^1 = a, \qquad u^2 = u^2 \text{ on } S.$$

Write equations of C in the form $u^1 = a$, $u^2 = s/a$, where s is the arc parameter and show with the aid of (71.6) that $\varkappa_g = \sqrt{2}/2a$. Verify this result by (71.10).

3. Show that the parallels $u^1 =$ constant on a sufficiently smooth surface of revolution

$$y^1 = u^1 \cos u^2, \quad y^2 = u^1 \sin u^2, \quad y^3 = f(u^1)$$

are curves of constant geodesic curvature.

4. A curve C on a surface S is called an *asymptotic line* if $b_{\alpha\beta}\lambda^\alpha\lambda^\beta = 0$ along C. Show that the principal normal $\boldsymbol{\mu}$ to the asymptotic line is tangent to S and the binormal $\boldsymbol{\nu}$ is normal to S.

5. Show that the normal curvatures in the directions of the coordinate curves are b_{11}/a_{11} and b_{22}/a_{22}.

6. Prove the theorem: If a curve is a geodesic on the surface, then either it is a straight line or its principal normal is orthogonal to the surface at every point, and conversely.

72. The Principal Curvatures of a Surface

We will be concerned in this section with the determination of directions $\lambda^\alpha = du^\alpha/ds$ on the surface such that the normal curvature $\varkappa_{(n)}$, given by the formula

$$\varkappa_{(n)} = b_{\alpha\beta}\lambda^\alpha\lambda^\beta, \tag{71.12}$$

assumes an extreme value.

Since the vector λ^α is a unit vector, $\varkappa_{(n)}$ in (71.12) has to be maximized subject to the constraining relation

$$a_{\alpha\beta}\lambda^\alpha\lambda^\beta = 1. \tag{72.1}$$

Following the usual procedure of determining constrained maxima and minima, we deduce that a necessary condition for an extremum is

$$b_{\alpha\beta}\lambda^\beta + \Lambda a_{\alpha\beta}\lambda^\beta = 0, \tag{72.2}$$

where Λ is the Lagrange multiplier. If equation 72.2 is multiplied by λ^α and account is taken of relations 71.12 and 72.1, it follows at once that $\Lambda = -\varkappa_{(n)}$. Thus equation 72.2, for the determination of directions yielding extreme values of $\varkappa_{(n)}$, can be written as

$$(b_{\alpha\beta} - \varkappa_{(n)}a_{\alpha\beta})\lambda^\beta = 0, \qquad (\alpha = 1, 2). \tag{72.3}$$

The set of homogeneous equations 72.3 will possess nontrivial solutions for λ^β if, and only if, the values of $\varkappa_{(n)}$ are the roots of the determinantal equation

$$|b_{\alpha\beta} - \vartheta a_{\alpha\beta}| = 0. \tag{72.4}$$

The quadratic equation 72.4, when written out in expanded form, is

$$\vartheta^2 - a^{\alpha\beta}b_{\alpha\beta}\vartheta + \frac{b}{a} = 0, \tag{72.5}$$

where $b = |b_{\alpha\beta}|$ and $a = |a_{\alpha\beta}|$.

Since the Gaussian curvature K is given by

$$K = \frac{b}{a} \tag*{[70.1]}$$

and the mean curvature H is

$$H = \tfrac{1}{2}a^{\alpha\beta}b_{\alpha\beta}, \tag*{[70.2]}$$

we see that equation 72.5 assumes the form

$$\vartheta^2 - 2H\vartheta + K = 0. \tag{72.6}$$

The roots $\vartheta = \varkappa_{(1)}$ and $\vartheta = \varkappa_{(2)}$ of (72.6) are called the *principal curvatures* of the surface, and the directions $\lambda^\alpha_{(1)}$ and $\lambda^\alpha_{(2)}$, corresponding to these extreme values of $\varkappa_{(n)}$, are the *principal directions on the surfaces.* We leave it for the reader to show that these directions are real.

From (72.6) it is clear that the principal curvatures $\varkappa_{(1)}$ and $\varkappa_{(2)}$ are related to the mean and Gaussian curvatures by the formulas

$$\begin{cases} \varkappa_{(1)} + \varkappa_{(2)} = 2H, \\ \varkappa_{(1)}\varkappa_{(2)} = K. \end{cases} \tag{72.7}$$

From equation 72.3 it follows that the principal directions are determined by

$$\begin{cases} (b_{\alpha\beta} - \varkappa_{(1)}a_{\alpha\beta})\lambda^\beta_{(1)} = 0, \\ (b_{\alpha\beta} - \varkappa_{(2)}a_{\alpha\beta})\lambda^\beta_{(2)} = 0. \end{cases}$$

If the first of these equations is multiplied by $\lambda^\alpha_{(2)}$, the second by $\lambda^\alpha_{(1)}$, and the results subtracted, we obtain

$$(\varkappa_{(2)} - \varkappa_{(1)})a_{\alpha\beta}\lambda^\alpha_{(1)}\lambda^\beta_{(2)} = 0. \tag{72.8}$$

If $\varkappa_{(1)} \neq \varkappa_{(2)}$, equation 72.8 tells us that

$$a_{\alpha\beta}\lambda^{\alpha}_{(1)}\lambda^{\beta}_{(2)} = 0, \tag{72.9}$$

that is, the *principal directions are orthogonal.* If the extreme values of $\varkappa_{(n)}$ are equal at a given point, then every direction is a principal direction.

We can summarize these results as a

THEOREM. *At each point of a surface there exist two mutually orthogonal directions for which the normal curvature attains its extreme values.*

A curve on a surface such that the tangent line to it at every point is directed along a principal direction is called a *line of curvature.* The differential equation for which the lines of curvature on S are the integral curves follows directly from equations 72.3. If we eliminate $\varkappa_{(n)}$ from these equations and set $\lambda^{\beta} = du^{\beta}/ds$, we get

$$\frac{b_{1\beta}\,du^{\beta}}{a_{1\beta}\,du^{\beta}} = \frac{b_{2\beta}\,du^{\beta}}{a_{2\beta}\,du^{\beta}},$$

or

$$(b_{11}a_{12} - b_{12}a_{11})(du^1)^2 + (b_{11}a_{22} - b_{22}a_{11})\,du^1\,du^2 + (b_{12}a_{22} - a_{12}b_{22})(du^2)^2 = 0. \tag{72.10}$$

At each point of S where either $b_{\alpha\beta}\,du^{\alpha}\,du^{\beta} \neq 0$ or $b_{\alpha\beta}\,du^{\alpha}\,du^{\beta}$ is not proportional to $a_{\alpha\beta}\,du^{\alpha}\,du^{\beta}$, equation 72.10 specifies two orthogonal directions

$$\frac{du^2}{du^1} = \psi_{\alpha}(u^1, u^2), \qquad (\alpha = 1, 2), \tag{72.11}$$

which coincide with directions of the principal curvatures.[27] Each equation in (72.11) determines a family of curves on S covering the surface without gaps. These two families of curves are orthogonal, and, if they are taken as a parametric net on S, the first fundamental form has the form

$$(ds)^2 = \bar{a}_{11}(d\bar{u}^1)^2 + \bar{a}_{22}(d\bar{u}^2)^2.$$

Accordingly, equation 72.10 in the coordinate system $\bar{u}^{\alpha}$ takes the form

$$-\bar{b}_{12}\bar{a}_{11}(d\bar{u}^1)^2 + (\bar{b}_{11}\bar{a}_{22} - \bar{b}_{22}\bar{a}_{11})\,d\bar{u}^1\,d\bar{u}^2 + \bar{b}_{12}\bar{a}_{22}(d\bar{u}^2)^2 = 0,$$

and its solutions are

$$\bar{u}^1 = \text{constant}, \qquad \bar{u}^2 = \text{constant}.$$

If we take $d\bar{u}^1 \neq 0$ and $d\bar{u}^2 = 0$, we see that $\bar{b}_{12} = 0$, since $\bar{a}_{11} \neq 0$. Thus a necessary condition for the net of lines of curvature to be orthogonal

[27] We exclude those points on S at which $\varkappa_n = 0$ or $\varkappa_{(1)} = \varkappa_{(2)}$. See concluding remarks in Sec. 71.

is that $a_{12} = b_{12} = 0$. Conversely, if $a_{12} = b_{12} = 0$, then (72.10) has solutions $u^1 =$ constant, $u^2 =$ constant, so that the coordinate lines are the lines of curvature. Hence a

THEOREM. *A necessary and sufficient condition for the coordinate net on a surface S (other than a plane or a sphere) to be the net of lines of curvature is that $a_{12} = b_{12} = 0$ at all points of S.*

We note that for *every* orthogonal net on a plane or a sphere $a_{12} = b_{12} = 0$.

Formula 71.9, for the normal curvature $\varkappa_{(n)}$, when the coordinate system is taken to be the net of lines of curvature becomes

$$\varkappa_{(n)} = \frac{b_{11}(du^1)^2 + b_{22}(du^2)^2}{a_{11}(du^1)^2 + a_{22}(du^2)^2}.$$

If we set $du^1 = 0$, $du^2 \neq 0$, and $du^2 = 0$, $du^1 \neq 0$, we get

$$\varkappa_1 = \frac{b_{11}}{a_{11}}, \qquad \varkappa_2 = \frac{b_{22}}{a_{22}}$$

for the curvatures of the coordinate lines $u_1 =$ constant and $u_2 =$ constant.

The lines of curvature on S should not be confused with the normal sections of S. The normal sections C_n are necessarily plane curves, whereas the lines of curvature ordinarily are not plane curves.

We conclude this section by giving several definitions.

A surface at all points of which the Gaussian curvature K is positive is called a *surface of positive curvature*. In this case (see equation 70.1), $b_{11}b_{22} - b_{12}{}^2 > 0$, and, since $\varkappa_{(n)} = b_{\alpha\beta}\lambda^\alpha\lambda^\beta$, we see that the *principal radii* $R_{(n)} = 1/\varkappa_{(n)}$ to all normal sections of a surface with positive curvature do not differ in sign. If $K < 0$, at a given point, the principal radii differ in sign. Then the equation

$$b_{\alpha\beta}\lambda^\alpha\lambda^\beta = 0 \tag{72.12}$$

defines two directions for which the radii of curvature are infinite. A surface at all points of which $K < 0$ is called a *surface of negative curvature*. If $K = 0$ at a given point, the directions given by (72.12) coincide, and for this direction R is infinite.

From geometrical considerations it is clear that ellipsoids, biparted hyperboloids, and elliptic paraboloids are surfaces of positive curvature. Hyperboloids of one sheet and hyperbolic paraboloids are surfaces of negative curvature.

A point on S is said to be *elliptic* if the signs of the principal curvatures $\varkappa_{(1)}$, $\varkappa_{(2)}$ are the same. It follows then that $\varkappa_{(n)}$ at an elliptic point does not change sign for any direction of the normal section. A point is *hyperbolic* if $\varkappa_{(1)}$ and $\varkappa_{(2)}$ have opposite signs. At a hyperbolic point there are two directions for which $\varkappa_{(n)} = 0$. A point is *parabolic* if one of the values $\varkappa_{(1)}$ or $\varkappa_{(2)}$ is zero. In the special case $\varkappa_{(1)} = \varkappa_{(2)}$, all values of $\varkappa_{(n)}$ are equal and such points are called *spherical*. In the neighborhood of a spherical point the surface looks like a sphere, and we can prove that if all points of S are spherical, then the surface S is a sphere. In some books spherical points are also called *umbilical*.

Problems

1. Given an ellipsoid of revolution, whose surface is determined by

$$\begin{cases} y^1 = a \cos u^1 \sin u^2, \\ y^2 = a \sin u^1 \sin u^2, \\ y^3 = c \cos u^2, \qquad a^2 > c^2, \end{cases}$$

show that

$$a_{11} = a^2 \sin^2 u^2, \qquad a_{12} = 0, \qquad a_{22} = a^2 \cos^2 u^2 + c^2 \sin^2 u^2,$$

$$b_{11} = \frac{ac \sin^2 u^2}{\sqrt{a^2 \cos^2 u^2 + c^2 \sin^2 u^2}}, \qquad b_{12} = 0, \qquad b_{22} = \frac{ac}{\sqrt{a^2 \cos^2 u^2 + c^2 \sin^2 u^2}},$$

and

$$K = \varkappa_{(1)}\varkappa_{(2)} = \frac{c^2}{(a^2 \cos^2 u^2 + c^2 \sin^2 u^2)^2}.$$

Discuss the lines of curvature on this surface.

2. Find the principal curvatures of the surface defined by

$$\begin{cases} y^1 = u^1, \\ y^2 = u^2, \\ y^3 = f(u^1, u^2). \end{cases}$$

3. Show that the helicoid

$$y^1 = u^1 \cos u^2,$$
$$y^2 = u^1 \sin u^2,$$
$$y^3 = au^2$$

is a surface of negative curvature.

4. Show that when at all points of the surface $b_{\alpha\beta}\, du^\alpha\, du^\beta$ is proportional to $a_{\alpha\beta}\, du^\alpha\, du^\beta$, then $b_{\alpha\beta} = ka_{\alpha\beta}$, $k =$ constant. Interpret this result geometrically.

5. Prove that if every point of the surface S is parabolic, then S is developable.

6. Given a surface of revolution S,

$$y^1 = r \cos \phi, \qquad y^2 = r \sin \phi, \qquad y^3 = f(r),$$

with $f(r)$ of class C^2. Prove that the lines of curvature on S are the meridians $\phi =$ constant and the parallels $r =$ constant.

7. Refer to Problem 6 and show that the points on a surface of revolution S for which $f'f'' > 0$ are elliptic; those for which $f'f'' < 0$ are hyperbolic; and if $f'' \equiv 0$, then S is a cone.

8. Let the vector equation of a curve C drawn on a smooth surface S be $\mathbf{r} = \mathbf{r}(s)$. If $\mathbf{n}(s)$ is a unit normal to S at a given point of C, and v is a parameter measuring the distance along $\mathbf{n}$, the vector equation $\mathbf{R}(s, v) = \mathbf{r}(s) + v\mathbf{n}(s)$ defines a ruled surface S'. Prove that S' is developable if C is a line of curvature, and conversely. Outline of solution: Denote the coefficients in the second fundamental form of S' by $d_{\alpha\beta}$. Compute the $d_{\alpha\beta}$ from the formula

$$d_{\alpha\beta} = \frac{\partial^2 \mathbf{R}}{\partial v^\alpha \, \partial v^\beta} \cdot \mathbf{N},$$

where $v^1 = s$, $v^2 = v$, and $\mathbf{N}$ is the unit normal to S'. Show that $d_{22} = 0$. The developability condition $d_{11}d_{22} - d_{12}{}^2 = 0$ implies then that $d_{12} = 0$. But, along C, $\mathbf{N} = d\mathbf{r}/ds \times \mathbf{n}$; hence

$$d_{12} = \frac{d\mathbf{n}}{ds} \cdot \frac{d\mathbf{r}}{ds} \times \mathbf{n} = 0.$$

If $d\mathbf{n}/ds = 0$ along C, then S' is a cylinder, which is a developable surface. If $d\mathbf{n}/ds$ and $d\mathbf{r}/ds$ are collinear, then $d\mathbf{n} = k\, d\mathbf{r}$, which leads to the set of equations $(b_{\alpha\beta} - ka_{\alpha\beta})\, du^\beta = 0$ of the type (72.3). Retrace steps to prove the converse.

9. Let C be a smooth curve defined by $\mathbf{r} = \mathbf{r}(s)$. The *tangent surface* S to C is defined by $\mathbf{R}(s, v) = \mathbf{r}(s) + v(d\mathbf{r}/ds)$, v is a parameter measured along the tangent $d\mathbf{r}/ds$. Prove that S is developable. The curve C is called the *edge of regression* for S.

10. Prove Dupins theorem. The coordinate surfaces of every triply orthogonal curvilinear coordinate system in E_3 intersect along the lines of curvature of coordinate surfaces. *Hint:* Consider the surface $x^3 =$ constant and take $x^1 = u^1$, $x^2 = u^2$ as surface coordinates on it. Show that along the coordinate lines $u^1 =$ constant, $u^2 =$ constant, $b_{12} = 0$ if $a_{12} = 0$. See Problem 4, Sec. 67.

73. Parallel Surfaces

Let S be a smooth surface defined by equations

$$y^i = y^i(u^1, u^2), \qquad (i = 1, 2, 3), \tag{73.1}$$

where the coordinates y^i are orthogonal cartesian. A surface $\bar{S}$ determined by equations

$$\bar{y}^i(u^1, u^2) = y^i(u^1, u^2) + hn^i(u^1, u^2), \tag{73.2}$$

where n^i is the unit normal to S and h is the distance measured along the normal $\mathbf{n}$, is called a *parallel surface* to S.

Parallel surfaces figure prominently in the theory of elastic plates and shells, where relations connecting the Gaussian curvature K and the mean curvature H of S with the corresponding invariants for the surface $\bar{S}$ are important.

We proceed to outline a derivation of such relations by recalling first that the base vectors $\mathbf{a}_\alpha$, along the curves u_α = constant, are related to the base vectors $\mathbf{b}_i$ along the y^i-axes by

[64.10] $$\mathbf{a}_\alpha = \mathbf{b}_i \frac{\partial y^i}{\partial u^\alpha}.$$

For simplicity in writing we introduce the notations (cf. Sec. 64)

$$\frac{\partial y^i}{\partial u^\alpha} = y_\alpha{}^i \qquad \frac{\partial \bar{y}^i}{\partial u^\alpha} = \bar{y}_\alpha{}^i,$$

and, on differentiating (73.2), we get

(73.3) $$\bar{y}_\alpha{}^i = y_\alpha{}^i + hn^i_{,\alpha}$$

so that

(73.4) $$\bar{y}_\alpha{}^i n_i = y_\alpha{}^i n_i + hn^i_{,\alpha} n_i.$$

But $y_\alpha{}^i n_i = 0$, for the $\mathbf{a}_\alpha$ are orthogonal to $\mathbf{n}$ and $n^i_{,\alpha} n_i = 0$ since $n^i n_i = 1$. Thus (73.4) reduces to

(73.5) $$\bar{y}_\alpha{}^i n_i = 0.$$

On the other hand, the unit normal $\bar{n}_i$ to $\bar{S}$ is orthogonal to the base vectors $\bar{y}^i_\alpha$ on $\bar{S}$, so that

(73.6) $$\bar{y}_\alpha{}^i \bar{n}_i = 0.$$

We conclude from (73.5) and (73.6) that the vectors n_i and $\bar{n}_i$ are collinear, and, since they are unit vectors, $n_i = \bar{n}_i$

The metric coefficients $\bar{a}_{\alpha\beta}$ of S are given by[28]

$$\bar{a}_{\alpha\beta} = \bar{y}_\alpha{}^i \bar{y}_\beta{}^i,$$

which, on making use of (73.3), yield

$$\begin{aligned}\bar{a}_{\alpha\beta} &= (y_\alpha{}^i + hn^i_{,\alpha})(y_\beta{}^i + hn^i_{,\beta}) \\ &= y_\alpha{}^i y_\beta{}^i + hn^i_{,\alpha} y_\beta{}^i + hn^i_{,\beta} y_\alpha{}^i + h^2 n^i_{,\alpha} n^i_{,\beta}.\end{aligned}$$

The substitution in this expression from the Weingarten formula,

[69.5] $$n^i_{,\alpha} = -a^{\delta\gamma} b_{\delta\alpha} y_\gamma{}^i,$$

gives

(73.7) $$\bar{a}_{\alpha\beta} = a_{\alpha\beta} - 2hb_{\alpha\beta} + h^2 n^i_{,\alpha} n^i_{,\beta},$$

since $y_\alpha{}^i y_\beta{}^i = a_{\alpha\beta}$.

[28] See (64.6) and recall that $g_{ij} = \delta_{ij}$ since the coordinates y^i are orthogonal cartesian.

The last term in the right-hand member of (73.7) can be expressed in terms of the Gaussian curvature K and the mean curvature H as follows. On making use of (69.5) we get

(73.8)
$$\begin{aligned} n^i_{,\alpha} n^i_{,\beta} &= a^{\gamma\delta} b_{\alpha\delta} y_\gamma^i a^{\lambda\mu} b_{\beta\lambda} y_\mu^i \\ &= a^{\lambda\mu} b_{\alpha\mu} b_{\beta\lambda}, \end{aligned}$$

since $y^i_\gamma y^i_\mu = a_{\gamma\mu}$. On the other hand, the Gauss equations 69.10 require that

[69.10]
$$R_{\alpha\beta\gamma\delta} = b_{\alpha\gamma} b_{\beta\delta} - b_{\alpha\delta} b_{\beta\gamma},$$

where

$$R_{\alpha\beta\gamma\delta} = K\epsilon_{\alpha\beta}\epsilon_{\gamma\delta}$$

by (62.5), so that

$$K\epsilon_{\alpha\beta}\epsilon_{\gamma\delta} = b_{\alpha\gamma} b_{\beta\delta} - b_{\alpha\delta} b_{\beta\gamma}.$$

We multiply both members of this relation by $a^{\alpha\delta}$, sum on δ, and note that (cf. Sec. 62)

$$a^{\alpha\delta}\epsilon_{\alpha\beta}\epsilon_{\gamma\delta} = -a_{\beta\gamma},$$

and find

(73.9)
$$-Ka_{\beta\gamma} = a^{\alpha\delta} b_{\alpha\gamma} b_{\beta\delta} - 2Hb_{\beta\gamma},$$

since

[70.2]
$$H = \tfrac{1}{2} a^{\alpha\delta} b_{\alpha\delta}.$$

The substitution in the first term in the right-hand member of (73.9) from (73.8) yields

(73.10)
$$n^i_{,\alpha} n^i_{,\beta} = -Ka_{\alpha\beta} + 2Hb_{\alpha\beta},$$

and we can thus write (73.7) in the form

(73.11)
$$\bar{a}_{\alpha\beta} = a_{\alpha\beta}(1 - h^2K) - 2hb_{\alpha\beta}(1 - hH).$$

The important formula 73.11 enables us to compute the coefficients $\bar{a}_{\alpha\beta}$ at a given point $\bar{P}(u_1, u_2)$ on $\bar{S}$ from the values of $a_{\alpha\beta}$, $b_{\alpha\beta}$, K, and H at the corresponding point $P(u_1, u_2)$ on S.

To compute the coefficients $\bar{b}_{\alpha\beta}$ in the second fundamental form of $\bar{S}$, we recall that

$$b_{\alpha\beta} = y^i_{\alpha,\beta} n_i$$

by (67.7). But from (73.3),

$$\bar{y}^i_{\alpha,\beta} = y^i_{\alpha,\beta} + hn^i_{,\alpha\beta},$$

so that

(73.12)
$$\bar{b}_{\alpha\beta} = b_{\alpha\beta} + hn^i_{,\alpha\beta} n_i.$$

Since the coordinates y^i are rectangular cartesian, $n^i = n_i$, $n^i n^i = 1$, and we conclude that

$$n^i_{,\alpha} n^i = 0.$$

On differentiating this orthogonality relation we find that $n^i_{,\alpha\beta} n^i = -n^i_{,\alpha} n^i_{,\beta}$, and hence (73.12) can be written as

$$\bar{b}_{\alpha\beta} = b_{\alpha\beta} - h n^i_{,\alpha} n^i_{,\beta}.$$

On making use of (73.10), we finally obtain

$$\bar{b}_{\alpha\beta} = (1 - 2hH) b_{\alpha\beta} + hK a_{\alpha\beta}. \tag{73.13}$$

Formulas 73.11 and 73.13 appear on pages 110–111 of a monograph by T. Y. Thomas, *Concepts from Tensor Analysis and Differential Geometry*, Academic Press (1961). They are closely related to formulas on page 272 of L. P. Eisenhart's *Differential Geometry*, Princeton Press (1940).

Once the coefficients $\bar{a}_{\alpha\beta}$, $\bar{b}_{\alpha\beta}$ are known, the Gaussian and mean curvatures K and H can be computed from formulas 70.1 and 70.2. The result of somewhat lengthy computations, which will be found on pages 111 to 113 of the mentioned monograph by T. Y. Thomas, is

$$\begin{aligned} \bar{K} &= \frac{K}{1 + h^2 K - 2hH}, \\ \bar{H} &= \frac{H - hK}{1 + h^2 K - 2hH}. \end{aligned} \tag{73.14}$$

From the first of these elegant formulas it follows that when S is a developable surface, then the parallel surfaces $\bar{S}$ are also developable.

Problems

1. If S is a surface of revolution, show that a parallel surface $\bar{S}$ is also a surface of revolution. *Hint:* Consider $y^1 = u^1 \cos u^2$, $y^2 = u^1 \sin u^2$, $u^3 = f(u^1)$.

2. Show that the principal radii $R_{(\alpha)}$ of normal curvature of $\bar{S}$ are related to the principal radii $\bar{R}_{(\alpha)}$ of a parallel surface $\bar{S}$ by

$$\bar{R}_{\alpha} = R_{(\alpha)} - h.$$

74. The Gauss-Bonnet Theorem

The description of surfaces with the aid of differential equations has local character, since relations among the derivatives describe properties of surfaces only in the neighborhood of a point. To obtain results valid for the entire surface one must perform integrations. Because of the complex structure of differential equations of the theory of surfaces, relatively few global results have been obtained, and the available results

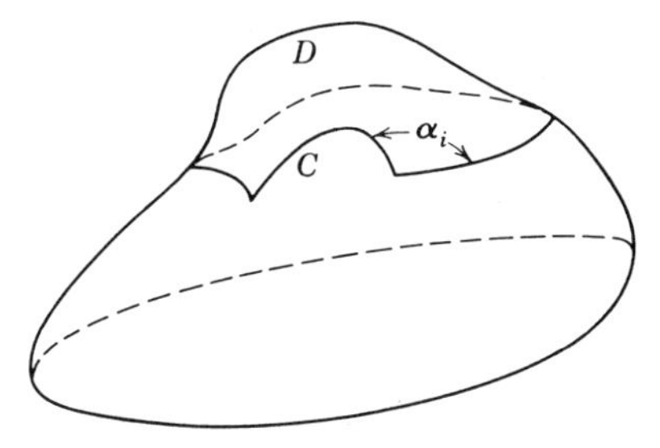

Fig. 31

in global geometry are largely concerned with a special class of convex surfaces. There is one important classical result, however, that relates the integral of the Gaussian curvature evaluated over the area of an arbitrary smooth surface to the line integral of the geodesic curvature computed over the curve that bounds the area. Gauss viewed this result as the most elegant theorem of geometry of surfaces in the large.

Let D be a region bounded by a closed piecewise smooth curve C drawn on a smooth surface S, shown in Fig. 31. We shall suppose that D is homeomorphic to a circular disk.[29] We saw in Sec. 63 that a unit tangent vector λ^α to a surface curve C is related to the unit vector η^α normal to λ^α by

$$\frac{\delta \lambda^\alpha}{\delta s} = \varkappa_g \eta^\alpha, \tag{63.4}$$

where $\varkappa_g$ is the geodesic curvature of C. Moreover, if C is a geodesic, then $\varkappa_g = 0$ at all points of C, and conversely.

Since η^α is orthogonal to λ^α, it follows from the concluding paragraph of Sec. 54 that $\epsilon_{\alpha\beta}\eta^\alpha\lambda^\beta = 1$, or

$$\eta_\alpha = \epsilon_{\alpha\beta}\lambda^\beta.$$

But from (63.4)

$$\varkappa_g = \eta_\alpha \frac{\delta \lambda^\alpha}{\delta s},$$

or

$$\varkappa_g = \epsilon_{\alpha\beta}\lambda^\beta \frac{\delta \lambda^\alpha}{\delta s}. \tag{74.1}$$

The integration of this expression over the curve C gives

$$\int_C \varkappa_g \, ds = \int_C \epsilon_{\alpha\beta}\lambda^\beta \frac{\delta \lambda^\alpha}{\delta s} \, ds, \tag{74.2}$$

[29] Two regions are said to be homeomorphic if they can be mapped into one another in a continuous one-to-one manner.

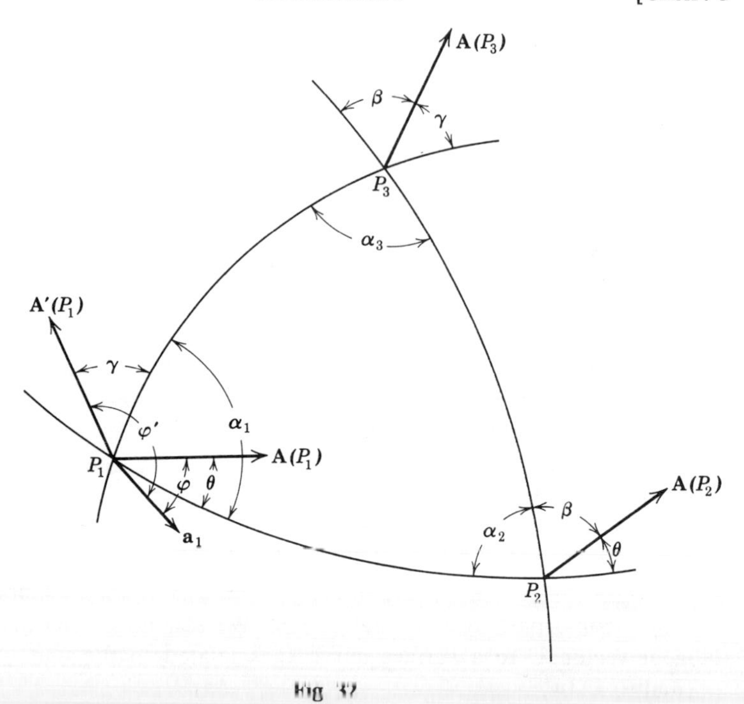

Fig. 32

and when the line integral in the right-hand member of (74.2) is transformed into a surface integral by Green's formula, we find that[30]

$$\int_C \varkappa_g \, ds = -\iint_D K \, d\sigma + 2\pi - \sum_i (\pi - \alpha_i), \tag{74.3}$$

where the α_i are interior angles of the contour C shown in Fig. 31, and $d\sigma = \sqrt{a}\, du^1 \, du^2$ is the element of the surface area of D. If C is smooth, the sum $\sum_i (\pi - \alpha_i) = 0$.

Formula (74.3) embodies the statement of the *Gauss-Bonnet theorem*. Instead of deducing this theorem with the aid of Green's formula, we give a geometric interpretation of formula (74.3) that suggests an alternative definition of Gaussian curvature.

Consider first a sphere S of radius R and a spherical triangle $P_1P_2P_3$ on S (Fig. 32) formed by the arcs P_1P_2, P_2P_3, P_3P_1 of three great circles. Denote the interior angles of this triangle at the vertices P_i by α_i and cover

[30] The details of easy calculations are found in the following books: A. V. Pogorelov, *Differential Geometry*, P. Noordhoff, N.V. (1959), p. 161; L. P. Eisenhart, *Differential Geometry*, Princeton (1940), p. 191; D. J. Struik, *Lectures on Classical Differential Geometry*, Addison-Wesley (1950), p. 154.

the sphere by some coordinate net (u_1, u_2). If the base vector along the u_1-coordinate line at P_1 is $\mathbf{a}_1$ and $\mathbf{A}(P_1)$ is an arbitrary surface vector at P_1, we denote the angle between $\mathbf{a}_1$ and $\mathbf{A}(P_1)$ by φ. Let θ be the angle made by $\mathbf{A}(P_1)$ with the geodesic arc P_1P_2. When $\mathbf{A}(P_1)$ is propagated in a parallel manner along the geodesic triangle $P_1P_2P_3$, it assumes the position $\mathbf{A}'(P_1)$ shown in Fig. 32. Our immediate object is to determine the angle φ' between $\mathbf{a}_1$ and $\mathbf{A}'(P_1)$.

During the parallel propagation of $\mathbf{A}(P_1)$ along P_1P_2, the angle θ is unchanged (see Sec. 60), and the vector $\mathbf{A}$ assumes the position $\mathbf{A}(P_2)$ with the geodesic arc P_2P_3, then

$$\beta = \pi - (\alpha_2 + \theta).$$

In the course of parallel propagation of $\mathbf{A}(P_2)$ along P_2P_3 the vector $\mathbf{A}$ continues making angle β with P_2P_3 and, on reaching the point P_3, it assumes the position $\mathbf{A}(P_3)$. Let γ be the angle between $\mathbf{A}(P_3)$ and the arc P_1P_3, then

$$\gamma = \alpha_3 - \beta = \alpha_3 - [\pi - (\alpha_2 + \theta)] = \alpha_2 + \alpha_3 + \theta - \pi.$$

On continuing propagations of $\mathbf{A}$ along P_3P_1, the vector $\mathbf{A}$ maintains the angle γ with P_1P_3 until it reaches the point P_1 when it assumes the position $\mathbf{A}'(P_1)$. Now, the angle φ' made by $\mathbf{A}'(P_1)$ with $\mathbf{a}_1$ is

$$\begin{aligned}\varphi' &= \gamma + \alpha_1 + \varphi - \theta \\ &= \alpha_1 + \alpha_2 + \alpha_3 + \varphi - \pi,\end{aligned}$$

so that the angle $\varphi' - \varphi$ between $\mathbf{A}(P')$ and $\mathbf{A}(P)$ is

$$\varphi' - \varphi = \alpha_1 + \alpha_2 + \alpha_3 - \pi. \tag{74.4}$$

The change $\varphi' - \varphi$ representing the difference between the sum of interior angles of the spherical triangle $P_1P_2P_3$ and the sum of interior angles of the rectilinear triangle is called the *spherical excess* of the spherical triangle $P_1P_2P_3$. If instead of interior angles α_i we introduce the exterior angles $\theta_i = \pi - \alpha_i$, formula 74.4 reads

$$\varphi' - \varphi = 2\pi - \sum_{i=1}^{n} \theta_i.$$

When the vector $\mathbf{A}$ is propagated along a geodesic polygon of n sides, entirely similar computations yield for the spherical excess of the polygon[31]

$$\varphi' - \varphi = \sum_{i=1}^{n} \alpha_i - (n - 2)\pi,$$

or

$$\varphi' - \varphi = 2\pi - \sum_{i=1}^{n} \theta_i,$$

[31] Note that the sum of interior angles of a rectilinear polygon of n sides is $(n - 2)\pi$ radians.

if we use the exterior angles $\theta_i = \pi - \alpha_i$. But it is known from spherical trigonometry that the spherical excess of a geodesic polygon is equal to σ/R^2, where σ is the area of the polygon and R is the radius of the sphere. Thus

$$\varphi' - \varphi = 2\pi - \sum_{i=1}^{n} \theta_i = \frac{\sigma}{R^2},$$

and since the Gaussian curvature K for the sphere is equal to $1/R^2$, we can write

$$K = \frac{2\pi - \Sigma\, \theta_i}{\sigma}. \tag{74.5}$$

This formula can be generalized to obtain the Gauss-Bonnet formula 74.3 for the case where C in Fig. 31 is a geodesic polygon. Thus, if the region D is subdivided by small geodesic polygons into subregions of areas $d\sigma_i$, the familiar procedures of integral calculus applied to (74.5) yield

$$\iint_D K\, d\sigma = 2\pi - \sum_{i=1}^{n} \theta_i. \tag{74.6}$$

This formula coincides with (74.3), since $\varkappa_g = 0$ when C is a geodesic polygon.

Formula 74.6, first obtained by Gauss, was generalized by Bonnet[32] to yield the result (74.3), which, as we have already noted follows directly from (74.2) on application of Green's formula.

The left-hand member $\iint_D K\, d\sigma$ in (74.6) is called the *integral curvature* of D. It turns out that the integral curvature is a topological invariant. Two surfaces are said to be topologically equivalent if they can be mapped into one another by a continuous one-to-one transformation. It can be shown by using (74.3) that $\iint_D K\, d\sigma = 4\pi$ for all regular surfaces topologically equivalent to a sphere and $\iint_D K\, d\sigma = 0$ for all regular surfaces topologically equivalent to a torus.[33]

75. The *n*-Dimensional Manifolds

It is the purpose of this section to introduce a few concepts from the geometry of n-dimensional metric manifolds which are of interest in applications to dynamics and relativity. Many of these concepts are straightforward generalizations of ideas introduced in this chapter in

[32] O. Bonnet, *Journal école polytechnique*, **19** (1848), pp. 1–146.
[33] See D. J. Struik, *op. cit.*, pp. 153–159.

connection with the study of surfaces imbedded in the three-dimensional Euclidean manifolds.

We shall suppose that the element of distance between two neighboring points in an n-dimensional manifold is given by the quadratic form

$$ds^2 = g_{ij}\, dx^i\, dx^j, \qquad (i, j = 1, \ldots, n), \qquad \text{with } |g_{ij}| \neq 0. \tag{75.1}$$

We extend the definition of Euclidean space, given in Sec. 29, by saying that the space is Euclidean if there exists a transformation of coordinates x^i such that the transform of ds^2 is a quadratic form with constant coefficients. Since every real quadratic form with constant coefficients can be reduced by a real linear transformation to the form

$$ds^2 = \lambda_i (dx^i)^2, \qquad (\lambda_i = \pm 1), \tag{75.2}$$

the form 75.2 can be used to define an Euclidean n-dimensional manifold.

If, in particular, the form 75.2 is *definite*, we shall say that the manifold is *purely Euclidean*, but if it is *indefinite*, the manifold will be called *pseudo-Euclidean.*

A linear manifold determined by a set of n equations

$$C: \quad x^i = x^i(t), \qquad t_1 \leq t \leq t_2,$$

with suitable differentiability properties, will be said to define a curve C in an n-dimensional manifold.

If the form 75.1 is positive definite, we shall say that the positive number

$$s = \int_{t_1}^{t_2} \sqrt{g_{ij}(dx^i/dt)(dx^j/dt)}\, dt$$

is the length of the curve C. There are definitions of metric manifolds which are not based on the expression for the element of arc in the form 75.1, but they need not concern us here (see Sec. 43).

The vector $\lambda^i = dx^i/ds$ will be said to define the direction of the curve, and it is clear that $g_{ij}\lambda^i\lambda^j = 1$, so that the vector λ^i is a unit vector. The length of any vector A^i is given by the formula

$$A = \sqrt{g_{ij}A^iA^j}$$

The notion of the angular metric in an n-dimensional manifold is a direct generalization of the definition of the angle in the three-dimensional case.

If λ^i and μ^i are two unit vectors, we define the cosine of the angle between them by the formula

$$\cos\theta = g_{ij}\lambda^i\mu^j. \tag{75.3}$$

It is not clear from this definition that the angle θ is necessarily real. We shall prove, however, that this is always so if the form $g_{ij}\,dx^i\,dx^j$ is positive definite. The proof follows at once from the Cauchy-Schwarz inequality

$$(g_{ij}x^iy^j)^2 \leq (g_{ij}x^ix^j)(g_{ij}y^iy^j), \tag{75.4}$$

where the form $g_{ij}x^ix^j \geq 0$.

We first establish the inequality 75.4. Let the form $Q(x) = g_{ij}x^ix^j$ be positive definite. If we replace in it x^i by $x^i + \lambda y^i$, where λ is an arbitrary scalar, we obtain

$$\begin{aligned} Q(x + \lambda y) &\equiv g_{ij}(x^i + \lambda y^i)(x^j + \lambda y^j) \\ &= g_{ij}x^ix^j + 2g_{ij}x^iy^j\lambda + g_{ij}y^iy^j\lambda^2 \\ &\equiv Q(x) + 2Q(x, y)\lambda + Q(y)\lambda^2. \end{aligned}$$

This is a quadratic expression in λ with real coefficients. By hypothesis $Q(x + \lambda y) \geq 0$, the sign of equality holding if, and only if, $x^i + \lambda y^i = 0$. Hence, the equation in λ,

$$f(\lambda) \equiv Q(y)\lambda^2 + 2Q(x, y)\lambda + Q(x) = 0,$$

possesses no distinct real roots. But a necessary and sufficient condition that this be true is that $[Q(x, y)]^2 - Q(y)Q(x) \leq 0$, that is,

$$(g_{ij}x^iy^j)^2 \leq (g_{ij}x^ix^j)(g_{ij}y^iy^j).$$

This is precisely the inequality 75.4.

If, now, in formula 75.4 we set $x^i = \lambda^i$ and $y^i = \mu^i$, we get

$$\frac{(g_{ij}\lambda^i\mu^j)^2}{(g_{ij}\lambda^i\lambda^j)(g_{ij}\mu^i\mu^j)} \leq 1,$$

and, since λ^i and μ^i are unit vectors, we have $(g_{ij}\lambda^i\mu^j)^2 \leq 1$, which states that the angle θ in formula 75.3 is real.

We define the volume element in R_n by the formula

$$d\tau = \sqrt{g}\,dx^1\,dx^2\cdots dx^n,$$

and the volume by the corresponding n-tuple integral.

A generalization of the concepts of curvature and torsion to curves imbedded in the n-dimensional Riemannian manifolds is direct and straightforward,[34] but matters become rapidly involved when one comes to consider hypersurfaces.

[34] See, for example, J. C. H. Gerretsen, *Lectures on Tensor Calculus and Differential Geometry*, P. Noordhoff, N.V. (1962), Chapter 6.

A set of n equations

$$x^i = x^i(u^1, u^2, \ldots, u^m), \qquad (i = 1, 2, \ldots, n), \qquad m \leq n, \tag{75.5}$$

is said to define an admissibly parametrized m-dimensional variety (or a hypersurface) over a neighborhood of the variables u^α if (*a*) the x^i in (75.5) are of class C^2 and (*b*) the Jacobian matrix $(\partial x^i/\partial u^\alpha)$, $(\alpha = 1, 2, \ldots, m)$, is of rank m at each point of the neighborhood.

In Secs. 64 to 73 we have studied two-dimensional Riemannian varieties (surfaces) imbedded in E_3. A question naturally arises: Under what circumstances an m-dimensional variety R_m, with a Riemannian metric

$$ds^2 = a_{\alpha\beta}\, du^\alpha\, du^\beta, \qquad (\alpha, \beta = 1, 2, \ldots, m), \tag{75.6}$$

can be imbedded in the n-dimensional Euclidean manifold with

$$ds^2 = dx^i\, dx^i\,? \tag{75.7}$$

Now equations 75.5 together with 75.6 and 75.7 require that

$$a_{\alpha\beta} = \frac{\partial x^i}{\partial u^\alpha}\frac{\partial x^i}{\partial u^\beta}, \qquad (i = 1, 2, \ldots, n). \tag{75.8}$$

The set of $\frac{1}{2}m(m + 1)$ partial differential equations 75.8 in n variables x^i will not be expected to possess a solution unless $n \geq \frac{1}{2}m(m + 1)$; thus, if $m = 2, n \geq 3$; if $m = 3, n \geq 6$, and so on. This estimate, however, does not constitute a proof that an m-dimensional variety can be imbedded in E_n whenever $n \geq \frac{1}{2}m(m + 1)$. It is possible, however, to prove that a neighborhood of R_m can be imbedded in E_n if $n \geq \frac{1}{2}m(m + 1)$. Concerning the global imbedding of the whole of R_m in E_n almost no general results are known. There are a few special theorems on the imbedding of two-dimensional varieties with special topological properties. These refer almost wholly to convex two-dimensional manifolds.[35] The problems on imbedding now lie at the forefront of researches in geometry.

[35] See L. Nirenberg, "The Weyl and Minkowski Problem," *Comm. on Pure and Appl. Math.*, **6,** 1948, and A. V. Pogorelov, *Some Questions of Geometry in the Large in a Riemannian space*, Moscow (1957).

4

ANALYTICAL MECHANICS

76. Basic Concepts. Kinematics

Analytical mechanics is concerned with a mathematical description of motion of material bodies subjected to the action of forces. Its development follows a familiar pattern. A material body is assumed to consist of a large number of minute bits of matter connected in some way with one another. The attention is first focused on a single particle, which is assumed to be free of constraints, and its behavior is analyzed when it is subjected to the action of external forces. The resulting body of knowledge constitutes the *mechanics of a particle*. To pass from mechanics of a single particle to mechanics of aggregates of particles composing a material body, we introduce the principle of superposition of effects and make specific assumptions concerning the nature of constraining forces, depending on whether the body under consideration is rigid, elastic, plastic, fluid, and so on.

We begin our study of mechanics of continua by analyzing the motion of a single particle. The particle is assumed to be an idealized entity having position and inertia, but no spatial extension. The measure of inertia is mass, and thus the particle is simply a *point-mass*. Another basic ingredient of mechanics is the concept of time, which arises in the assumption of causal connection between physical events. The hypothesis of causality implies the possibility of ordering events, and the time t, as it appears in the description of the physical universe, is an independent parameter whose range of variation is the real-number continuum.

We will suppose that physical events take place in the three-dimensional space whose metric is Euclidean, and we refer the position of a particle at a given time t to some curvilinear reference frame X. As in the study of geometry in Chapter 3, we denote the coordinates of the particle relative to a set of orthogonal cartesian axes by the symbols y^i. Clearly, the position of a particle is a relative concept depending on the selection of a reference frame. The reference system generally used in astronomy is that determined by the so-called fixed stars. It is termed the *primary inertial system*. Any system of axes moving relative to the primary inertial system with constant translational velocity is called a *secondary inertial system*. In many mechanical problems the motion of the earth relative to the primary inertial system is so nearly negligible that the Newtonian laws (Sec. 77), which are assumed to be valid only in the inertial reference frames, can be applied without modification to study the motion of particles referred to a system of axes fixed in the earth.

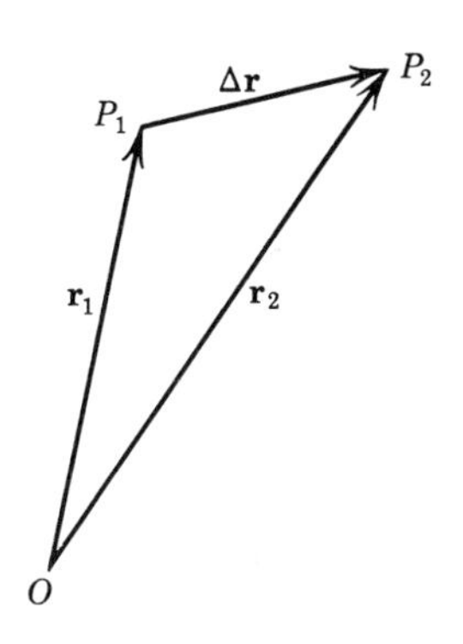

Fig. 33

When a particle changes its position in a given reference frame, it is said to undergo a *displacement*. Thus, suppose that the particle is at the point P_1 at time t. Its position at this time is given by the vector $\mathbf{r}_1$; at a later instant of time $t + \Delta t$ it is at P_2, determined by the position vector $\mathbf{r}_2$. We denote the displacement in the interval of time Δt by the vector $\overrightarrow{P_1P_2} = \Delta\mathbf{r}$ (Fig. 33) and suppose that the particle traverses a continuous path which is represented by the vector sum of the elementary displacements $d\mathbf{r}$.

We define the *average velocity* of the particle during the displacement $\Delta\mathbf{r}$ by the formula $\mathbf{v}_{\text{ave}} = \Delta\mathbf{r}/\Delta t$, and we assume that this ratio has a unique limit as $\Delta t \to 0$. Then the instantaneous velocity $\mathbf{v}$ is given by the formula $d\mathbf{r}/dt \equiv \dot{\mathbf{r}} = \mathbf{v}$. *Velocity* $\mathbf{v}$ *is*, of course, a vector.

The case in which $d\mathbf{r}/dt$ is constant is of relatively minor interest in mechanics, and generally we will be concerned with accelerated motions. We define the *average acceleration* of the particle, during the time interval Δt, by the formula $\mathbf{a}_{\text{ave}} = \Delta\mathbf{v}/\Delta t$, and the *instantaneous acceleration* by

$$\mathbf{a} = \lim_{\Delta t \to 0} \frac{\Delta\mathbf{v}}{\Delta t} = \frac{d\mathbf{v}}{dt} = \frac{d^2\mathbf{r}}{dt^2} = \ddot{\mathbf{r}}.$$

Hereafter, unless otherwise specified, the words *velocity* and *acceleration* are taken to mean the instantaneous values.

The velocity and acceleration are known as the *kinematical concepts* of mechanics to distinguish them from those concepts that utilize the idea of *force*. We consider this idea in the following section.

77. Newtonian Laws. Dynamics

In 1687, Sir Isaac Newton published three axioms or laws, the first of which was based on deductions from a set of remarkable experiments performed by Galileo (1564–1642) on bodies moving on inclined planes, and the other two represent a profound crystallization of the notions surrounding these experiments. These laws form the point of departure in all considerations in dynamics, and we give them here in a form that is almost a literal translation of Newton's Latin as it appears in the 1726 edition of the *Philosophia Naturalis Principia Mathematica.* The present-day formulation of analytical mechanics is essentially due to J. L. Lagrange (1736–1813), whose greatest work, *Mécanique analytique*, was written in 1788, and W. R. Hamilton (1805–1865), whose celebrated principle embraces the whole of mechanics.

NEWTONIAN LAWS. I. *Every body continues in its state of rest, or of uniform motion in a straight line, except insofar as it is compelled by impressed forces to change that state.*

II. *The change of motion is proportional to the impressed motive force, and takes place in the direction of the straight line in which that force is impressed.*

III. *To every action there is always an equal and contrary reaction; or the mutual actions of two bodies are always equal and oppositely directed along the same straight line.*

The first law depends for its meaning on the dynamical concept of force and on the kinematical idea of uniform rectilinear motion. It ascribes anthropomorphic attributes to a particle, which is bent on continuing its motion in a straight line but is somehow deflected from its intentions by a push or pull. Newton doubtless felt that the idea of force is intuitively known and requires no further explanation. We shall presently see that the first law is in reality a corollary of the second.

The second law of motion also introduces the kinematical concept of motion and the *dynamical* idea of force. To understand its meaning it should be noted that Newton uses the term *motion* in the sense of *momentum*, that is, the product of mass by velocity. Thus the "change of motion" means the *time rate of change of momentum*, and hence in vector notation the second law can be stated as a formula

$$\mathbf{F} = \frac{d(m\mathbf{v})}{dt}, \tag{77.1}$$

provided that our units are so chosen as to make the proportionality constant equal to one.

If we postulate the invariance of mass, then equation 77.1 can be written in the familiar form

$$\mathbf{F} = m\mathbf{a}. \tag{77.2}$$

We note from (77.1) that, if $\mathbf{F} = 0$, then $d(m\mathbf{v})/dt = 0$, so that $m\mathbf{v} =$ constant, and hence $\mathbf{v}$ is a constant vector. Thus the first law is a consequence of the second.

The concept of mass can obviously be defined with the aid of the second law in terms of force and acceleration. There were numerous attempts to define mass and force independently of one another. The most familiar of such definitions is due to Ernst Mach,[1] who formulated a definition of mass with the aid of Newton's third law of motion. In our opinion a fine-grained analysis of Mach's definition of mass reveals certain logical difficulties which cannot be resolved by appealing to the third law alone. For this reason it seems best to leave one of the fundamental building blocks of mechanics (mass or force) undefined and admit it in the science of mechanics on the same basis as the "God-given integers" in mathematics.

The third law of motion states that accelerations always occur in pairs. In terms of force we may say that, if a force acts on a given body, the body itself exerts an equal and oppositely directed force on some other body. Newton called the two aspects of the force *action* and *reaction*, whence the usual statements of the law.

The entity of mass entering in the formulation of Newtonian laws is sometimes called the *inertial mass* (or simply *inertia*) to distinguish it from the *gravitational mass* M entering in the Newtonian law of gravitation. This law states that the force of attraction between a pair of particles is proportional to the product of their masses, is inversely proportional to the square of the distance r between them, and is directed along the line joining the particles. In symbols,

$$\mathbf{F} = k \frac{M_1 M_2}{r^3} \mathbf{r}, \tag{77.3}$$

where k is a universal constant and $\mathbf{r}$ is a vector directed from mass M_1 to mass M_2.

If it is assumed (as it is usually done) that the gravitational and inertial masses are equal, the law 77.3 furnishes a practical means for comparing masses with the aid of beam balances.

In order to develop the science of mechanics of a universe consisting of more than two particles, it is necessary to adjoin to Newtonian laws the principle of superposition of effects and make further assumptions regarding the nature of constraints.

78. Equations of Motion of a Particle. Work. Energy

Let the position of a moving particle P be determined by a vector $\mathbf{r}$. If the curvilinear coordinates of the terminal point of $\mathbf{r}$ are denoted by

[1] E. Mach, *The Science of Mechanics*. An interesting survey of it is contained in R. B. Lindsay and H. Margenau, *Foundations of Physics*.

$x^i(t)$, then the equations of the path C of the particle can be written in the form

$$C\colon\ x^i = x^i(t), \tag{78.1}$$

and we call the curve C the *trajectory* of the particle.

The velocity of P is a vector $\mathbf{v} = d\mathbf{r}/dt$, whose components are

$$v^i = \frac{dx^i}{dt}. \tag{78.2}$$

The acceleration $\mathbf{a} = d\mathbf{v}/dt = d^2\mathbf{r}/dt^2$ has the components (see Secs. 46 and 47)

$$a^i = \frac{\delta v^i}{\delta t} \equiv \frac{d^2x^i}{dt^2} + \begin{Bmatrix} i \\ jk \end{Bmatrix} \frac{dx^j}{dt}\frac{dx^k}{dt}, \tag{78.3}$$

where $\delta v^i/\delta t$ is the intrinsic derivative and the $\begin{Bmatrix} i \\ jk \end{Bmatrix}$ are the Christoffel symbols calculated from the metric tensor g_{ij}, associated with the reference system X.

If the mass of P is m, Newton's second law of motion yields the equation $\mathbf{F} = m\, d^2\mathbf{r}/dt^2$, or

$$F^i = m\frac{\delta v^i}{\delta t} = ma^i. \tag{78.4}$$

In orthogonal cartesian coordinates, equation 78.4 assumes the familiar form $\bar{F}^i = m\, d^2y^i/dt^2$.

We introduce next the concept of energy, which will permit us to give a more elegant formulation of the theory. The germ of the energy concept can be traced back at least to Galileo, who remarked, "What is gained in power is lost in speed," but the first clear introduction of the idea of energy in mechanics as a quantity equal to the product of mass and the square of velocity of the particle (*vis viva*) was made by Huygens in the seventeenth century. The full use of this idea, however, and of its relation to the concept of work, did not come until the nineteenth century.

We define the *element of work* done by the force $\mathbf{F}$ in producing a displacement $d\mathbf{r}$ by the invariant $dW = \mathbf{F}\cdot d\mathbf{r}$, and, since the components of $\mathbf{F}$ and $d\mathbf{r}$ are, respectively, F^i and dx^i, this scalar product is equal to

$$\begin{aligned} dW &= g_{ij}F^i\, dx^j \\ &= F_j\, dx^j, \end{aligned} \tag{78.5}$$

where the $F_j = g_{ij}F^i$ are the covariant components of the vector $\mathbf{F}$. We shall suppose that, in general, the functions $F^i(x)$, defining the vector field

F, belong to class C^1. The work done in displacing a particle along the trajectory C, joining a pair of points P_1 and P_2, is the line integral

$$W = \int_{P_1}^{P_2} F_i \, dx^i. \tag{78.6}$$

Making use of Newton's second law of motion 78.4, we can write 78.6 in the form

$$\begin{aligned} W &= \int_{P_1}^{P_2} m g_{ij} \frac{\delta v^i}{\delta t} \, dx^j \\ &= \int_{t_1}^{t_2} m g_{ij} \frac{\delta v^i}{\delta t} v^j \, dt. \end{aligned} \tag{78.7}$$

But

$$\frac{\delta(g_{ij} v^i v^j)}{\delta t} = 2 g_{ij} \frac{\delta v^i}{\delta t} v^j,$$

and, since $g_{ij} v^i v^j$ is an invariant,

$$\frac{\delta(g_{ij} v^i v^j)}{\delta t} = \frac{d}{dt}(g_{ij} v^i v^j),$$

and hence

$$\frac{d}{dt}(g_{ij} v^i v^j) = 2 g_{ij} \frac{\delta v^i}{\delta t} v^j.$$

Inserting from this result in the integrand of (78.7) yields

$$\begin{aligned} W &= \int_{t_1}^{t_2} \frac{m}{2} \frac{d}{dt}(g_{ij} v^i v^j) \, dt \\ &= \frac{m}{2} g_{ij} v^i v^j \Big|_{P_1}^{P_2} \\ &= T_2 - T_1, \end{aligned} \tag{78.8}$$

where

$$T \equiv \frac{m}{2} g_{ij} v^i v^j = \frac{mv^2}{2}.$$

We have the result that the work done by the force F_i in displacing the particle from the point P_1 to the point P_2 is equal to the difference of the values of the quantity $T = \frac{1}{2}mv^2$ at the end and at the beginning of the displacement. We define the quantity $T = \frac{1}{2}mv^2$, which is exactly one-half of the *vis viva* of Huygens, as the *kinetic energy of the particle.*

The statement embodied in the formula 78.8 can be enunciated as a

THEOREM. *The work done in displacing a particle along its trajectory is equal to the change in the kinetic energy of the particle.*

It may happen that the force field F_i is such that the integral 78.6 is independent of the path. In this event the integrand $F_i\,dx^i$ is an exact differential,

$$dW = F_i\,dx^i, \tag{78.9}$$

of the *work function W*. The negative of the work function W is called the *force potential* or *potential energy*. We denote the potential energy by the symbol V, and conclude from (78.9) that

$$F_i = -\frac{\partial V}{\partial x^i}. \tag{78.10}$$

The fields of force for which potential functions exist are called *conservative*. There is a simple criterion for a field of force F_i to be conservative. We state it as a

THEOREM. *A necessary and sufficient condition that a force field F_i, defined in a simply connected region, be conservative is that $F_{i,j} = F_{j,i}$.*

The proof of this theorem follows immediately from the observation that a necessary and sufficient condition for the expression $F_i\,dx^i$ to be an exact differential of a single-valued function V is that

$$\frac{\partial F_i}{\partial x^j} = \frac{\partial F_j}{\partial x^i}, \tag{78.11}$$

since these derivatives are assumed to be continuous functions.[2] But

$$F_{i,j} = \frac{\partial F_i}{\partial x^j} - \begin{Bmatrix} k \\ ij \end{Bmatrix} F_k,$$

and, since $\begin{Bmatrix} k \\ ij \end{Bmatrix}$ is symmetric in i and j, we conclude that the condition 78.11 is completely equivalent to the one stated in the theorem.

As a corollary we observe that a parallel force field (Sec. 48) is necessarily conservative, since the condition for a vector field F_i to be parallel is $F_{i,j} = 0$.

79. Lagrangean Equations of Motion

An alternative formulation of the Newtonian law 78.4, phrased in terms of the kinetic energy of the particle, was obtained by Lagrange from the principle discussed in Sec. 84. We derive these equations in this section by a direct calculation which makes use of Newton's second law of motion.

[2] If the region is multiply connected, the conditions 78.11 still guarantee the existence of potential V related to F_i by formula 78.10, but, in this case, the function V, in general, is multiple valued.

The kinetic energy $T = \frac{1}{2}mv^2$ can be written as

$$T = \frac{m}{2} g_{ij}\dot{x}^i\dot{x}^j, \tag{79.1}$$

since $\dot{x}^i = v^i$. If we differentiate (79.1) with respect to $\dot{x}^i$ we obtain $\partial T/\partial \dot{x}^i = mg_{ij}\dot{x}^j$. The derivative of this expression with respect to t is

$$\frac{d}{dt}\left(\frac{\partial T}{\partial \dot{x}^i}\right) = m\left(g_{ij}\ddot{x}^j + \frac{\partial g_{ij}}{\partial x^k}\dot{x}^k\dot{x}^j\right).$$

If we subtract from this the derivative of (79.1) with respect to x^i, namely,

$$\frac{\partial T}{\partial x^i} = \frac{m}{2}\frac{\partial g_{jk}}{\partial x^i}\dot{x}^j\dot{x}^k,$$

we get

$$\begin{aligned}\frac{d}{dt}\left(\frac{\partial T}{\partial \dot{x}^i}\right) - \frac{\partial T}{\partial x^i} &= m\left[g_{ij}\ddot{x}^j + \frac{1}{2}\left(\frac{\partial g_{ij}}{\partial x^k} + \frac{\partial g_{ik}}{\partial x^j} - \frac{\partial g_{jk}}{\partial x^i}\right)\dot{x}^j\dot{x}^k\right] \\ &= m\{g_{ij}\ddot{x}^j + [jk, i]\dot{x}^j\dot{x}^k\} \\ &= mg_{il}\left(\ddot{x}^l + \begin{Bmatrix} l \\ jk \end{Bmatrix}\dot{x}^j\dot{x}^k\right).\end{aligned}$$

But by (78.3) the expression in parentheses on the right is the acceleration a^l, and, since $mg_{il}a^l = ma_i = F_i$, we can write

$$\frac{d}{dt}\left(\frac{\partial T}{\partial \dot{x}^i}\right) - \frac{\partial T}{\partial x^i} = F_i. \tag{79.2}$$

Equations 79.2 give the statement of Newton's second law in the form used by Lagrange.

For a conservative system, $F_i = -\partial V/\partial x^i$, and equations 79.2 becomes

$$\frac{d}{dt}\left(\frac{\partial T}{\partial \dot{x}^i}\right) - \frac{\partial T}{\partial x^i} = -\frac{\partial V}{\partial x^i}, \tag{79.3}$$

or

$$\frac{d}{dt}\left(\frac{\partial T}{\partial \dot{x}^i}\right) - \frac{\partial (T - V)}{\partial x^i} = 0. \tag{79.4}$$

We recall that the potential energy V is a function of the coordinates x^i alone; hence, if we introduce the *Lagrangean function*

$$L \equiv T - V,$$

we can write equation 79.4 in the form

$$\frac{d}{dt}\left(\frac{\partial L}{\partial \dot{x}^i}\right) - \frac{\partial L}{\partial x^i} = 0. \tag{79.5}$$

In the application of Lagrangean equations to specific problems one frequently deals with the *physical* components $\bar{F}^i$ of the force vector $\mathbf{F}$ instead of the *tensor* components F^i. The physical components of $\mathbf{F}$, we recall, are the coefficients in the representation

$$\mathbf{F} = \bar{F}^i \mathbf{e}_i,$$

where the $\mathbf{e}_i$'s are unit vectors codirectional with the base vectors $\mathbf{a}_i$. (See Sec. 45.) Since $\mathbf{F} = F^i \mathbf{a}_i$ and $\mathbf{a}_i \cdot \mathbf{a}_j = g_{ij}$, the physical components $\bar{F}^i$ are related to the tensor components F^i by the formula

$$\bar{F}^i = \sqrt{g_{ii}}\, F^i, \qquad \text{(no sum)}.$$

Problems

1. Show that the covariant components of the acceleration vector in a spherical coordinate system with $ds^2 = (dx^1)^2 + (x^1\, dx^2)^2 + (x^1)^2 \sin^2 x^2 (dx^3)^2$ are

$$a_1 = \ddot{x}^1 - x^1(\dot{x}^2)^2 - x^1(\dot{x}^3 \sin x^2)^2,$$

$$a_2 = \frac{d}{dt}\,[(x^1)^2 \dot{x}^2] - (x^1)^2 \sin x^2 \cos x^2\, (\dot{x}^3)^2,$$

$$a_3 = \frac{d}{dt}\,[(x^1 \sin x^2)^2 \dot{x}^3].$$

Deduce these expressions from formula 78.3 and also from Lagrangean equations 79.2. *Hint:* $F_i = ma_i$ and $T = \dfrac{m}{2}\left(\dfrac{ds}{dt}\right)^2 = \dfrac{mg_{ij}}{2}\, \dot{x}^i \dot{x}^j$.

2. Use Lagrangean equations to show that, if a particle is not subjected to the action of forces, then its trajectory is given by $y^i = a^i t + b^i$, where the a^i and b^i are constants and the y^i are orthogonal cartesian coordinates.

3. Find, with the aid of Lagrangean equations, the trajectory of a particle moving in a uniform gravitational field. *Hint:* $T = \frac{1}{2} m \dot{y}^i \dot{y}^i$ and $V = mgy^3$, where y^3 is normal to the plane of the earth.

4. Deduce from Newtonian equations the equation of energy, $T + V = h$, where h is a constant. *Hint:* Show that $dT/dt = ma_j v^j = -dV/dt$.

5. Prove that, if a particle moves so that its velocity is constant in magnitude, then its acceleration vector is either orthogonal to the velocity vector, or it is zero. *Hint:* Compute the intrinsic derivatives of $v^2 = g_{ij} v^i v^j$.

6. We have shown in Sec. 79 that $\dfrac{d}{dt}\dfrac{\partial T}{\partial \dot{x}^i} - \dfrac{\partial T}{\partial x^i}$ is a covariant vector F_i whenever $T(x, \dot{x})$ is an invariant defined by (79.1). Prove more generally that if $W(x, \dot{x})$ is an invariant, then both $\partial W / \partial \dot{x}^i$ and $\dfrac{d}{dt}\dfrac{\partial W}{\partial \dot{x}^i} - \dfrac{\partial W}{\partial x^i}$ are covariant vectors. *Hint:* Let $x^i = x^i(q^1, q^2, q^3)$ be an admissible transformation of coordinates. Compute $\dot{x}^i$, show that $\partial \dot{x}^i / \partial \dot{q}^j = \partial x^i / \partial q^j$, and observe that the invariance of $W(x, \dot{x})$ requires that $W(x, \dot{x}) = W[x(q), \dot{x}(q)] \equiv \bar{W}(q, \dot{q})$.

80. Applications of Lagrangean Equations

As an illustration of the application of Lagrangean equations to the determination of trajectories, we consider several examples, which include the important cases of particles moving on smooth curves and surfaces.

1. *Free-Moving Particle.* If a particle is not subjected to the action of forces, the right-hand member of equation 79.2 vanishes, and we have

$$\frac{d}{dt}\left(\frac{\partial T}{\partial \dot{x}^i}\right) - \frac{\partial T}{\partial x^i} = 0. \tag{80.1}$$

If the coordinates x^i are chosen to be rectangular cartesian, then $T = (m/2)\dot{y}^i\dot{y}^i$, and hence equation 80.1 yields $m\ddot{y}^i = 0$. Integration of this equation gives $y^i = a^i t + b^i$, which represents a straight line.

2. *Constant Gravitational Field.* Again we choose a cartesian reference frame and take the Y^3-axis to be normal to the plane of the earth. The potential V of the constant gravitational field is $V = mgy^3$, if the positive Y^3-axis is directed upward. In this case equations 79.2 give

$$\ddot{y}^1 = 0, \qquad \ddot{y}^2 = 0, \qquad \ddot{y}^3 = -g,$$

so that the trajectory is determined by

$$y^\alpha = a^\alpha t + b^\alpha, \qquad (\alpha = 1, 2),$$
$$y^3 = -\tfrac{1}{2}gt^2 + at + b.$$

Thus the trajectory is a parabola whose axis is parallel to the Y^3-axis.

3. *Motion of a Particle on a Curve.* Let a particle be constrained to move on a curve C whose equations are

$$x^i = x^i(s), \qquad (i = 1, 2, 3), \tag{80.2}$$

s being the arc parameter. We shall suppose that C has a continuously turning tangent, so that the $x^i(s)$ are of class C^2.

The components v^i of the velocity vector $\mathbf{v}$ of the particle are

$$v^i = \frac{dx^i}{dt} = \frac{dx^i}{ds}\frac{ds}{dt} = v\lambda^i, \tag{80.3}$$

where $\lambda^i = dx^i/ds$ is the unit tangent vector to C and $v = ds/dt$ is the magnitude of $\mathbf{v}$.

The components a^i of the acceleration vector $\mathbf{a}$ are determined by computing the intrinsic derivative of (80.3) with respect to t

$$a^i = \frac{dv}{dt}\lambda^i + v\frac{\delta\lambda^i}{\delta t}, \tag{80.4}$$

where we wrote $\delta v/\delta t = dv/dt$, since v is a scalar. However,

(80.5) $$\frac{\delta\lambda^i}{\delta t} = \frac{\delta\lambda^i}{\delta s}\frac{ds}{dt} = v\frac{\delta\lambda^i}{\delta s} = v\varkappa\mu^i,$$

where we recalled the Frenet formula

[50.1] $$\frac{\delta\lambda^i}{\delta s} = \varkappa\mu^i, \qquad \varkappa > 0,$$

defining the curvature $\varkappa$ and the principal normal unit vector $\boldsymbol{\mu}$.

On substituting from (80.5) in (80.4), we get

(80.6) $$a^i = \frac{dv}{dt}\lambda^i + \varkappa v^2\mu^i,$$

which states that the acceleration vector **a** lies in the osculating plane of the curve. Moreover, the component of **a** in the tangential direction is equal to the time rate of change of speed v, whereas the component in the direction of the principal normal is v^2/R, where $R = 1/\varkappa$ is the radius of curvature of C.

The force $\mathbf{F} = m\mathbf{a}$ acting on a particle of mass m moving along C is determined by

(80.7) $$F^i = m\frac{dv}{dt}\lambda^i + m\varkappa v^2\mu^i.$$

It should be observed that F^i is the resultant of all external forces that act on the particle and thus **F** includes the reaction **R** of the curve on the particle. Since **F** lies in the osculating plane of the curve, the component of all external forces normal to this plane is zero. This condition enables us to compute reaction **R** in the general case. In mechanics the curve C is said to be smooth if the reaction **R** is normal[3] to C, that is, if $R^i\lambda_i = 0$: If $\mathbf{R} = \mathbf{0}$ the curve C is called the *natural trajectory* of the particle.

As an illustration, let a bead of mass m slide under gravity along a smooth curve C lying in the vertical Y^1Y^2-plane (Fig. 34).

The force **F** acting on m is

$$\mathbf{F} = m\mathbf{g} + \mathbf{R},$$

where **R** is the pressure exerted by the curve on the particle and $m\mathbf{g}$ is the gravitational force. Since the curve is smooth, **R** is normal to C. If α is the angle between the direction of **R** and the positive Y^2-axis, the components of **F** in the directions of the tangent $\boldsymbol{\lambda}$ and the principal normal $\boldsymbol{\mu}$ are

$$F_{(\lambda)} = -mg\sin\alpha, \qquad F_{(\mu)} = -mg\cos\alpha + R.$$

[3] This is equivalent to saying that the frictional force is zero. The term "smooth" employed in mechanics is different from the term "smooth" used in geometry, where a "smooth curve" is one with a continuously turning tangent.

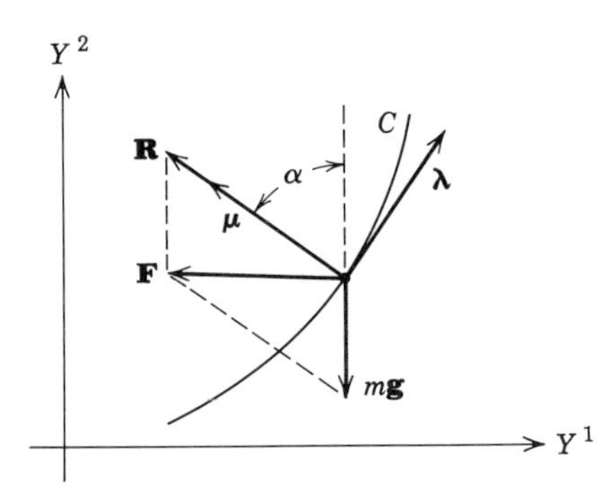

Fig. 34

On referring to (80.7), we conclude that

$$(80.8)\qquad m\frac{dv}{dt} = -mg\sin\alpha, \qquad m\varkappa v^2 = -mg\cos\alpha + R.$$

But $\cos\alpha = dy^1/ds$, $\sin\alpha = dy^2/ds$, and $dv/dt = (dv/ds)(ds/dt)$. Accordingly, the first of equations 80.8 yields

$$mv\frac{dv}{ds} = -mg\frac{dy^2}{ds},$$

so that

$$(80.9)\qquad \tfrac{1}{2}mv^2 = -mgy^2 + \text{constant}.$$

Since in this case the component of **R** in the direction of the path is zero, we could have written equations 80.9 directly from the energy equation $T + V = \text{constant}$.

Equation 80.9 determines the speed v along C as a function of y^2. The second equation in (80.8) then serves to determine R as a function of the curvature $\varkappa$. If the curve is rough, **R** is no longer normal to C and the angle α depends on the coefficient of friction.

As a concrete example, let a particle of mass m move under gravity along a smooth cycloid

$$(a)\qquad \begin{aligned} y^1 &= a(\theta - \sin\theta) \\ y^2 &= a(1 + \cos\theta), \qquad 0 \le \theta \le 2\pi, \end{aligned}$$

shown in Fig. 35. Then the first of equations 80.8 yields

$$(b) \qquad m\frac{d^2s}{dt^2} = -mg\frac{dy^2}{ds},$$

where

$$s = \int_0^\theta \sqrt{(dy^1)^2 + (dy^2)^2} = a\int_0^\theta \sqrt{2(1 - \cos\theta)}\, d\theta$$

$$= 2a\int_0^\theta \sin\frac{\theta}{2}\, d\theta = 4a\left(1 - \cos\frac{\theta}{2}\right).$$

Since $\cos^2(\theta/2) = \frac{1}{2}(1 + \cos\theta)$ we deduce, on noting the second of equations (a), that

$$y^2 = \frac{(s - 4a)^2}{8a}.$$

Accordingly, (b) yields the equation

$$\ddot{s} + \frac{g}{4a}s = g,$$

the general solution of which is

$$(c) \qquad s = c_1 \cos(\sqrt{g/4a}\, t + c_2) + 4a.$$

The integration constants c_1 and c_2 are determined by the initial position and initial velocity of m on the cycloid.

It is clear from (c) that the period of motion is independent of the amplitude c_1 and is equal to $2\pi/\sqrt{g/4a}$. This fact was discovered by Christian Huygens, about 300 years ago. Huygens proposed the use of cycloidal pendulum in the construction of isochronous clocks. Calculations, making use of the second equation 80.8 show that $R = 2\,mg\cos\alpha$.

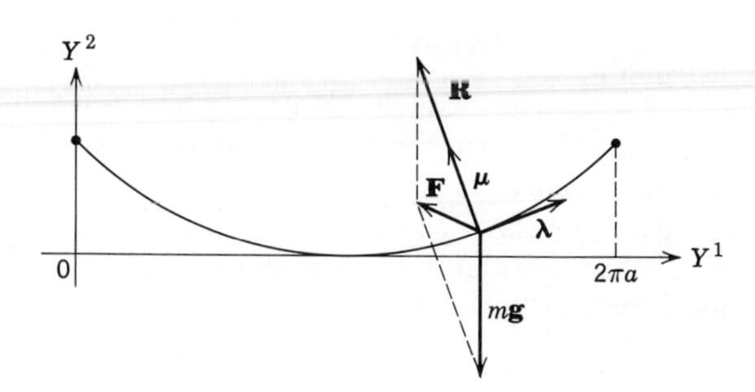

Fig. 35

Problems

1. Deduce the differential equations for a simple pendulum of length l, and show that for small oscillations the period is $2\pi/\sqrt{g/l}$.

2. Derive the equations of motion for a particle moving under gravity on a smooth helix:

$$y^1 = a\cos\theta, \qquad y^2 = a\sin\theta, \qquad y^3 = k\theta.$$

Note that since the helix is smooth, the reaction **R** is normal to the helix and hence the component of the resultant force **F** in the tangential direction is equal to the component of the gravitational force mg in that direction. The latter component can be computed from gravitational potential $V = mgy^3$. Furthermore, by (80.9) the energy equation in this case yields $\frac{1}{2}m(v^2 - v_0^2) = mg(y^3 - y_0^3)$.

3. If a particle of mass m moves on a smooth parabola with its axis vertical and concavity downwards, show that the reaction $R = \varkappa m(v^2 - v'^2)$, where v' is the velocity of m for which this parabola is the natural trajectory and v is the velocity in constrained motion.

4. *Motion of a Particle on a Surface.* Let the equations of a regular surface S be given in a parametric form as

$$S:\quad x^i = x^i(u^1, u^2), \qquad (i = 1, 2, 3), \tag{80.10}$$

and let a particle of mass m be constrained to move on S under the action of the force **F**. The force **F** is the resultant of all external forces acting on the particle and thus includes the reaction **R** of the surface on the particle. When the surface is smooth, **R** is normal to S and represents pressure that constrains the particle to remain on S.

The space components v^i of the velocity vector **v** of the particle are related to the surface components v^α by the formula[4]

$$v^i = \frac{dx^i}{dt} = \frac{\partial x^i}{\partial u^\alpha}\frac{du^\alpha}{dt} = x_\alpha^i \dot{u}^\alpha, \qquad (\alpha = 1, 2),$$

or

$$v^i = x_\alpha^i v^\alpha, \tag{80.11}$$

where $v^\alpha = \dot{u}^\alpha$.

The acceleration $a^i = \delta v^i/\delta t$; hence equation 80.11 yields

$$a^i = x_\alpha^i \frac{\delta v^\alpha}{\delta t} + \frac{\delta x_\alpha^i}{\delta t} v^\alpha,$$

[4] See equation 64.5. The reader should take care not to confuse the base vectors $\mathbf{a}^\alpha$ used in Chapter 3 with the acceleration components a^α used in this section.

or

(80.12) $$a^i = x_\alpha{}^i a^\alpha + x^i_{\alpha,\beta} v^\alpha v^\beta,$$

where $a^\alpha \equiv \delta v^\alpha/\delta t$.

If we make use of the Gauss formula

[67.7] $$x^i_{\alpha,\beta} = b_{\alpha\beta} n^i,$$

equation 80.12 reads

(80.13) $$a^i = x_\alpha{}^i a^\alpha + b_{\alpha\beta} v^\alpha v^\beta n^i.$$

Thus

$$a^i = x_\alpha{}^i a^\alpha + b_{\alpha\beta} v^2 \lambda^\alpha \lambda^\beta n^i,$$

and, since the normal curvature $\varkappa_{(n)} = b_{\alpha\beta}\lambda^\alpha\lambda^\beta$, we have

$$a^i = x_\alpha{}^i a^\alpha + v^2 \varkappa_{(n)} n^i.$$

Since $F^i = ma^i$, we have

(80.14) $$F^i = m x_\alpha{}^i a^\alpha + m v^2 \varkappa_{(n)} n^i.$$

The first term in the right-hand member of (80.14) is the component of **F** in the tangent plane to S, whereas the second is the component of **F** along the normal **n**. For, the component of **F** in the direction of the normal **n** is

(80.15) $$\begin{aligned} F^i n_i &= m x_\alpha{}^i n_i a^\alpha + m v^2 \varkappa_{(n)} n^i n_i, \\ &= 0 + m v^2 \varkappa_{(n)}, \end{aligned}$$

since the surface vectors $x_\alpha{}^i$ are orthogonal to n_i and $n^i n_i = 1$. The components of **F** in the plane tangent to S, on the other hand, are given by

$$\begin{aligned} g_{ij} x_\gamma{}^j F^i &= m g_{ij} x_\gamma{}^j x_\alpha{}^i a^\alpha + m v^2 \varkappa_{(n)} g_{ij} x_\gamma{}^j n^i \\ &= m a_{\gamma\alpha} a^\alpha + 0, \end{aligned}$$

since $g_{ij} x_\gamma{}^i x_\alpha{}^i = a_{\gamma\alpha}$ by (64.6), and $g_{ij} x_\gamma{}^j n^i = 0$, because the surface vectors $x_\gamma{}^j$ are orthogonal to n_j. If we rewrite this relation as

$$x_\gamma{}^j F_j = m a_\gamma,$$

and set $F_\gamma \equiv x_\gamma{}^j F_j$, we obtain a pair of Newtonian equations

(80.16) $$F_\gamma = m a_\gamma,$$

relating the surface force vector F_γ to the surface acceleration vector a_γ.

Equations 80.16 can be recast into equivalent Lagrangean form by noting that the kinetic energy $T = \frac{1}{2}mv^2$ is

$$T = \frac{m}{2} a_{\alpha\beta} v^\alpha v^\beta = \frac{m}{2} a_{\alpha\beta} \dot{u}^\alpha \dot{u}^\beta.$$

We obtain, as in Sec. 79,

$$\frac{d}{dt}\left(\frac{\partial T}{\partial \dot{u}^\alpha}\right) - \frac{\partial T}{\partial u^\alpha} = F_\alpha, \tag{80.17}$$

where F_α is defined by (80.16). When the force field is conservative, $F_\alpha = -\partial V/\partial u^\alpha$, where V is the potential.

We can deduce the equation analogous to (80.6) for the acceleration along the trajectory of the particle moving on S. The velocity v^α of the particle, along the trajectory, is $v^\alpha = v\lambda^\alpha$, hence

$$\begin{aligned} a^\alpha = \frac{\delta v^\alpha}{\delta t} &= \frac{dv}{dt}\lambda^\alpha + v\frac{\delta \lambda^\alpha}{\delta t} \\ &= \frac{dv}{dt}\lambda^\alpha + v^2\frac{\delta\lambda^\alpha}{\delta s}. \end{aligned}$$

If we recall that

$$\frac{\delta \lambda^\alpha}{\delta s} = \varkappa_g \eta^\alpha, \tag{71.6}$$

where η^α is the unit normal to the trajectory in the tangent plane, and $\varkappa_g$ is the geodesic curvature, we can write

$$\begin{aligned} a^\alpha &= \frac{dv}{dt}\lambda^\alpha + v^2\varkappa_g\eta^\alpha \\ &= v\frac{dv}{ds}\lambda^\alpha + v^2\varkappa_g\eta^\alpha, \end{aligned}$$

so that

$$a^\alpha = \frac{1}{2}\frac{dv^2}{ds}\lambda^\alpha + \varkappa_g v^2 \eta^\alpha.$$

If follows from this result (cf. equation 80.7) that

$$F^\alpha = \frac{dT}{ds}\lambda^\alpha + 2T\varkappa_g\eta^\alpha,$$

where $T = mv^2/2$. If the vector F^α vanishes identically, then $dT/ds = 0$ and $\varkappa_g = 0$ along the trajectory. The first of these equations states that $v =$ constant, and, if $v \neq 0$, then the trajectory is a geodesic by the theorem of Sec. 63.

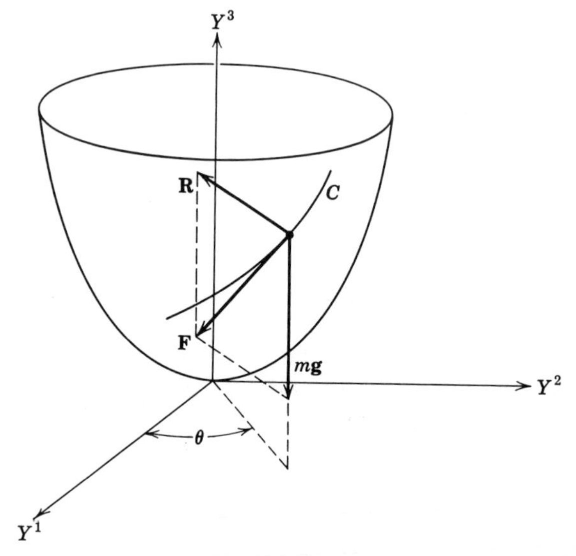

Fig. 36

As an illustration of the use of equations 80.17, we consider a particle of mass m constrained to move under gravity on a smooth paraboloid of revolution (Fig. 36),

$$y^3 = \frac{1}{4a}[(y^1)^2 + (y^2)^2], \qquad a = \text{constant}. \tag{80.18}$$

If we introduce cylindrical coordinates (r, θ, z) by setting

$$y^1 = r\cos\theta, \qquad y^2 = r\sin\theta, \qquad y^3 = z,$$

equation 80.18 becomes

$$z = \frac{r^2}{4a}, \tag{80.19}$$

and the kinetic energy $T = \frac{1}{2}m\dot{y}^i\dot{y}^i$ takes the form

$$T = \frac{m}{2}\left[\left(1 + \frac{r^2}{4a^2}\right)\dot{r}^2 + r^2\dot{\theta}^2\right].$$

The potential energy of the gravitational field is $V = mgy^3$, which, on noting (80.19), takes the form $V = mgr^2/4a$. Since the surface is smooth, the reaction $\mathbf{R}$ is normal to S and we can use equations 80.17 with $F_\alpha = -\partial V/\partial u^\alpha$, since the components of $\mathbf{R}$ in the tangent plane to S are zero.

We parametrize the surface by setting $u^1 = r$, $u^2 = \theta$, substitute in (80.17) for T and $F_\alpha = -\partial V/\partial u^\alpha$ and obtain two equations

$$\left(1 + \frac{r^2}{4a^2}\right)\ddot{r} + \frac{r\dot{r}^2}{4a^2} - r\,\dot{\theta}^2 = -\frac{gr}{2a},$$

(80.20)

$$\frac{d}{dt}(r^2\dot{\theta}) = 0.$$

The second of equations 80.20 gives on integration the equation of angular momentum

(80.21) $$r^2\dot{\theta} = h, \qquad h = \text{constant}.$$

The elimination of $\dot{\theta}$ from the first of equations 80.20 with the aid of (80.21) gives

(80.22) $$\left(1 + \frac{r^2}{4a^2}\right)\ddot{r} + \frac{r\dot{r}^2}{4a^2} - \frac{h^2}{r^3} = -\frac{gr}{2a},$$

which has a unique solution when the initial position $r = r_0$ and the initial velocity $\dot{r} = v_0$ of the particle are specified.

If our particle is constrained to move on a horizontal circle $r =$ constant, (80.22) requires that $h^2 = gr^4/2a$, and equation 80.21 then shows that $\dot{\theta}^2 = g/2a$, so that the angular velocity $\dot{\theta}$ is independent of the radius of the circle. When the path of the particle is the meridional line $\theta =$ constant, we get from 80.20 the equation

$$\left(1 + \frac{r^2}{4a^2}\right)\ddot{r} + \frac{r\dot{r}^2}{4a^2} + \frac{gr}{2a} = 0.$$

The integration of equation 80.22 and the calculation of the reaction **R** required to constrain the motion to the paraboloid is tedious. To compute the magnitude of reaction, we need equation 80.15, in which $\mathbf{F} = m\mathbf{g} + \mathbf{R}$.

If we replace the surface of the paraboloid in this illustration by the surface of the sphere, we have the problem of a spherical pendulum. The solution of equations of motion for the spherical pendulum can be obtained with the aid of elliptic functions. When the surface is a cylinder $r = a$, the integration of equations of motion is easy.[5]

[5] Interested readers are referred to pp. 99–109 in E. T. Whittaker's *Analytical Mechanics*, Cambridge Press (1917), where the motion of the particle on a surface is analyzed in a different way.

Problem

Let a particle of mass m be constrained to move on the surface of a sphere of radius a. Relate the orthogonal cartesian coordinates y^i to the surface coordinates u^α by the formulas

$$\begin{cases} y^1 = a \sin u^1 \cos u^2, \\ y^2 = a \sin u^1 \sin u^2, \\ y^3 = a \cos u^1. \end{cases}$$

Show that equations 80.17 yield

$$\begin{cases} \ddot{u}^1 - (\dot{u}^2)^2 \sin u^1 \cos u^1 = \dfrac{F_1}{ma^2}, \\ \ddot{u}^2 \sin^2 u^1 + 2\dot{u}^1\dot{u}^2 \sin u^1 \cos u^1 = \dfrac{F_2}{ma^2}. \end{cases}$$

Solve these equations for the case when $F^\alpha = 0$, and show that the trajectory is an arc of a great circle, and the speed $v = \text{const}$.

Hint: The first integral of the second equation is $\dot{u}^2 \sin^2 u^1 = \text{constant}$. Use this result in the first equation and observe that $v^2 = a^2[(\dot{u}^1)^2 + (\dot{u}^2)^2 \sin^2 u^1]$.

81. The Symbol of Variation

In this section we recall the definition of the variational symbol δ, first introduced in Sec. 56, and record several of its properties. The notation introduced here permits one to give a concise formulation of Hamilton's principle and Lagrange's principle of least action. Either of these principles (rather than the Newtonian laws) can serve as a starting point in the development of analytical dynamics.

Let $F(x^1, x^2, \ldots, x^n)$ be a function of n independent variables x^i of class C^2 in some region R of an n-dimensional manifold. We shall be concerned with the behavior of the function F in a certain neighborhood of the curve C, defined by the parametric equations

$$C: \ x^i = x^i(t), \qquad t_1 \le t \le t_2,$$

where we assume that the $x^i(t)$ are of class C^2.

Consider an h-neighborhood of the curve C, defined by the inequalities

$$x^i - h < \bar{x}^i < x^i + h, \qquad (i = 1, \ldots, n),$$

where h is a small positive number, and the x^i are the coordinates of a point on C. We introduce a class of functions

$$C': \ \bar{x}^i(t, \epsilon) = x^i(t) + \epsilon\xi^i(t), \qquad (i = 1, \ldots, n),$$

where $-1 \leq \epsilon \leq 1$ and the $\xi^i(t)$ are single-valued functions of class C^2 in $t_1 \leq t \leq t_2$, such that

$$\xi^i(t_1) = \xi^i(t_2) = 0$$

and $|\xi^i(t)| < h$, uniformly in $t_1 \leq t \leq t_2$.

A set of n functions $\bar{x}^i(t, \epsilon)$ constitutes a *varied path*, and it is clear that the curves C' so defined can be made to belong to the h-neighborhood of C. In the space of two dimensions the curves C' all lie in a band of width $2h$ about the curve C and coincide with C at the end points of the interval (t_1, t_2).

The variation δx^i was defined in Sec. 56 by the formula

$$\delta x^i = \left.\frac{\partial \bar{x}^i}{\delta \epsilon}\right|_{\epsilon=0} \cdot \epsilon = \xi^i(t)\epsilon, \tag{81.1}$$

and the variation δF of the function $F(x^1, \ldots, x^n)$ is

$$\delta F = \epsilon\left(\frac{\partial F}{\partial \epsilon}\right)_0,$$

where

$$\left(\frac{\partial F}{\partial \epsilon}\right)_0 = \left.\frac{\partial F(x^1 + \epsilon\xi^1, \ldots, x^n + \epsilon\xi^n)}{\partial \epsilon}\right|_{\epsilon=0} = \frac{\partial F}{\partial x^i}\,\xi^i.$$

Thus

$$\partial F = \frac{\partial F}{\partial x^i}\,\delta x^i. \tag{81.2}$$

Consider next the function $\dot{x}^i(t) \equiv dx^i/dt$. We form

$$\dot{\bar{x}}^i(t, \epsilon) \equiv \dot{x}^i(t) + \epsilon\dot{\xi}^i(t),$$

and conclude from definition 81.1 that

$$\delta \dot{x}^i = \epsilon\left(\frac{\partial \dot{\bar{x}}^i(t, \epsilon)}{\partial \epsilon}\right)_0 = \epsilon\dot{\xi}^i(t) = \frac{d}{dt}\,\delta x^i.$$

Hence

$$\delta\,\frac{dx^i}{dt} = \frac{d}{dt}\,\delta x^i, \tag{81.3}$$

so that *the variation of the derivative is the derivative of the variation.*

Clearly, if we have a function $F(x^1, \ldots, x^n, \dot{x}^1, \ldots, \dot{x}^n, t)$ of $2n + 1$ variables x^i, $\dot{x}^i \equiv dx^i/dt$, and t, which is of class C^2, we can write

$$\delta F = \frac{\partial F}{\partial x^i}\,\delta x^i + \frac{\partial F}{\partial \dot{x}^i}\,\delta \dot{x}^i. \tag{81.4}$$

A simple calculation, analogous to that used in deducing formula 81.3, leads to the conclusion that

$$\delta \frac{dF}{dt} = \frac{d}{dt} \delta F, \tag{81.5}$$

and one can readily show with the aid of (81.4) that

$$\delta(F + \Phi) = \delta F + \delta\Phi,$$
$$\delta(F\Phi) = F\delta\Phi + \Phi\delta F,$$

where F and Φ are any functions satisfying the conditions laid down above, and the variational symbol δ refers to the same varied path C'.

In Sec. 57 we considered the functional

$$J = \int_{t_1}^{t_2} F(t, x^1, \ldots, x^n, \dot{x}^1, \ldots, \dot{x}^n)\, dt,$$

where the functional arguments $x^i(t)$, $t_1 \leq t \leq t_2$, belonged to the h-neighborhood of an extremal of J. That is, we considered the behavior of the integral J along the varied paths $\bar{x}^i(t, \epsilon) = x^i + \epsilon\xi^i(t)$. Making use of equation 81.4 of this section and referring to formula 57.6, we see that formula 57.6 can be written

$$\delta J = \int_{t_1}^{t_2} \left(\frac{\partial F}{\partial x^i} \delta x^i + \frac{\partial F}{\partial \dot{x}^i} \delta \dot{x}^i \right) dt,$$

so that, for a pair of *fixed* limits t_1 and t_2,

$$\delta J = \int_{t_1}^{t_2} \delta F\, dt = \delta \int_{t_1}^{t_2} F\, dt.$$

When stated in words the foregoing equation reads: *The variation of the integral with fixed limits is equal to the integral of the variation of the integrand.*

We shall make use of the symbolism introduced in this section to formulate *Hamilton's principle.*

82. Hamilton's Principle

Consider a particle of mass m moving in a three-dimensional Euclidean manifold, referred to a curvilinear system of coordinates X. The particle is in motion under the influence of force $\mathbf{F}$, and our problem is to determine the trajectory

$$C: \quad x^i = x^i(t), \qquad (i = 1, 2, 3), \qquad t_1 \leq t \leq t_2,$$

where t denotes the time.

The kinetic energy T of the particle (which has a physical meaning only along the trajectory C) is given by the formula $T = \frac{1}{2}mg_{ij}\dot{x}^i\dot{x}^j$. If we define a family of varied paths

$$C': \quad \bar{x}^i(\epsilon, t) = x^i(t) + \delta x^i(t),$$

with $\delta x^i(t) = \epsilon\xi^i(t)$ and $\xi^i(t_1) = \xi^i(t_2) = 0$, belonging to the h-neighborhood of C, we can speak of the variation of T, namely,

$$\delta T = \frac{\partial T}{\partial \dot{x}^i}\,\delta\dot{x}^i + \frac{\partial T}{\partial x^i}\,\delta x^i, \tag{82.1}$$

and we can phrase Hamilton's principle as follows:

HAMILTON'S PRINCIPLE. *If a particle is at the point P_1 at the time t_1 and at the point P_2 at the time t_2, then the motion of the particle takes place in such a way that*

$$\int_{t_1}^{t_2} (\delta T + F_i\,\delta x^i)\,dt = 0, \tag{82.2}$$

where $x^i = x^i(t)$ are the coordinates of the particle along the trajectory and $x^i + \delta x^i$ are the coordinates along a varied path beginning at P_1 at time t_1 and ending at P_2 at time t_2.

It will be shown next that this principle is equivalent to Lagrangean equations of motion 79.2, and hence to Newtonian laws. The proof is simple. Substituting (82.1) in (82.2) yields

$$\int_{t_1}^{t_2} \left(\frac{\partial T}{\partial \dot{x}^i}\,\delta\dot{x}^i + \frac{\partial T}{\partial x^i}\,\delta x^i + F_i\,\delta x^i\right) dt = 0. \tag{82.3}$$

Integrating the first term under the integral sign of (82.3) by parts,

$$\int_{t_1}^{t_2} \frac{\partial T}{\partial \dot{x}^i}\,\delta\dot{x}^i\,dt = \frac{\partial T}{\partial \dot{x}^i}\,\delta x^i \Big|_{t_1}^{t_2} - \int_{t_1}^{t_2} \frac{d}{dt}\left(\frac{\partial T}{\partial \dot{x}^i}\right)\delta x^i\,dt,$$

and, since $\delta x^i(t_2) = \delta x^i(t_1) = 0$ by virtue of $\xi^i(t_2)$ and $\xi^i(t_1)$ vanishing, equation 82.3 becomes

$$\int_{t_1}^{t_2} \left(F_i + \frac{\partial T}{\partial x^i} - \frac{d}{dt}\frac{\partial T}{\partial \dot{x}^i}\right)\delta x^i\,dt = 0. \tag{82.4}$$

Since this integral vanishes for arbitrary δx^i, the argument used in Sec. 57 shows that

$$\frac{d}{dt}\frac{\partial T}{\partial \dot{x}^i} - \frac{\partial T}{\partial x^i} = F_i, \qquad (i = 1, 2, 3). \tag{82.5}$$

Conversely, if Lagrangean equations 82.5 hold, then equation 82.4, and hence equation 82.2, is valid.

In the foregoing formulation of Hamilton's principle no reference is made to the nature of the force field F_i. If, in particular, this field is conservative, then there exists a potential function $V(x^1, x^2, x^3)$ such that $\partial V/\partial x^i = -F_i$. In this case equation 82.2 reads

$$\int_{t_1}^{t_2} \left(\delta T - \frac{\partial V}{\partial x^i} \delta x^i \right) dt = 0,$$

and, since $\delta V = (\partial V/\partial x^i)\, \delta x^i$, we have

$$\int_{1t}^{t_2} \delta(T - V)\, dt = 0. \tag{82.6}$$

But in Sec. 79 we defined the Lagrangean function $L = T - V$, so that equation 82.6 can be written as $\int_{t_1}^{t_2} \delta L\, dt = 0$, and, since the limits of integration are fixed, we have a concise formulation of Hamilton's principle for a conservative field in the form

$$\delta \int_{t_1}^{t_2} L\, dt = 0. \tag{82.7}$$

We can state equation 82.7, in words, as follows: *In a conservative field of force a particle moves so that the integral* $\int_{t_1}^{t_2} L\, dt$, *evaluated along the trajectory* $x^i = x^i(t)$, $t_1 \leq t \leq t_2$, *has a stationary value in comparison with its values for all neighboring paths beginning at the point* P_1 *at* $t = t_1$ *and ending at point* P_2 *at* $t = t_2$.

Equations of motion in form 79.5, namely,

$$\frac{d}{dt}\left(\frac{\partial L}{\partial \dot{x}^i}\right) - \frac{\partial L}{\partial x^i} = 0,$$

follow at once from the formulation 82.7.

83. Integral of Energy

We establish in this section an important general

THEOREM. *The motion of a particle in a conservative field of force is such that the sum of its kinetic and potential energies is a constant.*

The proof of this theorem follows from an identity which will be established next.

Since the kinetic energy $T = \frac{1}{2}mg_{ij}\dot{x}^i\dot{x}^j$ is an invariant,

$$\frac{dT}{dt} = \frac{\delta T}{\delta t} = \frac{\delta}{\delta t}\left[\frac{m}{2}(g_{ij}\dot{x}^i\dot{x}^j)\right]$$

$$= \frac{m}{2}g_{ij}\left(\frac{\delta\dot{x}^i}{\delta t}\dot{x}^j + \dot{x}^i\frac{\delta\dot{x}^j}{\delta t}\right)$$

$$= mg_{ij}a^i v^j,$$

so that

$$\frac{dT}{dt} = ma_i v^i, \tag{83.1}$$

where v^i is the velocity and a_i is the acceleration of the particle.

For a conservative field of force, $ma_i = F_i = -\partial V/\partial x^i$, and we can write (83.1) as

$$\frac{dT}{dt} = -\frac{\partial V}{\partial x^i}\frac{dx^i}{dt},$$

or

$$\frac{dT}{dt} = -\frac{dV}{dt}. \tag{83.2}$$

Integrating (83.2) yields the result

$$T + V = h,$$

where h is a constant of integration.

84. Principle of Least Action

The history of science abounds in attempts to imbed the laws of nature in the structure of theology. Several of these, based on the minimal concepts, such as Heron's (100 B.C.) doctrine of the shortest path and Fermat's (1601–1665) principle of least time, had an innate esthetic appeal to mathematicians. The most celebrated of such attempts, in the domain of mechanics, is the doctrine of least action propounded by P. M. L. Maupertuis *circa* 1740. Maupertuis asserted that all activities of nature are performed with the least possible expenditure of "action," which he defined as the product of mass, velocity, and distance. In order to fit his principle to the known results of mechanics, Maupertuis was obliged to alter the definitions of the quantities entering in the product *mvs* so as to suit each problem under consideration. Thus, in the anlaysis of inelastic collision of two particles of masses m_1 amd m_2, moving with velocities v_1 and v_2, he minimized the product mvs, where s was the distance per unit time. This made the "action" proportional to the kinetic energy. Maupertuis obtained the known correct expression for the final common velocity, $v = (m_1v_1 + m_2v_2)/(m_1 + m_2)$. On the other hand, in the problem of refraction of light passing from one optical medium to another he used the actual distance s and got the constant (but incorrect) value for the ratio of the sines of the angles of the incident and refracted rays. The doctrine of Maupertuis, who believed that it furnished a scientific demonstration of the existence of God, excited the imaginations of Daniel Bernoulli and Euler and was defended by them. In 1744, Euler showed that the integral $\int mv\, ds$ has a stationary value along the trajectory of a particle moving in a central field of force. In 1760, Lagrange extended Euler's result by demonstrating that the integral $A = \int_{P_1}^{P_2} m\mathbf{v} \cdot d\mathbf{s}$ has a stationary value along the trajectories of particles moving in a conservative force field, provided that the constraints are not functions of the time. This led him to formulate the principle of least action. This formulation still left a great deal to be desired from the point of view of clarity, and Hamilton, in an attempt to understand Lagrange's formulation of the principle, deduced a broader and different principle (1827) discussed in Sec. 82. The proof of the Lagrangean principle, which put it on a secure basis, was supplied by Jacobi.

Let us consider the integral of Lagrange

$$A = \int_{P_1}^{P_2} m\mathbf{v} \cdot d\mathbf{s}, \tag{84.1}$$

evaluated over the path

$$C: \quad x^i = x^i(t), \qquad t_1 \leq t \leq t_2,$$

where C is the trajectory of the particle of mass m moving in a conservative field of force. We suppose that neither the kinetic energy T nor the potential energy V is a function of time.

In curvilinear coordinates the integral 84.1 assumes the form

$$\begin{aligned} A &= \int_{P_1}^{P_2} mg_{ij} \frac{dx^i}{dt} dx^j \\ &= \int_{t(P_1)}^{t(P_2)} mg_{ij} \frac{dx^i}{dt} \frac{dx^j}{dt} dt, \end{aligned}$$

and, since

$$T = \frac{m}{2} g_{ij} \frac{dx^i}{dt} \frac{dx^j}{dt},$$

we have

$$A = \int_{t(P_1)}^{t(P_2)} 2T\,dt. \tag{84.2}$$

This integral has a physical meaning only when evaluated over the trajectory C, but its value can be computed along any varied path joining the points P_1 and P_2. Let us consider a particular set of admissible paths C' along which the function $T + V$, for each value of parameter t, has the same constant value h. The functional A so determined is called the action integral, and concerning it we can formulate

THE PRINCIPLE OF LEAST ACTION.[6] *Of all curves C' passing through P_1 and P_2 in the neighborhood of the trajectory C, which are traversed at a rate such that, for each C', for every value of t, $T + V = h$, that one for which the action integral A is stationary is the trajectory of the particle.*

When stated in the form of the variational equation, this principle reads

$$\delta \int_{t(P_1)}^{t(P_2)} 2T\,dt = 0, \tag{84.3}$$

with the auxiliary condition

$$T + V - h = 0, \qquad \text{on } C'. \tag{84.4}$$

It is important to recognize that in this instance we cannot determine the extremals of the action integral by setting F in the Euler equations 57.7 equal to $2T$, because of the auxiliary condition (84.4). Since T is a function of the velocity v, and V is a function of position alone, the times $t(P_2) - t(P_1)$ required to traverse the varied paths C' will differ in general. Thus the upper limit $t(P_2)$ in the integral 84.4 is not fixed. In this case we have the problem in the calculus of variations with variable end points and with one auxiliary condition 84.4. The procedure employed in solving this problem makes use of Lagrange's method of multipliers for a problem with nonholonomic constraints, which we briefly indicate. (Compare Sec. 57.)

We construct a function $F = 2T + \lambda\phi$, where $\phi = T + V - h$, and determine the solution of the system of four equations

$$\begin{cases} \dfrac{\partial F}{\partial x^i} - \dfrac{d}{dt}\left(\dfrac{\partial F}{\partial \dot{x}^i}\right) = 0, & (i = 1, 2, 3), \\ T + V - h = 0. \end{cases}$$

[6] Strictly speaking this principle should be called the principle of stationary action.

An investigation of this system shows that[7] $\lambda(t) = -1$, and it follows from this fact that the trajectory C is determined by the solution of the system

$$\frac{d}{dt}\left(\frac{\partial T}{\partial \dot{x}^i}\right) - \frac{\partial T}{\partial x^i} = -\frac{\partial V}{\partial x^i}, \qquad (i = 1, 2, 3). \tag{84.5}$$

These are precisely the Lagrangean equations of motion.

A different and somewhat more illuminating mode of attack on this problem is to reduce it to a consideration of the variational problem with fixed end points by a change of variable. Since the kinetic energy

$$T = \frac{m}{2} g_{ij} \frac{dx^i}{dt} \frac{dx^j}{dt} = \frac{m}{2}\left(\frac{ds}{dt}\right)^2,$$

$$dt = \sqrt{\frac{m}{2T}}\, ds \tag{84.6}$$

$$= \sqrt{\frac{m}{2(h - V)}}\, ds.$$

Consequently the action integral 84.2 can be written[8]

$$A = \int_{s_1}^{s_2} \sqrt{2m(h - V)}\, ds, \tag{84.7}$$

since along all admissible paths $T = h - V$. The integrand in the preceding integral is clearly independent of t. We now parametrize our varied paths C', so that

$$C:\quad x^i = x^i(u), \qquad u_1 \leq u \leq u_2,$$

where P_1: $x^i(u_1)$ and P_2: $x^i(u_2)$, and write

$$ds = \sqrt{g_{ij} x'^i x'^j}\, du,$$

where $x'^i = dx^i/du$.

This permits us to write the action integral 84.7 in the form

$$A = \int_{u_1}^{u_2} \sqrt{2m(h - V) g_{ij} x'^i x'^j}\, du, \tag{84.8}$$

and, since the limits of integration in (84.8) are fixed, we see that the determination of the trajectory is equivalent to finding the geodesics in a three-dimensional Riemannian manifold with the arc element

$$dS^2 = 2m(h - V) g_{ij}\, dx^i\, dx^j. \tag{84.9}$$

[7] See Sec. 88 below and O. Bolza, *Vorlesungen über Variationsrechnung*, p. 586.

[8] The form 84.7 of the action integral was used by Jacobi. See a discussion of this integral and its generalizations in C. Carathéodory's *Variationsrechnung*, pp. 255, 290.

If we form Euler's equations

$$F_{x^i} - \frac{d}{du} F_{x'^i} = 0,$$

with $F = \sqrt{2m(h - V)g_{ij}x'^i x'^j}$, and take cognizance of equation 84.6 in the form

$$dt = \sqrt{\frac{mg_{ij}x'^i x'^j}{2(h - V)}}\, du,$$

we get the desired equations 84.5.

We see from formulas 84.8 and 84.9 that the action is equal numerically to the length of the curve in a Riemannian manifold with metric coefficients

$$h_{ij} \equiv 2m(h - V)g_{ij},$$

and that the trajectories in E_3 correspond to the geodesics in a Riemannian space metrized by the formula $dS^2 = h_{ij}\, dx^i\, dx^j$. This geometrization of dynamics had a far-reaching effect on the developments in relativistic dynamics.

85. Systems of Particles. Generalized Coordinates

We have already remarked (in Sec. 77) that the passage from mechanics of a single particle to mechanics of material bodies can be accomplished by introducing certain assumptions regarding the nature of constraining forces operating on particles making up the body. In some dynamical problems the change of shape of the body is so slight that one is justified in supposing that the particles remain at fixed distances from one another. This assumption leads to the *dynamics of rigid bodies*. If a body suffers nonnegligible deformations we can postulate, with varying degrees of realism, the nature of constraining forces and thus arrive at the dynamics of elastic bodies, ideal fluids, viscoplastic media, and so on. The assumptions concerning the nature of constitutive forces permit us to characterize the positions of a large number of material particles in terms of relatively few descriptive parameters. Thus a thin rigid rod of length l, moving in space, requires only five parameters for the determination of its position. These can be taken as space coordinates of its center of mass and two direction ratios of one of the ends relative to the center of mass. The choice of descriptive parameters is not unique, and they clearly need not have the dimensions of length. A bead sliding on a curved wire requires only one parameter for the description of its location, say the distance from some fixed point on the wire; a particle moving on the surface is located unambiguously by a pair of Gaussian coordinates. Whatever is the nature of descriptive parameters, they will be termed *the generalized coordinates.* Clearly, if the characterization of dynamical systems is to be complete, the generalized coordinates must be functionally connected with the space coordinates of particles making up the system.

Let there be N particles composing a system, and let $x^i_{(\alpha)}$, $(i = 1, 2, 3)$, $(\alpha = 1, 2, \ldots, N)$, be the positional coordinates of these particles referred to some convenient reference frame in E_3. The system of N free particles is described by $3N$ parameters. If the particles are constrained in some way, there will be certain relations among the coordinates $x^i_{(\alpha)}$, and we suppose that there are r such independent relations.

$$(85.1)\quad f^i(x^1_{(1)}, x^2_{(1)}, x^3_{(1)}; x^1_{(2)}, x^2_{(2)}, x^3_{(2)}; \ldots ; x^1_{(N)}, x^2_{(N)}, x^3_{(N)}) = 0, \qquad (i = 1, 2, \ldots, r).$$

If these r equations of constraints 83.1 can be solved for some r coordinates in terms of the remaining $3N - r$ coordinates, the latter can be viewed as the independent generalized coordinates q^i. It is more convenient, however, to assume that each of the $3N$ coordinates is expressed in terms of $3N - r \equiv n$ independent variables q^i, and write $3N$ equations

$$(85.2)\qquad x^i_{(\alpha)} = x^i_{(\alpha)}(q^1, \ldots, q^n, t),$$

where we introduced the time parameter t which may enter in the problem explicitly if one deals with moving constraints.[9] If t does not enter explicitly in equations 85.2, the dynamical system is called a *natural* system.

We will suppose that the functions $x^i_{(\alpha)} = x^i_{(\alpha)}(q, t)$ are of class C^2 in the region of definition of the variables q^i and t and that the Jacobian matrix $(\partial x^i/\partial q^j)$ is of rank n [cf. (75.5)].

The velocities of the particles are given by differentiating equations 85.2 with respect to time. Thus

$$(85.3)\qquad \dot{x}^i_{(\alpha)} = \frac{\partial x^i_{(\alpha)}}{\partial q^j}\dot{q}^j + \frac{\partial x^i_{(\alpha)}}{\partial t}.$$

We shall call the time derivatives $\dot{q}^i$ of generalized coordinates q^i the *generalized velocities.*

Occasionally, for symmetry reasons, it is desirable to introduce a number of superfluous coordinates q^i, and describe the system with the aid of $k > n$ coordinates $q^1, \ldots, q^k$. In this event there will exist certain relations of the form

$$(85.4)\qquad f^j(q^1, \ldots, q^k, t) = 0,$$

[9] For example, a bead sliding on a wire while the wire itself is moving with specified velocity.

so that the quantities q^i, and hence $\dot{q}^i$, are no longer independent. There will be relations among the $\dot{q}$'s of the type

$$\frac{\partial f^j}{\partial q^i}\dot{q}^i + \frac{\partial f^j}{\partial t} = 0, \tag{85.5}$$

when the f^i are differentiable.

Since equations 85.5 were obtained by differentiating equations 85.4, it is clear that they are integrable, so that one can deduce from them equations 85.4, and use them to eliminate the superfluous coordinates. In some problems, however, functional relations of the type

$$F^j(q^1, q^2, \ldots, q^k; \dot{q}^1, \ldots, \dot{q}^k, t) = 0, \qquad (j = 1, 2, \ldots, m), \tag{85.6}$$

arise which are *nonintegrable*, that is, it may be impossible[10] to deduce from these differential equations solutions of the type 85.4. The behavior of the system in such event cannot be described with the aid of fewer than k coordinates, so that all k coordinates are independent. If nonintegrable relations 85.6 occur in the problem, we shall say that the given system has $k - m$ *degrees of freedom*, where m is the number of independent nonintegrable relations 85.6 and k is the number of independent coordinates. The dynamical systems involving nonintegrable relations 85.6 are called *nonholonomic* to distinguish them from *holonomic* systems in which the number of degrees of freedom is equal to the number of independent generalized coordinates. In other words, a holonomic system is one in which there are no nonintegrable relations involving the generalized velocities.

In the following section we derive the Lagrangean equations for a holonomic system, and in Sec. 88 we treat briefly one important class of nonholonomic systems occurring frequently in applications.

86. Lagrangean Equations in Generalized Coordinates

For concreteness of presentation the definitions of Sec. 85 were introduced with reference to systems consisting of a finite but, perhaps, large number of particles. These definitions can be readily extended to apply

[10] A billiard ball rolling and spinning on a rough table is an example of this situation. To specify the position of the ball one needs five generalized coordinates; two of these may locate its center, and three the angles describing the orientation of the ball relative to the center. Since the table is rough, the ball cannot slip, so that both velocity components of the point of contact must vanish. This gives two constraining relations of the form (85.6), involving the velocity components. They are nonintegrable, since, at any position of the center, the orientation of the ball can be changed without violating the constraints.

to continuous bodies, the points of which have coordinates x^r relative to some reference system X.

The particles of a continuous body are subjected to constraints of various sorts, and we shall suppose throughout the remainder of this chapter that the bodies under consideration are rigid, so that the material points remain at invariable distances from one another. If the points of the body are uniquely determined by a finite number of generalized coordinates q^i, we will write

$$x^r = x^r(q^1, \ldots, q^n, t), \qquad (r = 1, 2, 3),$$

and assume, as in Sec. 85, that the functions $x^r(q, t)$ are of class C^2. The velocity $\dot{x}^r$ of any point of the body is given by

$$\dot{x}^r = \frac{\partial x^r}{\partial q^j}\frac{dq^j}{dt} + \frac{\partial x^r}{\partial t}$$

$$= \frac{\partial x^r}{\partial q^j}\dot{q}^j + \frac{\partial x^r}{\partial t}, \qquad (j = 1, \ldots, n),$$

where the $\dot{q}^i$ are generalized velocities.

Let the system in question be natural, holonomic, with n degrees of freedom, so that the relations

$$x^r = x^r(q^1, \ldots, q^n) \tag{86.1}$$

involve n independent parameters q^i. The velocities $\dot{x}^r$ in this case are given by [cf. (80.11)]

$$\dot{x}^r = \frac{\partial x^r}{\partial q^j}\dot{q}^j, \qquad (r = 1, 2, 3; j = 1, 2, \ldots, n), \tag{86.2}$$

where the $\dot{q}^i$ transform under any admissible transformation

$$\bar{q}^k = \bar{q}^k(q^1, \ldots, q^n), \qquad (k = 1, \ldots, n), \tag{86.3}$$

in accordance with the contravariant law.

The kinetic energy of the system is given by the expression of the form

$$T = \tfrac{1}{2}\sum_{\alpha} m_{(\alpha)} g_{rs}\dot{x}^r_{(\alpha)}\dot{x}^s_{(\alpha)}, \qquad (r, s = 1, 2, 3), \tag{86.4}$$

where m is the mass of the particle located at the point x^r and the summation (or integration) is carried over the entire region occupied by the body. The g_{rs} in (86.4) are the components of the metric tensor associated with the coordinate system X covering E_3.

If we insert in (86.4) the values of $\dot{x}^i$ from (86.2), we obtain[11]

$$T = \tfrac{1}{2} \sum_{\alpha} m g_{rs} \frac{\partial x^r}{\partial q^i} \frac{\partial x^s}{\partial q^j} \dot{q}^i \dot{q}^j$$
$$\equiv \tfrac{1}{2} a_{ij} \dot{q}^i \dot{q}^j,$$

where

$$a_{ij} \equiv \sum_{\alpha} m g_{rs} \frac{\partial x^r}{\partial q^i} \frac{\partial x^s}{\partial q^j}, \qquad (r, s = 1, 2, 3),$$
$$(i, j = 1, \ldots, n),$$

Since

(86.5) $$T = \tfrac{1}{2} a_{ij} \dot{q}^i \dot{q}^j$$

is an invariant, and the quantities a_{ij} are symmetric, we conclude that the a_{ij} are components of a covariant tensor of rank two with respect to a class of admissible transformations 86.3 of generalized coordinates. We note that, since the kinetic energy T is a positive form in the velocities $\dot{q}^i$, $|a_{ij}| > 0$, and we can construct the reciprocal tensor a^{ij}.

If we carry out a computation, in every detail identical with that of Sec. 79, by using the expression for the kinetic energy in the form 86.5, we obtain the formula

(86.6) $$\frac{d}{dt}\left(\frac{\partial T}{\partial \dot{q}^i}\right) - \frac{\partial T}{\partial q^i} = a_{il}\left(\ddot{q}^l + \left\{ {l \atop jk} \right\}_a \dot{q}^j \dot{q}^k\right),$$

where the Christoffel symbols $\left\{ {l \atop jk} \right\}_a$ are constructed from the tensor a_{kl}. We denote the expression appearing in the parentheses of the right-hand member of (86.6) by

$$Q^l \equiv \ddot{q}^l + \left\{ {l \atop jk} \right\}_a \dot{q}^j \dot{q}^k$$

and write equation 86.6 in the form

(86.7) $$\frac{d}{dt}\left(\frac{\partial T}{\partial \dot{q}^i}\right) - \frac{\partial T}{\partial q^i} = a_{il} Q^l$$
$$= Q_i, \qquad (i = 1, 2, \ldots, n).$$

The expression in the left-hand member of (86.7) can also be computed by starting with formula 86.4 and by taking cognizance of the dependence of the variables x^i on the parameters q^i. A straight-forward but somewhat lengthy computation making use of the formula $\partial \dot{x}^r / \partial \dot{q}^j = \partial x^r / \partial q^j$ and the

[11] For simplicity in writing we omit the subscripts α in terms affected by the symbol $\sum_{\alpha}$.

relations $\dfrac{\partial \dot{x}^r}{\partial q^i} = \dfrac{\partial^2 x^r}{\partial q^i\,\partial q^j}\dot{q}^j$ and $\dfrac{\partial \dot{x}^r}{\partial q^i} = \dfrac{d}{dt}\dfrac{\partial x^r}{\partial q^i}$, following from equation 86.2, leads to the result

$$\frac{d}{dt}\left(\frac{\partial T}{\partial \dot{q}^i}\right) - \frac{\partial T}{\partial q^i} = \sum_{\alpha} m a_r \frac{\partial x^r}{\partial q^i}, \tag{86.8}$$

in which $a_j = g_{ij}a^i$ is the acceleration of the point $P(x)$.

On the other hand, Newton's second law gives

$$m a_r = F_r, \tag{86.9}$$

where the F_r's are the components of force $\mathbf{F}$ acting on the particle located at the point $P(x)$. It follows from (86.9) that

$$\sum_{\alpha} m a_r \frac{\partial x^r}{\partial q^i} = \sum_{\alpha} F_r \frac{\partial x^r}{\partial q^i},$$

and hence equations 86.8 can be written

$$\frac{d}{dt}\left(\frac{\partial T}{\partial \dot{q}^i}\right) - \frac{\partial T}{\partial q^i} = \sum_{\alpha} F_r \frac{\partial x^r}{\partial q^i}. \tag{86.10}$$

Comparing (86.7) with (86.10), we conclude that

$$Q_i = \sum_{\alpha} F_r \frac{\partial x^r}{\partial q^i},$$

in which the vector Q_i is called *generalized force.*

The equations

$$\frac{d}{dt}\left(\frac{\partial T}{\partial \dot{q}^i}\right) - \frac{\partial T}{\partial q^i} = Q_i \tag{86.11}$$

are known as *Lagrangean equations in generalized coordinates.* They yield a system of n second-order ordinary differential equations for the generalized coordinates q^i. The solutions of these equations in the form

$$C{:}\quad q^i = q^i(t)$$

represent the *dynamical trajectory* of the system.

If there exists a function $V(q^1, \ldots, q^n)$, such that

$$\frac{\partial V}{\partial q^i} = -Q_i,$$

the system is said to be conservative, and for such systems equations 86.11 assume the form

$$\frac{d}{dt}\left(\frac{\partial L}{\partial \dot{q}^i}\right) - \frac{\partial L}{\partial q^i} = 0, \tag{86.12}$$

where $L \equiv T - V$ is the *kinetic potential.*

Since $L(q, \dot{q})$ is a function of both the generalized coordinates and velocities,

$$\frac{dL}{dt} = \frac{\partial L}{\partial \dot{q}^i}\ddot{q}^i + \frac{\partial L}{\partial q^i}\dot{q}^i.$$

Inserting in this expression from Lagrangean equations 86.12, we get

$$\frac{dL}{dt} = \frac{\partial L}{\partial \dot{q}^i}\ddot{q}^i + \frac{d}{dt}\left(\frac{\partial L}{\partial \dot{q}^i}\right)\dot{q}^i = \frac{d}{dt}\left(\frac{\partial L}{\partial \dot{q}^i}\dot{q}^i\right). \tag{86.13}$$

But, since $L = T - V$, and the potential energy V is not a function of the $\dot{q}^i$,

$$\frac{\partial L}{\partial \dot{q}^i}\dot{q}^i = \frac{\partial T}{\partial \dot{q}^i}\dot{q}^i = 2T,$$

since $T = \frac{1}{2}a_{ij}\dot{q}^i\dot{q}^j$. Thus equation 86.13 can be written in the form

$$\frac{d(L - 2T)}{dt} = -\frac{d(T + V)}{dt} = 0,$$

which implies that $T + V = h$ (constant). Thus, along the dynamical trajectory, the sum of the kinetic and potential energies is a constant.

It follows from this development that the study of natural holonomic dynamical systems with n degrees of freedom can be reduced to a study of motion of a single particle in the n-dimensional space.

We can phrase the problem of determining the dynamical trajectory of the system in the language of calculus of variations. Indeed, the statements of Hamilton's principle and of the least action principle, given in Secs. 82 and 84, can be repeated word-for-word if the "point" is interpreted to mean a set of n parameters $q^1, \ldots, q^n$, specifying the configuration of our dynamical system in a certain n-dimensional space.

In symbols the principle of Hamilton reads

$$\int_{t_1}^{t_2} (\delta T + Q_i\, \delta q^i)\, dt = 0, \tag{86.14}$$

and, if the force field Q_i is conservative, the principle can be stated in the form

$$\delta \int_{t_1}^{t_2} L\, dt = 0.$$

These variational equations imply the satisfaction of Lagrangean equations 86.11 and 86.12.

It follows at once from the formulation of the principle of least action in generalized coordinates (cf. equations 84.3 and 84.4) that dynamical trajectories in a conservative field are geodesics in the n-dimensional Riemannian manifold with the arc element dS given by

$$dS^2 = 2(h - V)a_{ij}\, dq^i\, dq^j.$$

The fact that the dynamical trajectory can be regarded as a geodesic permits one to geometrize dynamics.

Problems

Show that the dynamical equations in spherical coordinates with

$$ds^2 = (dr)^2 + r^2(d\theta)^2 + r^2 \sin^2\theta\, (d\phi)^2$$

assume the form

$$m(\ddot{r} - r\dot{\theta}^2 - r\dot{\phi}^2 \sin^2\theta) = -\frac{\partial V}{\partial r},$$

$$m\left[\frac{1}{r}\frac{d}{dt}(r^2\dot{\theta}) - r\dot{\phi}^2 \sin\theta \cos\theta\right] = -\frac{1}{r}\frac{\partial V}{\partial \theta},$$

$$m\left[\frac{1}{r\sin\theta}\frac{d}{dt}(r^2\dot{\phi}\sin^2\theta)\right] = -\frac{1}{r\sin\theta}\frac{\partial V}{\partial \phi},$$

whereas in cylindrical coordinates, with $ds^2 = (dr)^2 + r^2(d\theta)^2 + (dz)^2$, they are

$$m(\ddot{r} - r\dot{\theta}^2) = -\frac{\partial V}{\partial r},$$

$$m\left[\frac{1}{r}\frac{d}{dt}(r^2\dot{\theta})\right] = -\frac{1}{r}\frac{\partial V}{\partial \theta},$$

$$m\ddot{z} = -\frac{\partial V}{\partial z}.$$

87. Virtual Work and Generalized Forces

In the developments of the preceding sections no characterization of forces F_r acting at a point (x^r) of a rigid body was made. It is customary in the study of mechanics of continuous media to classify forces into three categories.[12]

(*a*) Internal constitutive forces.
(*b*) Reactive forces produced by constraints.
(*c*) External impressed forces.

[12] The reactive forces produced by constraints are also *external* forces.

We can visualize a material body as being composed of a vast number of particles which interact with one another in a rather complicated way. As long as the constitutive internal forces are of the action-reaction type, they need not be taken into account in the dynamical equations, since their resultant at any point P of the body vanishes. Thus the forces F_r, appearing in the formulas of Sec. 86, consist of reactive forces produced by constraints and external impressed forces.

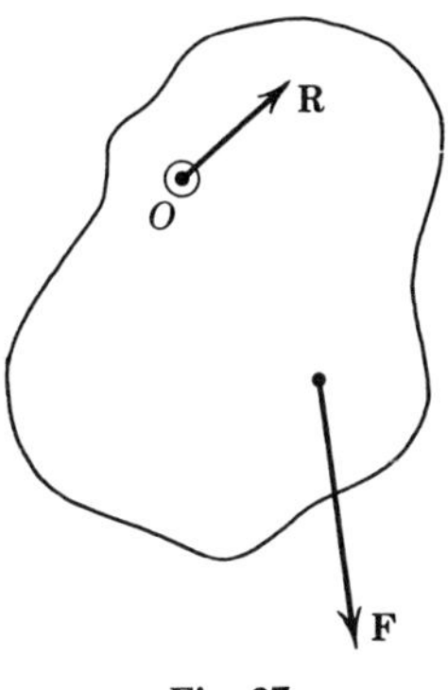

Fig. 37

To illustrate the meaning of this we can consider a rigid body fixed at some point O by a smooth pin, and subjected to the action of impressed force F_r (see Fig. 37). The pin at O constrains the motion of a body to that of rotation about the point O. The reactive force R_r acting at O does no work if the body is displaced so as not to violate the constraints at O. We shall term all reactive forces that do no work in an arbitrary displacement which does not violate the constraints *workless forces*. Any displacement of a point of a body that is consistent with imposed constraints is a *virtual displacement*,[13] and we denote such virtual displacements at a point x^r by δx^r.

The work done by the impressed forces F_r in a virtual displacement δx^r is

$$W_\delta = \Sigma\, F_r\, \delta x^r, \tag{87.1}$$

where the summation is carried over all particles of the body; this will be the total work *if the reactive forces are of the workless type*. We define W_δ to be the *virtual work* in producing a virtual displacement δx^r, provided that the reactions are workless. Otherwise, W_δ will also contain contributions from the working reactive forces.

It should be noted carefully that a virtual displacement δx^r is not necessarily the actual displacement dx^r that the point $P(x^r)$ undergoes under the action of specified forces. It is merely *any conceivable displacement* that a body can perform without violating the constraints.

If a given natural holonomic system with n degrees of freedom is described by the generalized coordinates q^i, then $x^r = x^r(q^1, \ldots, q^n)$, and the virtual displacements δx^r are related linearly to the *generalized virtual displacements* δq^i, namely,

$$\delta x^r = \frac{\partial x^r}{\partial q^j}\, \delta q^j. \tag{87.2}$$

[13] Virtual displacements that violate constraints are also used in dynamics, especially if one is concerned with the computation of reactive forces.

In formula 87.2 the δq^j's are arbitrary, and they are necessarily consistent with constraints imposed on the system, since the coordinates q^i are independent.[14]

If we insert expressions from (87.2) in (87.1), we get

$$
\begin{aligned}
W_\delta &= \Sigma F_r \frac{\partial x^r}{\partial q^j} \delta q^j \\
&= Q_j \, \delta q^j,
\end{aligned}
\tag{87.3}
$$

where the last step makes use of the definition of the generalized force Q_i. It follows from this formula that one can calculate the generalized forces Q_i, acting on the system, by computing the work W_δ produced by displacing the system through a virtual displacement $\delta q^j \neq 0$, (j fixed), and with $\delta q^i = 0$, $i \neq j$. Then $Q_i = W_\delta / \delta q^j$. We shall resort to this method of computing generalized forces in the illustrative examples of Sec. 89.

88. Nonholonomic Systems

The derivation of Lagrangean equations in Sec. 86 is based on the assumption that the dynamical system is holonomic and that its configuration is described by n independent generalized coordinates q^i. When the q^i are not independent, the derivation of appropriate dynamical equations from Hamilton's principle (86.14) hinges on general considerations presented in Sec. 57.

In dealing with nonholonomic dynamical systems it is customary to assume that the generalized velocities $\dot{q}^i$ enter in the constraining relations linearly. Accordingly, we shall suppose that n generalized coordinates q^i satisfy $m < n$ conditions of the type

$$
c_{ki}(q^1, \ldots, q^n)\dot{q}^i = 0, \qquad (k = 1, \ldots, m), \qquad (i = 1, \ldots, n),
\tag{88.1}
$$

in which the coefficients c_{ki} are continuously differentiable functions of the variables q^i.

The set of m equations 88.1 can be written in the form

$$
c_{ki}\dot{q}^i \, \delta t = 0,
$$

and, since $\dot{q}^i \, \delta t = \delta q^i$, we have m relations

$$
c_{ki} \, \delta q^i = 0,
\tag{88.2}
$$

in which the variations δq^i in general are not independent.

[14] We call attention to the distinction between the virtual displacements δq^i and the actual displacements dq^i taking place along the dynamical trajectory $q^i = q^i(t)$.

To deduce the dynamical equations from Hamilton's variational equation

(88.3) $$\int_{t_1}^{t_2} (\delta T + Q_i\,\delta q^i)\,dt = 0,$$

in which the δq^i are constrained by m relations 88.2, we introduce (cf. Sec. 57) m unknown functions $\lambda^k(q^1, \ldots, q^n)$ and form with the aid of (88.2) the sum

(88.4) $$\lambda^k c_{ki}\,\delta q^i = 0, \qquad (k = 1, \ldots, m), \qquad (i = 1, \ldots, n).$$

Since $\delta T = (\partial T/\partial \dot{q}^i)\,\delta\dot{q}^i + (\partial T/\partial q^i)\,\delta q^i$, equation 88.3 yields

(88.5) $$\int_{t_1}^{t_2} \left(\frac{\partial T}{\partial \dot{q}^i}\,\delta\dot{q}^i + \frac{\partial T}{\partial q^i}\,\delta q^i + Q_i\,\delta q^i\right) dt = 0.$$

But, $\delta\dot{q}^i = (d/dt)\,\delta q^i$, and the integration by parts of the first term in the integrand of (88.5) gives (cf. 82.3)

(88.6) $$\int_{t_1}^{t_2} \left(\frac{\partial T}{\partial q^i} - \frac{d}{dt}\frac{\partial T}{\partial \dot{q}^i} + Q_i\right)\delta q^i\,dt = 0,$$

when we recall that $\delta q^i(t_1) = \delta q^i(t_2) = 0$ along each varied path.

We rewrite (88.6) by inserting in the integrand the term $\lambda^k c_{ki}\,\delta q^i = 0$,

(88.7) $$\int_{t_1}^{t_2} \left(\frac{\partial T}{\partial q^i} - \frac{d}{dt}\frac{\partial T}{\partial \dot{q}^i} + Q_i + \lambda^k c_{ki}\right)\delta q^i\,dt = 0.$$

In formula 88.7, the q^i are constrained by m relations 88.4, and, if we agree to consider the first $n - m$ coordinates q^i as independent variables, and suppose that m functions $\lambda^k(q^1, \ldots, q^n)$ can be chosen so that

(88.8) $$\frac{\partial T}{\partial q^i} - \frac{d}{dt}\frac{\partial T}{\partial \dot{q}^i} + Q_i + \lambda^k c_{ki} = 0, \qquad \text{for } i = n - m + 1, \ldots, n,$$

then (88.7) reduces to

(88.9) $$\int_{t_1}^{t_2} \left(\frac{\partial T}{\partial q^i} - \frac{d}{dt}\frac{\partial T}{\partial \dot{q}^i} + Q_i + \lambda^k c_{ki}\right)\delta q^i\,dt = 0,$$
$$(i = 1, 2, \ldots, n - m).$$

Since the first $n - m$ variables q^i in the integrand of (88.9) are independent, the variations δq^i for $i = 1, 2, \ldots, n - m$ can be chosen arbitrarily, and we conclude that

(88.10) $$\frac{\partial T}{\partial q^i} - \frac{d}{dt}\frac{\partial T}{\partial \dot{q}^i} + Q_i + \lambda^k c_{ki} = 0, \qquad (i = 1, 2, \ldots, n - m).$$

The two sets of equations (88.8) and (88.10) involve n generalized coordinates q^i and m Lagrangean multipliers $\lambda^k(q^1, \ldots, q^n)$. By adjoining to these equations m equations 88.1, we get $n + m$ equations for the determination of the q's and λ's.

The circumstances under which the λ^k can be determined so as to satisfy (88.8) were detailed in Sec. 57; they relate to the rank of the Jacobian matrix for (88.1).

We note that when the equations in (88.8) and (88.10) are written as a single set

$$\text{(88.11)} \qquad \frac{d}{dt}\frac{\partial T}{\partial \dot{q}^i} - \frac{\partial T}{\partial q^i} = Q_i + \lambda_k c_{ki}, \qquad (i = 1, 2, \ldots, n),$$

the right-hand member of (88.11) differs from the right-hand member of (86.11) by the term $R_i = \lambda^k c_{ki}$. This term corresponds to the generalized reactive forces produced by constraints when the Q_i are generalized forces that act on the system in the absence of constraints.

In special situations the Q_i may be derived from potential $V(q^1, \ldots, q^n)$.

As an illustration of the use of equations 88.11, we consider a homogeneous circular cylinder rolling under gravity down a rough inclined plane.

Let the cylinder of radius a and mass m roll without slipping down the plane making a fixed angle ϕ with the horizontal. The position of the cylinder is determined by the angle of roll θ and by the distance x through which the center of mass of the cylinder moves down the plane. We shall take as our generalized coordinates $q^1 = \theta$, $q^2 = x$, and note that the kinetic energy T of the system is the sum of the kinetic energy of translation of the center of mass and the kinetic energy of rotation about the center of mass. Thus

$$\text{(88.12)} \qquad T = \tfrac{1}{2}m\dot{x}^2 + \tfrac{1}{2}mk^2\dot{\theta}^2,$$

where k is the radius of gyration of the cylinder.

Since the plane is rough, there is frictional force $\mathbf{F}$ acting in the plane, and we suppose that this force is just sufficient to prevent slipping. In this event x and θ are related by

$$\text{(88.13)} \qquad a\, d\theta = dx,$$

where a is the radius of the cylinder.

The constraint (88.13) is actually holonomic since (88.13) can be integrated to yield $x = a\theta$, so that the problem can be reduced to the consideration of one independent variable, say x. However, to illustrate the theory of this section, we write (88.13) in the form $c_{ki}\dot{q}^i = 0$ [cf. (88.1)],

$$\text{(88.14)} \qquad a\frac{d\theta}{dt} - \frac{dx}{dt} = 0,$$

so that $c_{11} = a$, $c_{12} = -1$.

Equations (88.11) then yield

$$(88.15)\qquad \begin{aligned} \frac{\partial T}{\partial \theta} - \frac{d}{dt}\frac{\partial T}{\partial \dot{\theta}} + Q_1 + \lambda a = 0, \\ \frac{\partial T}{\partial x} - \frac{d}{dt}\frac{\partial T}{\partial \dot{x}} + Q_2 - \lambda = 0. \end{aligned}$$

Now, the work W done by the gravitational force alone, when the center of mass moves through a distance x is $W = xmg \sin \phi$. Hence $V = -xmg \sin \phi$, and

$$Q_1 = -\frac{\partial V}{\partial \theta} = 0, \qquad Q_2 = -\frac{\partial V}{\partial x} = mg \sin \phi.$$

On inserting these expressions in (88.15) and using T in the form 88.12, we get a pair of equations

$$(88.16)\qquad m\ddot{\theta} = \frac{\lambda a}{k^2}, \qquad m\ddot{x} = mg \sin \phi - \lambda,$$

which, when compared with (88.11), show that the generalized reactions R_i are

$$R_1 = \lambda a, \qquad R_2 = -\lambda.$$

To compute λ, observe that $a\theta = x$, so that $\ddot{\theta} = \ddot{x}/a$, and use this relation to eliminate $\ddot{x}$ and $\ddot{\theta}$ in (88.16). The result is

$$\lambda = \frac{mg \sin \phi}{1 + a^2/k^2},$$

and hence equations 88.16 yield

$$(88.17)\quad m\ddot{\theta} = \frac{mga \sin \phi}{a^2 + k^2}, \qquad m\ddot{x} = mg \sin \phi - \frac{k^2 mg \sin \phi}{a^2 + k^2}.$$

The term $k^2 mg \sin \phi/(a^2 + k^2)$, in the second of equations 88.17, represents the frictional force F opposing the component $mg \sin \phi$ of the gravitational force along the plane. If the cylinder is solid, $k^2 = a^2/2$, and $F = \frac{1}{3} mg \sin \phi$. The magnitude of frictional force $F = \mu N$, where μ is the coefficient of friction and N is the pressure of the cylinder on the plane. Since $N = mg \cos \phi$, we conclude that $\mu = F/N = \frac{1}{3} \tan \phi$.

As another illustration of the use of equations 88.11 we consider the brachistochrone problem in a resisting medium.[15]

[15] See G. A. Bliss, "The Problem of Lagrange in the Calculus of Variations," *American Journal of Mathematics*, **52** (1930) and L. A. Pars, *Calculus of Variations* (1962), pp. 241–243.

Let it be required to determine an arc of a continuously differentiable curve

$$C = y = y(x), \qquad y(x_1) = y_1, \qquad y(x_2) = y_2, \tag{88.18}$$

such that the time of descent of a bead of unit mass, moving on C under gravity, is as short as possible. We suppose that the motion is opposed by a force $R(v)$ per unit mass, where $R(v)$ is a continuously differentiable function of the speed v.

We choose the positive Y-axis in the direction of gravity. Since the work done by gravity on the particle less work done by the resisting force $R(v)$ is equal to the change in kinetic energy, we have

$$\frac{dv^2}{2} = g\,dy - R(v)\,ds.$$

If we take x as our independent variable, this relation gives the constraining condition in the form

$$\phi(y, v, y', v') = vv' - gy' + R(v)\sqrt{1 + (y')^2} = 0, \tag{88.19}$$

where primes denote derivatives with respect to x.

The integral to be minimized under the condition 88.19 is

$$J = \int_{t_1}^{t_2} dt = \int_{x_1}^{x_2} \frac{ds}{v} = \int_{x_1}^{x_2} \frac{\sqrt{1 + (y')^2}}{v}\,dx. \tag{88.20}$$

We denote the integrand in (88.20) by $F = \sqrt{1 + (y')^2}/v$, and construct the function

$$\begin{aligned} G &= F + \lambda\phi \\ &= \frac{\sqrt{1 + (y')^2}}{v} + \lambda(x)[vv' - gy' + R(v)\sqrt{1 + (y')^2}]. \end{aligned}$$

If we define

$$H = \frac{1}{v} + \lambda R(v), \tag{88.21}$$

we can write G as

$$G = H\sqrt{1 + (y')^2} + \lambda(vv' - gy'). \tag{88.22}$$

The equations of C are determined from Euler's equations

$$\frac{dG_{y'}}{dx} - G_y = 0, \qquad \frac{dG_{v'}}{dx} - G_v = 0, \tag{88.23}$$

and, since G does not contain y, we conclude from the first of equations 88.23 that $G_{y'} = a$ or

$$\frac{Hy'}{\sqrt{1 + (y')^2}} - \lambda g = a, \tag{88.24}$$

where a is a constant.

The second of equations 88.23 yields

$$\frac{d}{dx}(\lambda v) - \sqrt{1 + (y')^2} H_v - \lambda v' = 0,$$

or

$$\frac{v\lambda'(x)}{\sqrt{1 + (y')^2}} = H_v. \tag{88.25}$$

We thus have the system of three equations 88.19, 88.24, 88.25 for the determination of $y(x)$, $v(x)$, and $\lambda(x)$. We can rewrite them as

$$\begin{aligned} v\frac{dv}{ds} &= g\frac{dy}{ds} - R, \\ H\frac{dy}{ds} &= \lambda g + a, \\ v\frac{d\lambda}{ds} &= H_v \end{aligned} \tag{88.26}$$

by setting $ds = \sqrt{1 + (y')^2}\,dx$. On eliminating dy and ds from (88.26), we get the equation

$$H(H_v\,dv + R\,d\lambda) = (g\lambda + a)g\,d\lambda,$$

and since $R = H_\lambda$ by (88.21), we can write it as

$$H(H_v\,dv + H_\lambda\,d\lambda) = (g\lambda + a)g\,d\lambda$$

or

$$H\,dH = (g\lambda + a)g\,d\lambda.$$

The integration of this equation yields

$$H^2 = (g\lambda + a)^2 + b^2, \tag{88.27}$$

where b^2 is the constant of integration.

It follows from (88.21) that (88.27) is a quadratic equation in λ so that λ can be regarded as a *known function of v and the integration constants a, b*. This suggests that the equation of C be sought in the parametric form

$$C\colon\quad x = x(v), \qquad y = y(v). \tag{88.28}$$

Since $dy/dv = dy/ds \cdot ds/dv$, we find with the aid of the first two equations in (88.26) that

(88.29)
$$\frac{dy}{dv} = \frac{v(\lambda g + a)}{g(\lambda g + a) - RH},$$

the right-hand member of which is a known function of v. On performing quadrature we then get

(88.30)
$$y = f_1(v, a, b) + c,$$

where c is a constant. Equation 88.30 is one of the desired equations in (88.28). To obtain $x = x(v)$, we note that $dx/dv = dx/ds \cdot ds/dv$ and, since $dx/ds = \sqrt{1 - (dy/ds)^2}$ and both dy/ds and ds/dv are determined by the first two equations in (88.26), we see that dx/ds is also a known function of v. The reader will check that

$$\frac{dx}{dv} = \frac{bv}{g(\lambda g + a) - RH},$$

so that

(88.31)
$$x = f_2(v, a, b) + d,$$

where d is a constant. The constants of integration in (88.30) and (88.31) must be determined for the initial conditions. To make the problem physically meaningful we must impose some restrictions on the relative magnitudes of $R(v)$ and the gravitational force g, as, for example, $R < g$ for all relevant values of v.

Problems

1. A hollow cylindrical drum of mass m rolls under gravity down a rough inclined plane making an angle ϕ with the horizontal. What must the coefficient of friction μ be to prevent slipping? (*Answer:* $\mu \geq \frac{1}{2} \tan \phi$.)

2. A bead of mass m slides on a smooth rod rotating in a vertical plane about one end with constant angular velocity ω. Show that the equation of motion is $\ddot{r} - \omega^2 r = g \sin \omega t$, and solve it.

3. A bead slides on a smooth circular wire of radius a, which is rotating with constant angular velocity about the vertical diameter of the wire. Show that $\ddot{\theta} - \omega^2 \sin \theta \cos \theta = (g/a) \sin \theta$, where θ is the angle made by the radius to the particle with the diameter.

89. Illustrative Examples

We give next three examples illustrating the use of generalized co-ordinates.

Consider first the problem of a simple pendulum, consisting of a bob of mass m supported by a light inextensible cord of length l. We shall

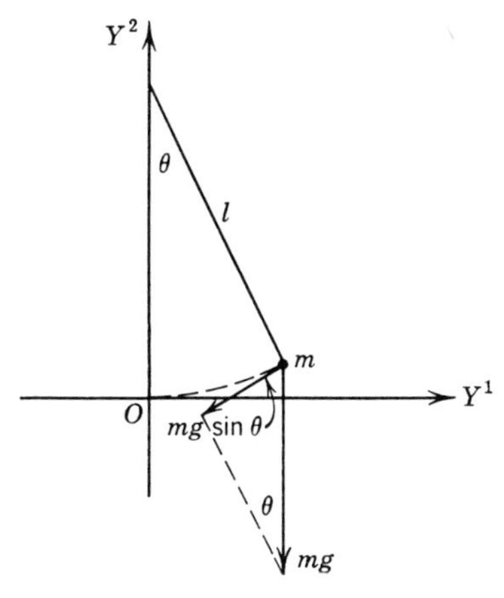

Fig. 38

suppose that the pendulum is set in vibration in some plane which we take as the Y^1Y^2-plane. (See Fig. 38.)

In order to form Lagrangean equations

$$\frac{d}{dt}\left(\frac{\partial T}{\partial \dot{q}^i}\right) - \frac{\partial T}{\partial q^i} = Q_i, \tag{89.1}$$

we need the expression for the kinetic energy

$$T = \tfrac{1}{2}m\dot{y}^i\dot{y}^i. \tag{89.2}$$

However,

$$\begin{cases} y^1 = l \sin \theta \equiv l \sin \dfrac{q}{l}, \\ y^2 = l(1 - \cos \theta) = l\left(1 - \cos \dfrac{q}{l}\right), \end{cases} \tag{89.3}$$

where we take the arc-length $q = l\theta$ as our generalized coordinate. Since $\dot{y}^2 = \dot{q} \sin q/l$ and $\dot{y}^1 = \dot{q} \cos q/l$, equation 89.2 becomes $T = \frac{1}{2}m(\dot{q})^2$.

The work W_δ done in producing a virtual displacement δq is

$$\begin{aligned} W_\delta &= -mg \sin \theta \, \delta q \\ &= -mg \sin \frac{q}{l} \, \delta q, \end{aligned}$$

and hence the generalized force $Q = -mg \sin q/l$. Thus equation 89.1 yields

$$\ddot{q} + g \sin \frac{q}{l} = 0, \tag{89.4}$$

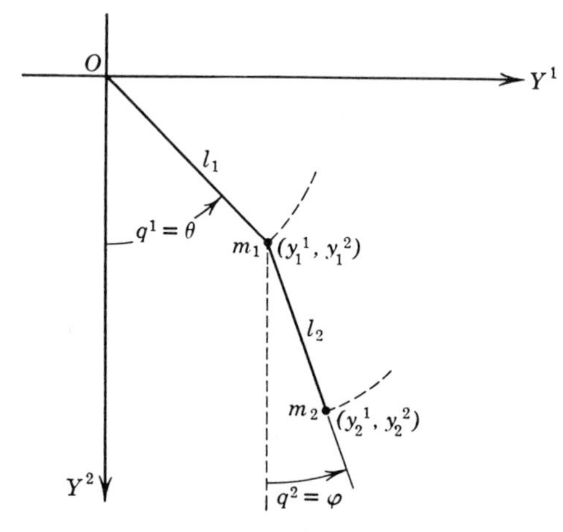

Fig. 39

and, since for small displacements $\sin\theta \doteq \theta$, for small vibrations we have

$$\ddot{q} + k^2 q = 0,$$

where $k^2 = g/l$. The solution of this equation is $q = a\cos(kt + \alpha)$. The solution of (89.4) can be expressed in terms of elliptic integrals of the first kind.

We turn next to a more interesting problem of a double pendulum. Consider an arrangement of particles shown in Fig. 39, where we suppose that the masses m_1 and m_2 are supported by inextensible light cords of lengths l_1 and l_2, respectively. The pendulum is assumed to vibrate in one plane, and we take as our generalized coordinates the quantities θ and ϕ, which give the angular deviations of the cords of lengths l_1 and l_2 from the vertical.

The equations connecting the coordinates (y_1^1, y_1^2) and (y_2^1, y_2^2), of the masses m_1 and m_2, with generalized coordinates $q^1 = \theta$ and $q^2 = \phi$ are

$$\begin{cases} y_1^{\,1} = l_1 \sin q^1, \\ y_1^{\,2} = l_1 \cos q^1, \\ y_2^{\,1} = l_1 \sin q^1 + l_2 \sin q^2, \\ y_2^{\,2} = l_1 \cos q^1 + l_2 \cos q^2. \end{cases}$$

Since

$$T = \tfrac{1}{2} m_1 \dot{y}_1^{\,i} \dot{y}_1^{\,i} + \tfrac{1}{2} m_2 \dot{y}_2^{\,i} \dot{y}_2^{\,i}, \qquad (i = 1, 2),$$

an easy calculation gives

$$T = \tfrac{1}{2}\{m_1(l_1\dot{q}^1)^2 + m_2[(l_1\dot{q}^1)^2 + 2l_1 l_2 \dot{q}^1 \dot{q}^2 \cos(q^2 - q^1) + (l_2\dot{q}^2)^2]\}.$$

Now, the work done in a small virtual displacement δq^2 when $\delta q^1 = 0$ is

$$W_\delta^{(2)} = -m_2 g l_2 \sin q^2 \, \delta q^2;$$

so that

$$Q_2 = -m_2 l_2 g \sin q^2.$$

Also the work done in a displacement δq^1 when $\delta q^2 = 0$ is

$$W_\delta^{(1)} = -(m_1 + m_2) g l_1 \sin q^1 \, \delta q^1.$$

Thus

$$Q_1 = -(m_1 + m_2) g l_1 \sin q^1.$$

Making use of equations 89.1, we find a pair of simultaneous ordinary differential equations

$$(89.5) \quad \begin{cases} \dfrac{d}{dt}\{(m_1 + m_2)(l_1)^2 \dot{q}^1 + m_2 l_1 l_2 \dot{q}^2 \cos(q^2 - q^1)\} \\ \qquad - m_2 l_1 l_2 \dot{q}^1 \dot{q}^2 \sin(q^2 - q^1) = -(m_1 + m_2) g l_1 \sin q^1, \\ \dfrac{d}{dt}\{m_2 l_1 l_2 \dot{q}^1 \cos(q^2 - q^1) + m_2 (l_2)^2 \dot{q}^2\} \\ \qquad + m_2 l_1 l_2 \dot{q}^1 \dot{q}^2 \sin(q^2 - q^1) = -m_2 g l_2 \sin q^2, \end{cases}$$

for the determination of the dynamical trajectory.

Instead of determining the generalized forces Q_1 and Q_2 directly, we could have made use of the potential energy V, which is

$$V = m_1 g l_1 (1 - \cos q^1) + m_2 g (l_1 + l_2 - l_1 \cos q^1 - l_2 \cos q^2),$$

if we assume $V = 0$ when $q^1 = q^2 = 0$.

For a detailed discussion of the solution of the system of differential equations 89.5 we refer to standard treatises on analytical dynamics.

As our final example we consider the problem of small oscillations of a conservative dynamical system about the position of stable equilibrium.

We suppose that the system is natural, holonomic, with n degrees of freedom, and select the generalized coordinates q^i so that the equilibrium position is given by $q^i = 0$, $(i = 1, \ldots, n)$. Since the equilibrium is stable, the potential energy $V(q^1, \ldots, q^n)$ has a minimum value at $q^i = 0$, and hence $\left.\dfrac{\partial V}{\partial q^i}\right|_0 = 0$. If we choose the potential level to be zero at $q^i = 0$, then the expansion of $V(q^1, \ldots, q^n)$ in Taylor's series about $q^i = 0$ has the form $V = \frac{1}{2} b_{ij} q^i q^j + O(q^3)$, where $O(q^3)$ denotes the remainder after the second-degree terms in the q^i. Since we are concerned with small oscillations about the point $q^i = 0$, we shall suppose that the potential energy is represented with sufficient accuracy by the quadratic form

$$(89.6) \qquad V = \tfrac{1}{2} b_{ij} q^i q^j, \qquad (b_{ij} = b_{ji}).$$

The kinetic energy T of the system is

$$T = \tfrac{1}{2}a_{ij}\dot{q}^i\dot{q}^j, \qquad (a_{ji} = a_{ji}), \tag{89.7}$$

and we suppose that, in the neighborhood of the point $q^i = 0$, the a^{ij}'s do do not vary appreciably, so that they can be regarded as constants.

The Lagrangean equations 86.12 now yield the system of n simultaneous second-order ordinary differential equations with constant coefficients

$$a_{ij}\ddot{q}^j + b_{ij}q^j = 0.$$

Instead of integrating this coupled system directly we can simplify the problem by introducing a new set of independent variables q'^i, the so-called *normal coordinates*, which are related linearly to the coordinates q^i in such a way that the quadratic forms 89.6 and 89.7 reduce simultaneously[16] to a sum of squares. We then have

$$\begin{cases} T = (\dot{q}'^1)^2 + (\dot{q}'^2)^2 + \cdots + (\dot{q}'^n)^2 \\ V = \lambda_1^2(q'^1)^2 + \cdots + \lambda_n^2(q'^n)^2. \end{cases} \tag{89.8}$$

All the coefficients of the q''s in (89.8) are nonnegative since the quadratic form 89.6 is necessarily nonnegative if the potential energy V has a minimum at $q^i = 0$.

The Lagrangean equations now become

$$\ddot{q}'^i + \lambda_i^2 q'^i = 0, \qquad (\text{no sum on } i),$$

and their solutions obviously are

$$q'^i = c_1 (\cos \lambda_i t + c_2), \qquad (i = 1, \ldots, n).$$

Thus the oscillation of the system, in terms of the normal coordinates, is simple harmonic with normal modes of vibration determined by the characteristic values λ_i which satisfy the *frequency equation*

$$|b_{ij} - \lambda^2 a_{ij}| = 0. \tag{89.9}$$

If the roots λ_i are distinct, the normal coordinates q'^i are determined essentially uniquely. For multiple roots, the choice of normal coordinates is not unique. This follows from the analysis given in Sec. 16.

The problems of small oscillations are of great technical interest, and there is an extensive literature concerned with the study of oscillating systems with finite and infinite number of degrees of freedom.[17]

[16] This algebraic problem was considered in detail in Sec. 16.

[17] See, for some interesting examples, Frazer, Duncan, and Collar, *Elementary Matrices and Some Applications to Dynamics and Differential Equations*, Cambridge University Press, 1938.

As a concrete illustration of our general discussion of oscillation of dynamical systems about the position of stable equilibrium consider the double pendulum in Fig. 39 with $l_1 = l_2 = l$ and $m_1 = m_2 = m$. The expressions for T and V given on pages 250–251 in this case reduce to

$$T = \tfrac{1}{2}l^2m[2(\dot{q}^1)^2 + 2\dot{q}^1\dot{q}^2 \cos(q^2 - q^1) + (\dot{q}^2)^2],$$
$$V = mgl[(1 - \cos q^1) + (2 - \cos q^1 - \cos q^2)].$$

If we expand T and V in powers of q^i and $\dot{q}^i$ and retain only the second-degree terms in these variables, we get

$$T = \frac{ml^2}{2}[2(\dot{q}^1)^2 + 2\dot{q}^1\dot{q}^2 + (\dot{q}^2)^2],$$
$$V = \frac{mgl}{2}[2(q^1)^2 + (q^2)^2]. \tag{89.10}$$

To reduce (89.10) to the form 89.8 we introduce the normal coordinates $x = q'^1$, $y = q'^2$ by a linear transformation (cf. Sec. 16)

$$q^1 = a_1x + a_2y,$$
$$q^2 = b_1x + b_2y. \tag{89.11}$$

The coefficients a_i and b_i in (89.11) must be chosen so that T and V in (89.10) reduce to

$$T = \tfrac{1}{2}(\dot{x}^2 + \dot{y}^2),$$
$$V = \lambda_1^2x^2 + \lambda_2^2y^2. \tag{89.12}$$

The substitution from (89.11) in (89.10) yields two quadratic forms in which the cross-product terms must vanish. Thus

$$2b_1a_2 + 2b_2a_1 = 0,$$
$$4a_1a_2 + 2b_1b_2 = 0.$$

Solving these, we get

$$\frac{b_1}{a_1} = \sqrt{2}, \qquad \frac{b_2}{a_2} = -\sqrt{2}.$$

Furthermore, the comparison of coefficients of x^2 and y^2 shows that

$$a_1^{\ 2} = \frac{2 - \sqrt{2}}{4ml^2}, \qquad a_2^{\ 2} = \frac{2 + \sqrt{2}}{4ml^2}.$$

Thus the desired transformation 89.11 is

$$q^1 = \frac{1}{2l\sqrt{m}}(\sqrt{2 - \sqrt{2}}\, x + \sqrt{2 + \sqrt{2}}\, y),$$
$$q^2 = \frac{\sqrt{2}}{2l\sqrt{m}}(\sqrt{2 - \sqrt{2}}\, x - \sqrt{2 + \sqrt{2}}\, y), \tag{89.13}$$

under which V assumes the form

$$V = \frac{g}{2l}[(2 - \sqrt{2})x^2 + (2 + \sqrt{2})y^2].$$

Accordingly, the Lagrangean equations in normal coordinates are

$$\ddot{x} + \frac{g}{l}(2 - \sqrt{2})x = 0, \qquad \ddot{y} + \frac{g}{l}(2 + \sqrt{2})y = 0.$$

Solving these, we get

(89.14) $$x = c_1 \cos(\lambda_1 t + c_2), \qquad y = c_3 \cos(\lambda_2 t + c_4),$$

where

$$\lambda_1^2 = \frac{g}{l}(2 - \sqrt{2}), \qquad \lambda_2^2 = \frac{g}{l}(2 + \sqrt{2}).$$

The independent oscillations in (89.14) have periods $T_1 = 2\pi/\lambda_1$ and $T_2 = 2\pi/\lambda_2$. The vibration with the larger period is that of x; it is called the *grave mode*. The *rapid mode* is that of y. If we set $y = 0$ in (89.13) and consider the grave mode, we see that

$$q_2 = \sqrt{2}\, q_1.$$

The performance of the pendulum in this case is illustrated in Fig. 40*a*. On setting $x = 0$, we get the motion of the rapid mode for which $q_2 = -\sqrt{2}\, q_1$. This is shown in Fig. 40*b*. The angles shown in these figures

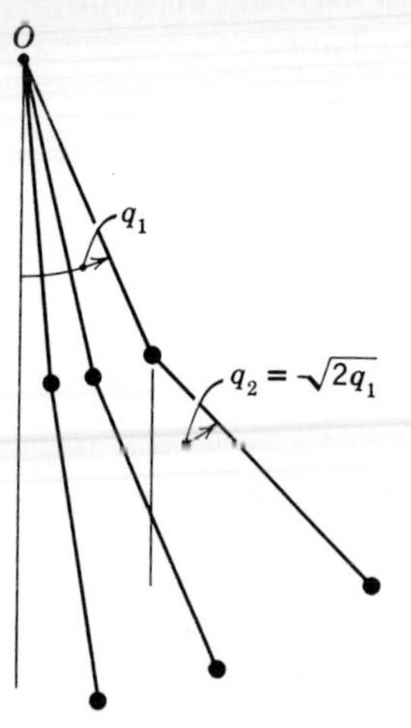

(*a*) Grave mode $q_2 = \sqrt{2}q_1$

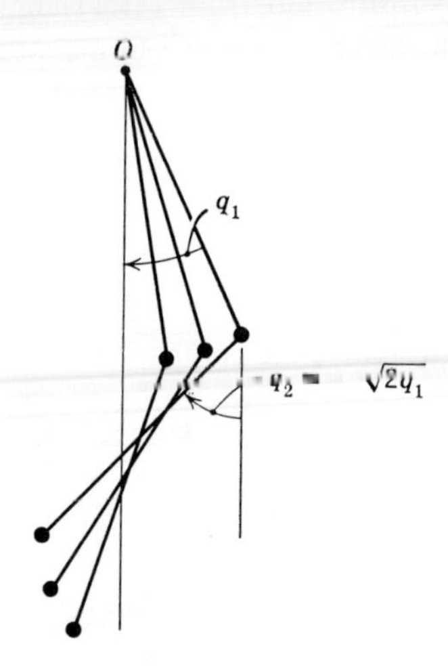

(*b*) Rapid mode $q_2 = -\sqrt{2}q_1$

Fig. 40

are exaggerated. The general motion given by (89.13) is a combination of motions of the two characteristic modes.

One can, of course, get the normal frequencies λ_1, λ_2 directly from the frequency equation (89.9).

If we substitute T and V from (89.10) in the Lagrangean equations 86.12, we get a pair of equations

$$2\ddot{q}^1 + \ddot{q}^2 + \frac{2g}{l} q^1 = 0, \tag{89.15}$$
$$\ddot{q}^1 + \ddot{q}^2 + \frac{g}{l} q^2 = 0,$$

in which the variables $q_1{}^1$ and $q_2{}^2$ are coupled. We assume solutions of (89.15) in the form

$$q^1 = a_1 e^{i\lambda t}, \qquad q^2 = a_2 e^{i\lambda t} \tag{89.16}$$

and determine λ so that equations 89.15 are satisfied. On substituting (89.16) in (89.15) we get two homogeneous equations

$$a_1\left(2\frac{g}{l} - 2\lambda^2\right) + a_2(-\lambda^2) = 0,$$

$$a_1(-\lambda^2) + a_2\left(\frac{g}{l} - \lambda^2\right) = 0,$$

which will have nontrivial solutions for a_1 and a_2 if, and only if,

$$\begin{vmatrix} 2g/i - 2\lambda^2 & -\lambda^2 \\ -\lambda^2 & g/l - \lambda^2 \end{vmatrix} = 0.$$

On expanding this determinant, we find that

$$\lambda^2 = \frac{g}{l}(2 \pm \sqrt{2}),$$

which yields two values $\lambda_1{}^2 = (g/l)(2 - \sqrt{2})$, $\lambda_2{}^2 = (g/l)(2 + \sqrt{2})$ corresponding to the grave and rapid modes found previously. Thus the solution (89.16) can be written

$$q^1 = c_1 e^{i\lambda_2 t} + c_2 e^{i\lambda_1 t},$$
$$q^2 = -\sqrt{2}\, c_1 e^{i\lambda_2 t} + \sqrt{2}\, c_2 e^{i\lambda_1 t},$$

as in (89.13).

Problems

1. Find the normal modes of vibration for the double pendulum in Fig. 39, assuming that $l_1 = l_2$, but $m_1 \neq m_2$.

2. A particle of mass m oscillates about the lowest point of a smooth surface $z = \frac{1}{2}(ax^2 + 2hxy + by^2)$, where the coordinates are orthogonal cartesian and the z-axis is directed vertically up. We suppose that the vertical component of the velocity is small, so that $T = \frac{1}{2}m(\dot{x}^2 + \dot{y}^2)$. The potential $V = mgz = (mg/2)(ax^2 + 2hxy + by^2)$. Obtain equations of motion, determine their solutions in the form $x = a_1 e^{i\lambda t}$, $y = a_2 e^{i\lambda t}$, and conclude that if $V = \min$ at $x = 0$, $y = 0$, then $a > 0$, $b > 0$, $ab - h^2 > 0$.

3. Let the particle in the problem at the end of Sec. 80 be acted on by the force of gravity, so that $F_1 = mga \sin u^1$, $F_2 = 0$. (Note that the work δW done in a small displacement δy^3 is $\delta W = -mg\,\delta y^3 = mga \sin u^1\,\delta u^1$.) Show that the motion, when the particle passes through the highest and lowest points on the sphere, is along an arc of a great circle. A complete discussion of this problem is involved. See P. Appell, *Mécanique rationelle*, **1,** Chapter 13, especially Sec. 277. See also a discussion of the spherical pendulum in J. L. Synge and B. A. Griffith, *Principles of Mechanics*.

4. Let the particle in the preceding problem execute small oscillations about the lower pole of the sphere. Consider projection of this motion on the plane tangent to the pole and discuss the motion.

Hint: Set $u^1 = \pi - (r/a)$, and deduce equations

$$\begin{cases} \ddot{r} + r\dot{u}^2 = -g\dfrac{r}{a}, \\ r\ddot{u} + 2\dot{r}\dot{u} = 0. \end{cases}$$

90. Hamilton's Canonical Equations

Consider a conservative holonomic dynamical system with n degrees of freedom and the integral

$$J = \int_{t_1}^{t_2} L(q, \dot{q})\,dt, \tag{90.1}$$

where $L = T - V$ is the kinetic potential. We saw in Sec. 86 that the system of Euler's equations associated with the variational problem $J =$ extremum consists of a set of n simultaneous second-order ordinary differential equations 86.12, which we write in the form

$$\frac{dL_{\dot{q}^i}}{dt} - L_{q^i} = 0, \qquad (i = 1, 2, \ldots, n), \tag{90.2}$$

by using the subscript notation for partial derivatives of $L(q, \dot{q})$. In a variety of considerations it is convenient to rewrite the system of n Lagrangean equations 90.2 in the form of an equivalent set of $2n$ first-order equations, known as Hamilton's equations.

The function $L(q, \dot{q}) = T(q, \dot{q}) - V(q)$ depends on n generalized coordinates q^i and n generalized velocities $\dot{q}^i$. Instead of the variables $\dot{q}^i$ we can introduce a set of n new variables p_i defined by the relations

$$p_i = L_{\dot{q}^i}(q, \dot{q}), \qquad (i = 1, 2, \ldots, n), \tag{90.3}$$

where we suppose that the system 90.3 is solvable for the $\dot{q}^i$ in terms of the p_i and q^i. This, surely, will be the case if the Jacobian determinant $\left|\dfrac{\partial L_{\dot{q}^i}}{\partial \dot{q}^j}\right| \neq 0$. We next construct a function $H(p, q)$ of the independent variables q and p,

$$H(p, q) = \dot{q}^i p_i - L(q, \dot{q}), \tag{90.4}$$

by expressing the $\dot{q}^i = \dot{q}^i(q, p)$ in the right-hand member of (90.4) in terms of the q^i and p_i with the aid of (90.3).

On differentiating (90.4) with respect to q^j, we get

$$H_{q^j} = \frac{\partial \dot{q}^i}{\partial q^j} p_i - L_{q^j} - L_{\dot{q}^i} \frac{\partial \dot{q}^i}{\partial q^j},$$

and since $p_i = L_{\dot{q}^i}$ by (90.3),

$$H_{q^j} = -L_{q^j}. \tag{90.5}$$

Similarly, we compute

$$H_{p_j} = \dot{q}^j + \frac{\partial \dot{q}^i}{\partial p_j} p_i - L_{\dot{q}^i} \frac{\partial \dot{q}^i}{\partial p_j},$$

which on using (90.3) reduces to

$$H_{p_j} = \dot{q}^j. \tag{90.6}$$

But the Lagrangean equations 90.2 state that

$$\frac{dL_{\dot{q}^i}}{dt} = L_{q^i}$$

and, if we recall the definition 90.3 and formula 90.5, we obtain a set of n first-order equations,

$$\frac{dp_i}{dt} = -H_{q^i}, \qquad (i = 1, \ldots, n), \tag{90.7}$$

which together with the n equations 90.6,

$$\frac{dq^i}{dt} = H_{p_i}, \qquad (i = 1, \ldots, n), \tag*{[90.6]}$$

constitute the system of $2n$ first-order *Hamilton's canonical equations.*

The function $H(p, q)$, known as the *Hamiltonian function*, has an important physical meaning. Since $L = T - V$ and V is a function of the q^i alone, we can rewrite (90.4) as

$$H = \dot{q}^i \frac{\partial L}{\partial \dot{q}^i} - L = \dot{q}^i \frac{\partial T}{\partial \dot{q}^i} - T + V. \tag{90.8}$$

However, $$T = \tfrac{1}{2} a_{ij} \dot{q}^i \dot{q}^j, \; \partial T / \partial \dot{q}^i = a_{ij} \dot{q}^j,$$

so that

$$\dot{q}^i \frac{\partial T}{\partial \dot{q}^i} = a_{ij}\dot{q}^i\dot{q}^j = 2T,$$

and hence (90.8) reduces to

$$H = T + V.$$

Thus H is the total energy of the system.

The variables

$$p_i \equiv \frac{\partial L}{\partial \dot{q}^i} = a_{ij}\dot{q}^j \tag{90.9}$$

are called the *generalized momenta*, and we note that the square of the magnitude of the vector p_i is

$$\begin{aligned} p^2 = a^{ij}p_ip_j &= a^{ij}a_{ik}a_{jl}\dot{q}^k\dot{q}^l \\ &= a_{kl}\dot{q}^k\dot{q}^l = 2T. \end{aligned} \tag{90.10}$$

As an illustration of a simple use of Hamilton's equation, consider a particle of mass m moving under the influence of a central force field with the potential $V(r)$, r being the distance of the particle from the center of attraction. If we choose polar coordinates $r = q^1$, $\theta = q^2$ as our generalized coordinates, then

$$T = \frac{m}{2}[\dot{r}^2 + (r\dot{\theta})^2] = \tfrac{1}{2}a_{ij}\dot{q}^i\dot{q}^j,$$

where

$$(a_{ij}) = \begin{pmatrix} m & 0 \\ 0 & mr^2 \end{pmatrix}.$$

But $H = T + V = \frac{1}{2}a^{ij}p_ip_j + V$, by (90.10), which yields on inserting the values of the a^{ij},

$$H = \frac{p_1^{\,2}}{2m} + \frac{p_2^{\,2}}{2mr^2} + V(r).$$

Thus

$$\frac{\partial H}{\partial r} = \frac{-p_2^{\,2}}{mr^3} + V'(r), \qquad \frac{\partial H}{\partial \theta} = 0, \qquad \frac{\partial H}{\partial p_i} = \frac{p_1}{m}, \qquad \frac{\partial H}{\partial p_2} = \frac{p_2}{mr^2},$$

and hence Hamilton's equations (90.6), (90.7) in this problem are

$$\frac{dr}{dt} = \frac{p_1}{m}, \qquad \frac{d\theta}{dt} = \frac{p_2}{mr^2}, \qquad \frac{dp_1}{dt} = \frac{p_2^{\,2}}{mr^3} - V'(r), \qquad \frac{dp_2}{dt} = 0. \tag{90.11}$$

The last of these equations, combined with the second, yields

$$\frac{d}{dt}(mr^2\dot{\theta}) = 0,$$

which is a statement of Kepler's second law of planetary motion. It is not difficult to show by using the remaining equations in (90.11) that if $V = -m/r$, the orbit is a conic section (cf. Sec. 97).

Problems

1. If a particle of mass m is constrained to move on a smooth surface, show that the system of Hamilton's equations is

$$\frac{du^\alpha}{dt} = \frac{\partial H}{\partial p^\alpha}, \qquad \frac{dp^\alpha}{dt} = -\frac{\partial H}{\partial u^\alpha}, \qquad (\alpha = 1, 2),$$

with $p_\alpha = m a_{\alpha\beta} \dot{u}^\beta$ and $H = (1/2m) a^{\alpha\beta} p_\alpha p_\beta + V$.

2. Show that along the dynamical trajectory $dH/dt = 0$, so that $H = \text{constant}$ is an integral of Hamilton's equations.

3. Show that $\partial L/\partial q^i + \partial H/\partial q^i = 0$.

4. Write Hamilton's canonical equations for Problem 1, Sec. 89.

5. If $T = \frac{1}{2} m(\dot{q})^2$ and $V = k(q)^2$, $k > 0$, show that $H = p^2/2m + m\omega^2(q)^2/2$, where $\omega^2 = k/m$. Deduce that $q = \sqrt{2h/m\omega^2} \sin(\omega t + \alpha)$.

6. Deduce Hamilton's equations from the variational principle $\delta \int L\, dt = 0$. *Hint:* Write L in the form $L = p_i(dq^i/dt) - H(p, q)$, treat the variations of p and q as independent, and show that

$$\int_{t^1}^{t_2} \left[\left(\dot{q}^i - \frac{\partial H}{\partial p_i} \right) \delta p_i - \left(\dot{p}_i + \frac{\partial H}{\partial q^i} \right) \delta q^i \right] dt = 0.$$

91. Newtonian Law of Gravitation

The general formulation of dynamical equations, outlined in the preceding sections, imposes no specific restrictions on the functional form of the fields of force. In various applications of dynamics, including those of astronomy and atomic physics, we are concerned with the behavior of dynamical systems subjected to the action of central fields of force and, in particular, those fields whose intensity varies inversely as the square of the distance of the particles from the center of attraction. The inverse square law of attraction had its origin in Newton's studies of motion of planetary bodies in what he termed[18] the "eccentric conic sections." We state this law as follows:

Two material particles attract each other with a force which is directly proportional to the product of their masses and inversely proportional to the square of the distance between them. The line of action of the force is along the line joining the particles.

Thus the law, when stated in the form of a vector equation, reads

$$\mathbf{F} = \gamma \frac{m_1 m_2}{r_{12}^{\ 3}} \mathbf{r}_{12},$$

[18] Newton's *Principia*, Book I, Sec. III, Propositions 1–17.

where m_1 and m_2 are the masses of the particles and $\mathbf{r}_{12}$ is the vector from P_1 to P_2. The constant of proportionality γ depends on the choice of units; in the cgs system its value is found to be 6.664×10^{-8}, and its physical dimensions are $M^{-1} L^3 T^{-2}$. In our work we shall make $\gamma = 1$, by a suitable choice of units of measure, so that

$$\mathbf{F} = \frac{m_1 m_2}{r_{12}^{\ 3}} \mathbf{r}_{12}. \tag{91.1}$$

We observe first that the law of gravitation 91.1 refers to two particles, and, since in dynamics one usually deals with continuous distributions of matter, it is necessary to generalize it. Thus one can subdivide the bodies into small parts, replace each part by an equivalent material particle, add the forces corresponding to discrete particles, and pass to the limit as the number of subdivisions is increased indefinitely. This procedure for two bodies τ_1 and τ_2 leads to the formula

$$\mathbf{F} = \int_{\tau_1} \int_{\tau_2} \frac{\rho_1 \rho_2}{r_{12}^{\ 3}} \mathbf{r}_{12} \, d\tau_1 \, d\tau_2, \tag{91.2}$$

where $d\tau_1$ and $d\tau_2$ are the volume elements of bodies τ_1 and τ_2, ρ_1 and ρ_2 their density functions, and $\mathbf{r}_{12}$ is the position vector of $d\tau_2$ relative to $d\tau_1$. We shall assume that ρ_1 and ρ_2 are piecewise continuous.

Since two interacting bodies ordinarily give rise not only to resultant forces but also to resultant moments, it is necessary to verify that the generalized law of gravitation 91.2 reduces to the parent law 91.1 and yields no nonvanishing couples when the bodies τ_1 and τ_2 are allowed to shrink to a point.

To show that this is indeed so, we introduce an orthogonal cartesian reference frame Y, and denote the coordinates of points of the bodies τ_1 and τ_2 by $(y_1^{\ i})$ and $(y_2^{\ i})$, respectively (Fig. 41). We replace the distributed mass $\rho_1 \Delta\tau_1$ by the concentrated mass m_1 at $P_1(y_1^{\ 1}, y_1^{\ 2}, y_1^{\ 3})$, and the mass $\rho_2 \Delta\tau_2$ by m_2 at $P(y_2^{\ 1}, y_2^{\ 2}, y_2^{\ 3})$.

In accordance with the law 91.1 we have, for the components of force ΔF^i due to these masses,

$$\Delta F^i = \rho_1 \rho_2 \Delta\tau_1 \Delta\tau_2 \frac{y_2^{\ i} - y_1^{\ i}}{r^3},$$

and for the components of moments[19] ΔL_i, relative to the origin O,

$$\begin{aligned} \Delta L_i &= e_{ijk} y_1^{\ j} \Delta F^k \\ &= e_{ijk} y_1^{\ j} \rho_1 \rho_2 \Delta\tau_1 \Delta\tau_2 \frac{y_2^{\ k} - y_1^{\ k}}{r^3} \end{aligned}$$

[19] We recall that the moment of force $\mathbf{F}$ relative to the origin, acting at a point determined by the position vector $\mathbf{r}$, is $\mathbf{L} = \mathbf{r} \times \mathbf{F}$ or, in terms of components, $L_i = e_{ijk} y^j F^k$.

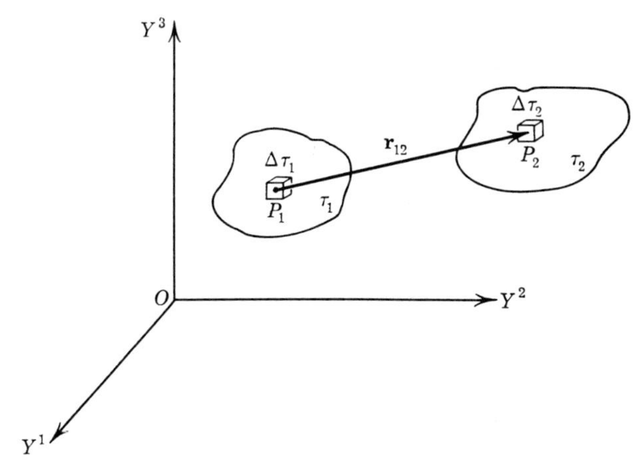

Fig. 41

Adding these vectorially gives the resultant force

$$F^i = \int_{\tau_1}\int_{\tau_2} \frac{\rho_1\rho_2(y_2^{\ i} - y_1^{\ i})}{r^3}\, d\tau_1\, d\tau_2, \tag{91.3}$$

and the resultant moment

$$L_i = \int_{\tau_1}\int_{\tau_2} \rho_1\rho_2 e_{ijk} y_1^{\ j} \frac{y_2^{\ k} - y_1^{\ k}}{r^3}\, d\tau_1\, d\tau_2. \tag{91·4}$$

We prove next that, as τ_1 and τ_2 are allowed to shrink toward P_1 and P_2, respectively (or, even if τ_1 alone is allowed to shrink to zero) the resultant moment L_i tends to zero and equation 91.3 specializes to the law in the form 91.1.

We choose the origin O of the coordinate system at P_1, and let τ_1 shrink toward O and τ_2 toward $P_2(y_2^{\ 1}, y_2^{\ 2}, y_2^{\ 3})$. Since ρ_1 and ρ_2 in equations 91.3 and 91.4 are nonnegative functions, the first mean value theorem for integrals is applicable and we obtain

$$F^i = \left[\frac{y_2^{\ i} - y_1^{\ i}}{r^3}\right]\int_{\tau_1}\int_{\tau_2} \rho_1\rho_2\, d\tau_1\, d\tau_2,$$

and

$$L_i = \left[e_{ijk}\frac{y_1^{\ j}(y_2^{\ k} - y_1^{\ k})}{r^3}\right]\int_{\tau_1}\int_{\tau_2} \rho_1\rho_2\, d\tau_1\, d\tau_2,$$

where brackets denote the values of affected quantities evaluated at certain points in τ_1 and τ_2. As the dimensions of τ_1 are allowed to approach zero, $y_1^{\,i} \to 0$, and hence $L_i \to 0$, whereas the first of the above integrals reduces to

$$F^i = \frac{y_2^{\,i}}{r^3} m_1 m_2.$$

This is precisely the law of gravitation 91.1 for two particles located at (0, 0, 0) and $(y_2^{\,1}, y_2^{\,2}, y_2^{\,3})$.

It follows from the foregoing that a material body interacting with a point mass produces no resultant moment **L**. Moreover, direct calculations show that this is also true when the point mass is replaced by a sphere τ whose density ρ is a continuous function of the radius alone. The resultant force **F**, exerted by the body on the sphere, turns out to be the same as that produced by the body acting on a point mass $m = \int_\tau \rho \, d\tau$, located at the center of the sphere.[20]

Consider next a body τ with piecewise continuous density ρ and let $P(y^1, y^2, y^3)$ be a fixed point either within or outside τ. The *gravitational potential* $V(P)$ at the point P due to τ is defined by the integral

$$V(P) = \int_\tau \frac{\rho(\xi^1, \xi^2, \xi^3)}{r} \, d\tau(\xi), \tag{91.5}$$

where $r = \sqrt{(y^1 - \xi^1)^2 + (y^2 - \xi^2)^2 + (y^3 - \xi^3)^2}$ is the distance between $P(y^1, y^2, y^3)$ and the variable point (ξ^1, ξ^2, ξ^3) associated with the volume element $d\tau(\xi)$ of τ. The integral 91.5, as we shall presently see, defines a differentiable function $V(y^1, y^2, y^3)$ for all locations of P.

If P is outside the body, the integral (91.5) is proper and we can compute as many derivatives of V as desired by differentiating (91.5) under the integral sign with respect to the parameters y^i. In particular,

$$\frac{\partial V}{\partial y^i} = -F_i, \tag{91.6}$$

where the F_i are components of the gravitational force

$$\mathbf{F}(P) = \int_\tau \frac{\rho(\xi)\mathbf{r}}{r^3} \, d\tau \tag{91.7}$$

exerted by the body τ on a particle of unit mass located at $P(y)$.

[20] See, for example, I. S. Sokolnikoff and R. M. Redheffer, *Mathematics of Physics and Modern Engineering*, McGraw-Hill Book Co. (1958), pp. 410–411.

If $P(y)$ is within τ, the integral 91.5 is improper, since $r = 0$ when the variable point (ξ^1, ξ^2, ξ^3) coincides with (y^1, y^2, y^3). However, an improper integral may still be differentiated under the sign when the derived integral is uniformly convergent. In our case the uniform convergence of (91.7) follows from the familiar test on convergence of improper integrals.[21] Moreover, it follows from the uniform convergence of (91.7) that $\mathbf{F}(P)$ is continuous throughout all space.

Although $V(P)$ is of class C^∞ whenever P is *exterior* to τ, more stringent restrictions must be imposed on the continuity of ρ to ensure the existence of second derivatives of $V(P)$ at points *within* τ. It is a fact that if ρ is of class C^1, then the second derivatives of $V(P)$ exist at all interior points of τ. A careful analysis of the difference quotients of the function $\mathbf{F}(P)$ shows, moreover, that[22] $V(P)$ satisfies the Poisson equation

$$\nabla^2 V = -4\pi\rho \tag{91.8}$$

at all points within τ and Laplace's equation

$$\nabla^2 V = 0, \tag{91.9}$$

at points exterior to τ.

Equations 91.8 and 91.9 imply that the second derivatives of $V(P)$ in general suffer discontinuities whenever P crosses the surface Σ of τ. In Sec. 93, we establish the validity of (91.8) and (91.9) with the aid of Gauss' flux theorem. A treatment based on Gauss' flux theorem has the advantages of physical suggestiveness that do not appear in a purely analytic discussion based on the aforementioned study of the difference quotients. However, it imposes quite severe restrictions on the character of regions and surfaces that bound the regions. The Gauss flux theorem is a theorem in the large, and it need not be used to deduce the local results 91.8 and 91.9, which concern the properties of potentials in the neighborhood of a given point.

92. Integral Transformation Theorems

To provide analytic tools for our further study, we translate the well-known integral transformation theorems of Gauss, Green, and Stokes in the language of tensor calculus.

[21] Since, for all values of (ξ^i) in the neighborhood of (y^i), $|r^n \rho(\xi)/r^2| < A$, if $2 < n < 3$, where A is a constant independent of (ξ^i). For a discussion of this test see I. S. Sokolnikoff, *Advanced Calculus*, McGraw-Hill Book Co. (1939), pp. 367–372, or O. D. Kellogg, *Foundations of Potential Theory*, Springer-Verlag (1929), pp. 146–156.

[22] See O. D. Kellogg, *op. cit.*, Chapter 6, pp. 146–156.

Let $\mathbf{F}$ be a vector point function of class C^1 in an open region τ bounded by the regular[23] surface Σ and continuous in the closed region $\Sigma + \tau$. We denote by $\mathbf{n}$ the exterior unit normal to Σ and state the *divergence theorem* in the form

$$\int_\tau \operatorname{div} \mathbf{F}\, d\tau = \int_\Sigma \mathbf{F} \cdot \mathbf{n}\, d\sigma. \tag{92.1}$$

The integral with the subscript τ is evaluated over the volume τ, whereas the integral in the right-hand member of (92.1) measures the flux of the vector quantity $\mathbf{F}$ over the surface Σ.

We recall from elementary vector analysis that, in orthogonal cartesian coordinates, the divergence of $\mathbf{F}$ is given by the formula

$$\operatorname{div} \mathbf{F} = \frac{\partial F^1}{\partial y^1} + \frac{\partial F^2}{\partial y^2} + \frac{\partial F^3}{\partial y^3}. \tag{92.2}$$

If the components of $\mathbf{F}$ relative to an arbitrary curvilinear coordinate system X are denoted by F^i, then the covariant derivative of F^i is

$$F^i_{,j} = \frac{\partial F^i}{\partial x^j} + \begin{Bmatrix} i \\ kj \end{Bmatrix} F^k,$$

and we observe that the invariant $F^i_{,i}$ in cartesian coordinates reduces to the right-hand member of (92.2), and hence it represents the divergence of the vector field $\mathbf{F}$. In addition,

$$\mathbf{F} \cdot \mathbf{n} = g_{ij} F^i n^j = F^i n_i,$$

and hence we can rewrite equation 92.1 in the form

$$\int_\tau F^i_{,i}\, d\tau = \int_\Sigma F^i n_i\, d\sigma. \tag{92.3}$$

From this theorem two other theorems (usually attributed to Green) can be derived easily.

Let $u(x^1, x^2, x^3)$ and $v(x^1, x^2, x^3)$ be two scalar functions of class C^2 in τ and of class C^1 in the closed region $\Sigma + \tau$. We denote the gradients of u and v by u_i and v_i, respectively, so that

$$u_i = \frac{\partial u}{\partial x^i} \quad \text{and} \quad v_i = \frac{\partial v}{\partial x^i}.$$

If we set

$$F_i = u v_i$$

[23] We omit a rather involved discussion of the properties of surfaces to which the divergence theorem is applicable. For a detailed treatment of this consult O. D. Kellogg, *Foundations of Potential Theory*, pp. 97–121.

and form the divergence of F^i, we get

$$F^i_{,i} = g^{ij}F_{i,j} = g^{ij}(uv_{i,j} + v_i u_j).$$

We insert this in equation 92.3 and obtain the desired formula

$$\int_\tau g^{ij}(uv_{i,j} + v_i u_j)\, d\tau = \int_\Sigma uv_i n^i\, d\sigma. \tag{92.4}$$

The invariant $g^{ij}v_{i,j}$ appearing in the left-hand member of equation 92.4, when expressed in cartesian coordinates, is the *Laplacian* of v, $\partial^2 v/\partial y^i\, \partial y^i$, and if we denote the Laplacian operator by the symbol ∇^2, we can write

$$g^{ij}v_{i,j} = \nabla^2 v.$$

Also the inner product $g^{ij}v_i u_j$ can be written as

$$g^{ij}v_i u_j = \nabla u \cdot \nabla v,$$

where we use the customary operator ∇ to denote the gradient.

Hence formula 92.4 can be written in the familiar form

$$\int_\tau u\nabla^2 v\, d\tau = \int_\Sigma u\mathbf{n} \cdot \nabla v\, d\sigma - \int_\tau \nabla u \cdot \nabla v\, d\tau, \tag{92.5}$$

where

$$\mathbf{n} \cdot \nabla v = v_i n^i = \frac{\partial v}{\partial n}.$$

Interchanging u and v in equation 92.5 and subtracting the resulting formula from equation 92.5 yields a *symmetrical form of Green's theorem*

$$\int_\tau (u\nabla^2 v - v\nabla^2 u)\, d\tau = \int_\Sigma \left(u\frac{\partial v}{\partial n} - v\frac{\partial u}{\partial n}\right) d\sigma. \tag{92.6}$$

Theorems stated in equations 92.3, 92.4, 92.5, and 92.6 are, perhaps, the ones most frequently used in mathematical physics.

The Laplacian of v,

$$\nabla^2 v = g^{ij}v_{i,j}, \tag{92.7}$$

when written out explicitly in terms of the Christoffel symbols associated with the curvilinear coordinates x^i covering E_3, is

$$\nabla^2 v = g^{ij}\left(\frac{\partial^2 v}{\partial x^i\, \partial x^j} - \begin{Bmatrix} k \\ ij \end{Bmatrix}\frac{\partial v}{\partial x^k}\right), \tag{92.8}$$

and the divergence of the vector F^i is

$$F^i_{,i} = \frac{\partial F^i}{\partial x^i} + \begin{Bmatrix} i \\ ji \end{Bmatrix} F^j. \tag{92.9}$$

Formulas 92.8 and 92.9 can be written in different forms, which frequently are more convenient in computations. Equation 31.10 yields

[31.10] $$\begin{Bmatrix} i \\ ji \end{Bmatrix} = \frac{\partial}{\partial x^j} \log \sqrt{g},$$

and hence the divergence $F^i_{,i}$, in (92.9), can be written as

$$F^i_{,i} = \frac{\partial F^i}{\partial x^i} + \left(\frac{\partial}{\partial x^j} \log \sqrt{g}\right) F^j,$$

or

(92.10) $$F^i_{,i} = \frac{1}{\sqrt{g}} \frac{\partial(\sqrt{g}\, F^i)}{\partial x^i}.$$

If we set in this formula $F^i = g^{ij}(\partial v/\partial x^j)$, we get

(92.11) $$\nabla^2 v = g^{ij} v_{j,i} = \frac{1}{\sqrt{g}} \frac{\partial(\sqrt{g}\, g^{ij}\, \partial v/\partial x^j)}{\partial x}.$$

We turn next to a consideration of Stokes's theorem which permits us to express certain surface integrals in terms of line integrals.

Let a portion of regular surface Σ be bounded by a closed regular curve C, and let $\mathbf{F}$ be any vector function of class C^1 defined on Σ and on C. The theorem of Stokes states that

(92.12) $$\int_\Sigma \mathbf{n} \cdot \textbf{curl}\, \mathbf{F}\, d\sigma = \int_C \mathbf{F} \cdot \boldsymbol{\lambda}\, ds,$$

where $\boldsymbol{\lambda}$ is the unit tangent vector to C, and curl $\mathbf{F}$ is the vector whose components in orthogonal cartesian coordinates are determined from

(92.13) $$\textbf{curl}\, \mathbf{F} = \begin{vmatrix} \mathbf{e}_1 & \mathbf{e}_2 & \mathbf{e}_3 \\ \dfrac{\partial}{\partial y^1} & \dfrac{\partial}{\partial y^2} & \dfrac{\partial}{\partial y^3} \\ F^1 & F^2 & F^3 \end{vmatrix},$$

the $\mathbf{e}_i$ being the unit base vectors in a cartesian frame. The determinant in 92.13 can be written as a symbolic vector product $\nabla \times \mathbf{F}$.

We consider the covariant derivative $F_{i,j}$ of the vector F_i and form a contravariant vector

(92.14) $$G^i = -\epsilon^{ijk} F_{j,k}.$$

It is readily checked that in cartesian coordinates equation 92.14 reduces to 92.13, and we define the vector $\mathbf{G}$ to be the curl of $\mathbf{F}$.

Since $\mathbf{n} \cdot \mathbf{curl}\, \mathbf{F} = n_i G^i = -\epsilon^{ijk} F_{j,k} n_i$, and the components of the unit tangent vector $\boldsymbol{\lambda}$ are dx^i/ds, we may rewrite equation 92.12 as

$$(92.15) \qquad -\int_\Sigma \epsilon^{ijk} F_{j,k} n_i \, d\sigma = \int_C F_i \frac{dx^i}{ds}\, ds.$$

The integral $\int_C F_i\, dx^i$ is called the *circulation* of $\mathbf{F}$ along the contour C.

Problems

1. Prove that

$$\int_\Sigma v_{,i} n^i \, d\sigma = \int_\tau \nabla^2 v \, d\tau ,$$

where $v_{,i} = \partial v / \partial x^i$ is continuous on Σ and of class C^2 in τ.

2. Show that

(*a*) In plane polar coordinates with $ds^2 = (dr)^2 + r^2(d\theta)^2$,

$$\operatorname{div} \mathbf{F} = \frac{1}{r}\left[\frac{\partial(rF_r)}{\partial r} + \frac{\partial F_\theta}{\partial \theta}\right],$$

$$\nabla^2 v = \frac{1}{r}\left[\frac{\partial}{\partial r}\left(r \frac{\partial v}{\partial r}\right) + \frac{\partial}{\partial \theta}\left(\frac{1}{r}\frac{\partial v}{\partial \theta}\right)\right],$$

where F_r and F_θ are the *physical* components of the vector $\mathbf{F}$, that is,

$$\mathbf{F} = F_r \mathbf{r}_1 + F_\theta \boldsymbol{\theta}_1,$$

where $\mathbf{r}_1$ and $\boldsymbol{\theta}_1$ are unit vectors.

(*b*) In cylindrical coordinates with $ds^2 = (dr)^2 + r^2(d\theta)^2 + (dz)^2$,

$$\operatorname{div} \mathbf{F} = \frac{1}{r}\frac{\partial(rF_r)}{\partial r} + \frac{1}{r}\frac{\partial F_\theta}{\partial \theta} + \frac{\partial F_z}{\partial z},$$

$$\nabla^2 v = \frac{1}{r}\frac{\partial\left(r\dfrac{\partial v}{\partial r}\right)}{\partial r} + \frac{1}{r^2}\frac{\partial^2 v}{\partial \theta^2} + \frac{\partial^2 V}{\partial z^2},$$

where $\mathbf{F} = F_r \mathbf{r}_1 + F_\theta \boldsymbol{\theta}_1 + F_z \mathbf{z}_1$, and $\mathbf{r}_1$, $\boldsymbol{\theta}_1$, $\mathbf{z}_1$ are unit vectors, so that F_r, F_θ, and F_z are the physical components of $\mathbf{F}$.

(*c*) In spherical coordinates with $ds^2 = (dr)^2 + r^2(d\theta)^2 + r^2 \sin^2\theta\, (d\phi)^2$,

$$\operatorname{div} \mathbf{F} = \frac{1}{r^2}\frac{\partial(r^2 F_r)}{\partial r} + \frac{1}{r \sin\theta}\frac{\partial(\sin\theta F_\theta)}{\partial \theta} + \frac{1}{r\sin\theta}\frac{\partial F_\phi}{\partial \phi},$$

$$\nabla^2 v = \frac{1}{r^2}\frac{\partial\left(r^2\dfrac{\partial v}{\partial r}\right)}{\partial r} + \frac{1}{r^2\sin\theta}\frac{\partial\left(\sin\theta\dfrac{\partial v}{\partial \theta}\right)}{\partial \theta} + \frac{1}{r^2\sin^2\theta}\frac{\partial^2 v}{\partial \phi^2},$$

where the *physical* components of $\mathbf{F}$ are F_r, F_θ, F_ϕ, so that $\mathbf{F} = r_1 F_r + \boldsymbol{\theta}_1 F_\theta + \boldsymbol{\phi}_1 F_\phi$, $\mathbf{r}_1$, $\boldsymbol{\theta}_1$, and $\boldsymbol{\phi}_1$ being the unit vectors.

3. Show that, in an orthogonal curvilinear frame X,

$$\operatorname{curl} \mathbf{F} = \frac{1}{\sqrt{g_{11}g_{22}g_{33}}} \begin{vmatrix} \sqrt{g_{11}}\,\mathbf{a}_1 & \sqrt{g_{22}}\,\mathbf{a}_2 & \sqrt{g_{33}}\,\mathbf{a}_3 \\ \dfrac{\partial}{\partial x^1} & \dfrac{\partial}{\partial x^2} & \dfrac{\partial}{\partial x^3} \\ \sqrt{g_{11}}\,F^1 & \sqrt{g_{22}}\,F^2 & \sqrt{g_{33}}\,F^3 \end{vmatrix},$$

where the $\mathbf{a}_i$ are the unit base vectors and $\mathbf{F} = F^1\mathbf{a}_1 + F^2\mathbf{a}_2 + F^3\mathbf{a}_3$.

4. Show that the contravariant components of the curl of a vector $\mathbf{F}$ are:

$$\frac{1}{\sqrt{g}}\left(\frac{\partial F_3}{\partial x^2} - \frac{\partial F_2}{\partial x^3}\right), \quad \frac{1}{\sqrt{g}}\left(\frac{\partial F_1}{\partial x^3} - \frac{\partial F_3}{\partial x^1}\right), \quad \frac{1}{\sqrt{g}}\left(\frac{\partial F_2}{\partial x^1} - \frac{\partial F_1}{\partial x^2}\right).$$

5. Prove that under suitable restrictions on continuity the curl of a gradient vector vanishes identically.

6. In orthogonal curvilinear coordinates,

$$g_{ij} = g^{ij} = 0, \quad i \neq j, \qquad \text{and} \qquad g_{11} = \frac{1}{g^{11}}, \quad g_{22} = \frac{1}{g^{22}}, \quad g_{33} = \frac{1}{g^{33}}.$$

If we set $ds^2 = e_1^2(dx^1)^2 + e_2^2(dx^2)^2 + e_3^2(dx^3)^2$, so that $g_{11} = e_1^2$, $g_{22} = e_2^2$, $g_{33} = e_3^2$, then

$$(a)\ [ij, k] = 0, \qquad \begin{Bmatrix} k \\ i\ j \end{Bmatrix} = 0, \qquad i, j, k \text{ distinct},$$

$$[ij, i] = -[ii, j] = e_i \frac{\partial e_i}{\partial x^j}, \qquad [ii, i] = e_i \frac{\partial e_i}{\partial x^i}, \qquad \begin{Bmatrix} i \\ i\ j \end{Bmatrix} = \frac{\partial \log e_i}{\partial x^j},$$

$$\begin{Bmatrix} j \\ i\ i \end{Bmatrix} = -\frac{e_i}{(e_j)^2}\frac{\partial e_i}{\partial x^j}, \qquad \begin{Bmatrix} i \\ i\ i \end{Bmatrix} = \frac{\partial \log e_i}{\partial x^i}, \qquad \text{(no sums)},$$

$$(b)\ \nabla^2 v = \frac{1}{e_1e_2e_3}\left[\frac{\partial}{\partial x^1}\left(\frac{e_2e_3}{e_1}\frac{\partial v}{\partial x^1}\right) + \frac{\partial}{\partial x^2}\left(\frac{e_3e_1}{e_2}\frac{\partial v}{\partial x^2}\right) + \frac{\partial}{\partial x^3}\left(\frac{e_1e_2}{e_3}\frac{\partial v}{\partial x^3}\right)\right].$$

93. Theorem of Gauss. Solution of Poisson's Equation

In accordance with Newton's law of gravitation, a particle P of mass m exerts on a particle P_1 of unit mass, located at a distance r from P, a force of magnitude $F = m/r^2$. Imagine a closed regular surface Σ drawn around the point P, and let θ denote the angle between the unit exterior normal $\mathbf{n}$ to Σ and the axis of a cone with its vertex at P. This cone subtends an element of surface $d\sigma$. (See Fig. 42.) The flux of the gravitational field produced by m is

$$\int_\Sigma \mathbf{F} \cdot \mathbf{n}\, d\sigma = \int_\Sigma \frac{m \cos\theta}{r^2} \frac{r^2\, d\omega}{\cos\theta},$$

where $d\sigma = r^2\, d\omega/\cos\theta$ and $d\omega$ is the solid angle subtended by $d\sigma$.

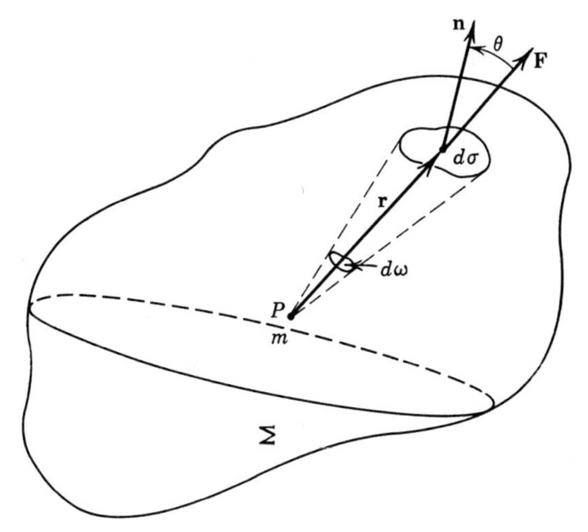

Fig. 42

We thus have

$$\int_\Sigma \mathbf{F} \cdot \mathbf{n} \, d\sigma = \int_\Sigma m \, d\omega = 4\pi m. \tag{93.1}$$

If there are n discrete particles of masses m_i located within Σ, then

$$\mathbf{F} \cdot \mathbf{n} = \sum_{i=1}^{n} \frac{m_i \cos \theta_i}{r_i^2}$$

and the total flux is

$$\int_\Sigma \mathbf{F} \cdot \mathbf{n} \, d\sigma = 4\pi \sum_{i=1}^{n} m_i. \tag{93.2}$$

The result embodied in formula 93.2 can be easily generalized to continuous distributions of matter whenever such distributions nowhere meet the surface Σ. The procedure is a standard one. The contribution to the flux integral from the mass element $\rho \, d\tau$, contained within τ, is

$$\int_\Sigma \mathbf{F} \cdot \mathbf{n} \, d\sigma = \int_\Sigma \frac{\cos \theta \rho \, d\tau}{r^2} \, d\sigma,$$

and the contribution from all masses contained entirely within Σ is

$$\int_\Sigma \mathbf{F} \cdot \mathbf{n} \, d\sigma = \int_\Sigma \left(\int_\tau \frac{\cos \theta \rho \, d\tau}{r^2} \right) d\sigma, \tag{93.3}$$

where $\int_\tau$ denotes the volume integral over all bodies interior to Σ. Since all masses are assumed to be interior to Σ, r never vanishes, so that the integrand in (93.3) is continuous, and hence one can interchange the order of integration to obtain

$$\int_\Sigma \mathbf{F}\cdot\mathbf{n}\,d\sigma = \int_\tau \rho\left(\int_\Sigma \frac{\cos\theta\,d\sigma}{r^2}\right)d\tau. \tag{93.4}$$

But the integral $\int_\Sigma \frac{\cos\theta\,d\sigma}{r^2} = 4\pi$, since it represents the flux due to a unit mass contained within Σ. Hence

$$\int_\Sigma \mathbf{F}\cdot\mathbf{n}\,d\sigma = 4\pi\int_\tau \rho\,d\tau = 4\pi m, \tag{93.5}$$

where m denotes the total mass contained within Σ.

We can now state

GAUSS'S THEOREM. *The integral of the normal component of the gravitational flux computed over a regular surface Σ containing gravitating masses wholly within it is equal to $4\pi m$, where m is the total mass enclosed by Σ.*

This theorem can be extended to situations where Σ intersects the distributed masses with sufficiently smooth density ρ. Let a regular closed surface Σ intersect a distribution of mass with continuously differentiable density ρ. We construct two surfaces Σ' and Σ'' parallel to Σ (cf. Sec. 73) such that Σ' is interior to Σ and Σ'' encloses Σ (Fig. 43). The flux produced by the gravitating masses varies continuously across Σ' and Σ'' when these surfaces, while remaining parallel, are made to approach Σ.

Since Σ'' does not intersect Σ, Gauss's flux theorem can be applied to compute the total flux over Σ'' produced by the masses within Σ. Accordingly,

$$\int_{\Sigma''} (\mathbf{F}\cdot\mathbf{n})_i\,d\sigma = 4\pi m, \tag{93.6}$$

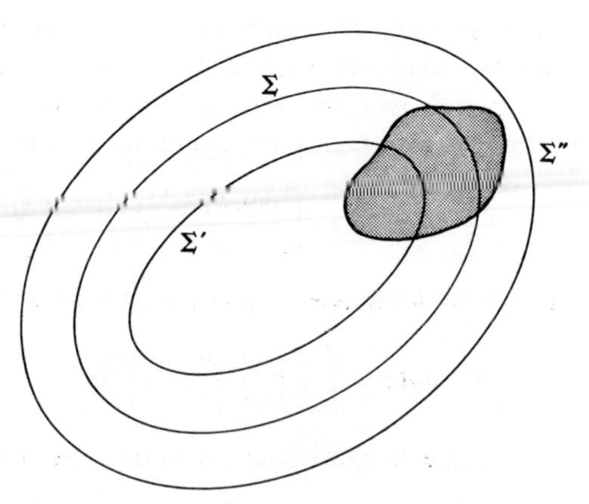

Fig. 43

where m is the total mass within Σ and the subscript i refers to the flux produced by the masses *inside* Σ. On the other hand, the *net* flux over Σ' produced by all masses *outside* Σ is

$$\int_{\Sigma'} (\mathbf{F} \cdot \mathbf{n})_0 \, d\sigma = 0, \tag{93.7}$$

for the flux cone from any point outside Σ cuts Σ' twice.

Now, if we let Σ' and Σ'' approach Σ, the right-hand members in (93.6) and (93.7) do not change, whereas the left-hand member of (93.6) becomes the flux integral over Σ produced by the masses within Σ and the left-hand member of (93.7) represents the flux over Σ due to all masses exterior to Σ. Thus the total flux produced by a distribution of masses within Σ is

$$\int_{\Sigma} (\mathbf{F} \cdot \mathbf{n}) \, d\sigma = 4\pi m = \int_{\tau} 4\pi\rho \, d\tau. \tag{93.8}$$

If we further suppose that $\mathbf{F}$ is continuously differentiable, we can apply the divergence theorem to the surface integral in (93.8), and get

$$\int_{\tau} (\operatorname{div} \mathbf{F} - 4\pi\rho) \, d\tau = 0. \tag{93.9}$$

This relation is true for an arbitrary region τ and, since the integrand in (93.8) is continuous, we conclude that

$$\operatorname{div} \mathbf{F} = 4\pi\rho \qquad \text{throughout } \tau. \tag{93.10}$$

However, formula 91.6 states that $\mathbf{F} = -\nabla V$, and thus (93.10) is equivalent to

$$\nabla^2 V = -4\pi\rho. \tag{93.11}$$

Thus at all points interior to the body τ, the gravitational potential satisfies the Poisson equation. We note in conclusion that formula

$$V(P) = \int_{\tau} \frac{\rho \, d\tau}{r} \tag*{[91.5]}$$

gives a solution of equation 93.11 at all points in τ.

94. Green's Third Identity. Harmonic Functions

Green's symmetrical formula

$$\int_{\tau} (u\nabla^2 v - v\nabla^2 u) \, d\tau = \int_{\Sigma} \left(u \frac{\partial v}{\partial n} - v \frac{\partial u}{\partial n} \right) d\sigma \tag*{[92.6]}$$

is applicable to any pair of functions u, v of class C^2 in the open region τ and of class C^1 in the closed region $\Sigma + \tau$. Let us set $u = 1/r$ and $v = V$, where r is the distance between the points $P(x^1, x^2, x^3)$ and $P_1(\xi^1, \xi^2, \xi^3)$, and V is the gravitational potential of a distribution of mass with continuously differentiable density ρ, so that V is of class C^2 in τ.

Since $1/r$ has a discontinuity at $(x^i) = (\xi^i)$, we delete $P(x)$ from τ by enclosing it by a sphere σ of radius δ and with center of σ at P. Functions $u = 1/r$ and $v = V$ then satisfy the conditions of theorem 92.6 in the region $\tau - \epsilon$ bounded by Σ and σ (Fig. 44). However, in the region $\tau - \epsilon$, $\nabla^2 u = \nabla^2(1/r) \equiv 0$, and formula 92.6 yields

$$(94.1) \quad \int_{\tau-\epsilon} \frac{1}{r} \nabla^2 V \, d\tau = \int_{\Sigma} \left(\frac{1}{r} \frac{\partial V}{\partial n} - V \frac{\partial (1/r)}{\partial n} \right) d\sigma + \int_{\sigma} \left(\frac{1}{r} \frac{\partial V}{\partial n} - V \frac{\partial (1/r)}{\partial n} \right) d\sigma,$$

where n is the unit exterior normal to the surface $\Sigma + \sigma$. Since, however, on σ the normal $\mathbf{n}$ is directed toward P,

$$\begin{aligned}(94.2) \quad \int_{\sigma} \left(\frac{1}{r} \frac{\partial V}{\partial n} - V \frac{\partial (1/r)}{\partial n} \right) d\sigma &= \int_{\sigma} \left(-\frac{1}{r} \frac{\partial V}{\partial r} + V \frac{\partial (1/r)}{\partial r} \right) d\sigma \\ &= \int_{\sigma} \left(-\frac{1}{r} \frac{\partial V}{\partial r} - \frac{V}{r^2} \right) r^2 \, d\omega \\ &= -\int_{\sigma} \left(r \frac{\partial V}{\partial r} + V \right) d\omega \\ &= -\delta \int_{\delta} \left(\frac{\partial V}{\partial r} \right)_{r=\delta} d\omega - 4\pi \bar{V},\end{aligned}$$

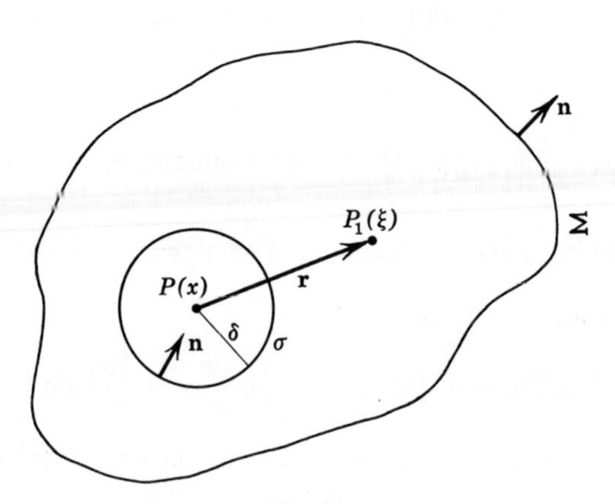

Fig. 44

where $\bar{V}$ is the mean value of V over the sphere σ, and ω denotes the solid angle.

On letting $\delta \to 0$, the right-hand member of (94.2) yields $-4\pi V(P)$, and it follows from (94.1) that

$$V(P) = \frac{1}{4\pi}\int_\tau -\frac{\nabla^2 V}{r}\,d\tau + \frac{1}{4\pi}\int_\Sigma \frac{\partial V}{\partial n}\frac{1}{r}\,d\sigma - \frac{1}{4\pi}\int_\Sigma V\frac{\partial(1/r)}{\partial n}\,d\sigma. \tag{94.3}$$

The important formula 94.3, known as *Green's third identity*, states that every function V of class C^1 in $\Sigma + \tau$ and of class C^2 in τ, can be represented as the sum of three integrals appearing in (94.3). If $V(P)$ is *regular at infinity*, that is, if for sufficiently large values of r, V is such that

$$|V| \le \frac{m}{r} \quad \text{and} \quad \left|\frac{\partial V}{\partial r}\right| \le \frac{m}{r^2}, \tag{94.4}$$

where m is a constant independent of r, then on extending the integration in (94.3) over all space we get

$$V(P) = \frac{1}{4\pi}\int_\infty -\frac{\nabla^2 V}{r}\,d\tau, \tag{94.5}$$

provided that this volume integral converges. The surface integrals in (94.3), when extended over all space, vanish by virtue of the regularity conditions 94.4.

At all points not occupied by matter (that is, where $\rho = 0$), the gravitational potential V satisfies Laplace's equation

$$\nabla^2 V = 0. \tag{94.6}$$

A function satisfying equation 94.6 in a given region is said to be *harmonic* in that region. If V is harmonic in the region τ, formula 94.3 reduces to

$$V(P) = \frac{1}{4\pi}\int_\Sigma \frac{1}{r}\frac{\partial V}{\partial n}\,d\sigma - \frac{1}{4\pi}\int_\Sigma V\frac{\partial(1/r)}{\partial n}\,d\sigma, \tag{94.7}$$

so that the values of V are completely determined in τ when the values of V and of its normal derivative $\partial V/\partial n$ are known on Σ. However, these surface values cannot be specified independently of one another, and we shall see that the specification of the values of V alone on Σ fully determines $V(P)$ at all points of τ. On the other hand, the specification of $\partial V/\partial n$ on Σ determines $V(P)$ in τ to within an arbitrary constant, provided that

$$\int_\Sigma \frac{\partial V}{\partial n}\,d\sigma = 0. \tag{94.8}$$

The condition 94.8 follows directly from the formula 92.6 on setting $u = 1$ and $v = V$. *It is a necessary condition satisfied by every harmonic function.*

If Σ in (94.7) is the surface of a sphere of radius R, with center at P, then $[\partial(1/r)]/\partial n = [\partial(1/r)]/\partial r = -1/r^2$ and (94.7) gives

$$V(P) = \frac{1}{4\pi R^2} \int_\Sigma V \, d\sigma, \tag{94.9}$$

when we note the condition 94.8. Formula 94.9 states an important property of harmonic functions: *The value of a harmonic function V at the center of a sphere is equal to the mean value of V over the surface of that sphere.* This property enables us to prove the following basic theorem on harmonic functions.

THEOREM. *A function V harmonic in a closed regular region $\Sigma + \tau$ assumes its maximum and minimum values on the boundary Σ of τ, with the single exception when $V =$ constant throughout τ.*

To prove this theorem assume that V takes on its maximum (or minimum) V_0 at some interior point P of τ. We construct a sphere S in τ with center at P and of radius R, then

$$V_0(P) = \frac{\int_S V \, d\sigma}{4\pi R^2}$$

by (94.9). But the right-hand member of this expression is the average value $\bar{V}$ of V over S, and the average value $\bar{V}$ can equal the maximum V_0 only if $V = V_0$ on S. Furthermore, since R is arbitrary, we conclude that $V = V_0$ at every interior point of S. To show that V has the same constant value V_0 at every point Q of τ, we connect P and Q by a curve C of finite length and cover it by a sequence of overlapping spheres with centers on C. Within each sphere of this sequence, V has the same constant value V_0 and hence $V(Q) = V_0$. Thus, unless $V =$ constant throughout τ, it takes on its extreme values on the boundary Σ.

The determination of a harmonic function V in τ from the specified values of V on the boundary Σ of τ is known as the *Problem of Dirichlet*. If τ is a finite region, we have an *interior problem* and when τ is an infinite region bounded by a closed surface Σ, we have an *exterior problem* of Dirichlet.

It is easy to prove that the interior problem of Dirichlet for a regular region $\Sigma + \tau$ does not have more than one solution. For, let there be two functions V_1 and V_2, harmonic in τ and which assume the same values on the boundary Σ. But $V \equiv V_1 - V_2$ is also harmonic, and it assumes zero values on Σ. This implies, however, that $V = 0$ throughout τ, since otherwise V would have to take on its positive maximum, or a negative minimum, in the interior. In the same way we can prove the uniqueness of solution of the exterior problem of Dirichlet if we suppose that V is regular at infinity.

The determination of a harmonic function V in τ which satisfies on the boundary Σ of τ, the condition

$$\frac{\partial V}{\partial n} = f(P), \quad \text{with} \quad \int_\Sigma f(P)\, d\sigma = 0, \tag{94.10}$$

is called the *Problem of Neumann.*

Since $V =$ constant is a harmonic function that satisfies the condition $\partial V/\partial n = 0$ on Σ, we conclude that the solution of the Neumann problem (if it exists) is determined to within an arbitrary constant. It is possible to prove, although the proof is by no means easy, that the Dirichlet and Neumann problems are solvable for finite regular regions when the specified values on the boundary are continuous.[24]

Problem

Show that formula 94.7 is valid in an infinite region τ exterior to a closed surface Σ whenever V is regular at infinity. *Hint:* Apply formula 94.7 to a finite region bounded by Σ and by a sphere S of radius R so large that S encloses Σ.

95. Functions of Green and Neumann

We have just shown that the solution of the interior problem of Dirichlet in Laplace's equation

$$\begin{aligned} \nabla^2 u &= 0 \qquad \text{in } \tau, \\ u &= f(P) \qquad \text{on } \Sigma, \end{aligned} \tag{95.1}$$

when it exists, is necessarily unique. Also the solution of the Neumann interior problem

$$\begin{aligned} \nabla^2 v &= 0 \qquad \text{in } \tau, \\ \frac{\partial v}{\partial n} &= g(P) \qquad \text{on } \Sigma, \end{aligned} \tag{95.2}$$

with $\int g(P)\, d\sigma = 0$, is determined to within an arbitrary constant when $g(P)$ is continuous. To make the solution of the Neumann problem unique, we adjoin to (95.2) the *normalizing condition*

$$\int_\Sigma v\, d\sigma = 0. \tag{95.3}$$

When Laplace's equations in (95.1) and (95.2) are replaced by the Poisson equations, we have the Dirichlet and Neumann problems in Poisson's equation.

[24] See O. D. Kellogg, *op. cit.*, p. 311.

Formula 94.7 is not directly applicable to the solution of problems 95.1 and 95.2, since it requires the knowledge of the values of the function *and* of its normal derivative on Σ. We show next how this difficulty can be avoided by introducing special functions that depend only on the shape of the region and not on the assigned boundary values $f(P)$ and $g(P)$. We begin with the Dirichlet problem.

Let $P(x)$ and $P'(\xi)$ be a fixed point and a variable point, respectively, in τ (Fig. 45). We construct a function $G(P, P')$ with the following properties:

$$(a) \qquad G(P, P') = \frac{1}{r} + w(P'),$$

where $r = \overline{PP'}$ and $W(P')$ is harmonic in τ.

$$(b) \qquad G(P, P') = 0 \qquad \text{on } \Sigma.$$

The condition (*b*) requires that

$$w(P') = -\frac{1}{r} \qquad \text{on } \Sigma,$$

so that $W(P')$, and hence $G(P, P')$, is uniquely determined by properties (*a*) and (*b*). We call $G(P, P')$ *Green's function for the region* τ.

We show next how Green's function can be used to construct an explicit integral formula solving the Dirichlet problem in Poisson's equation

$$(95.4) \qquad \begin{aligned} \nabla^2 V &= -4\pi\rho \qquad \text{in } \tau, \\ V &= f(P) \qquad \text{on } \Sigma. \end{aligned}$$

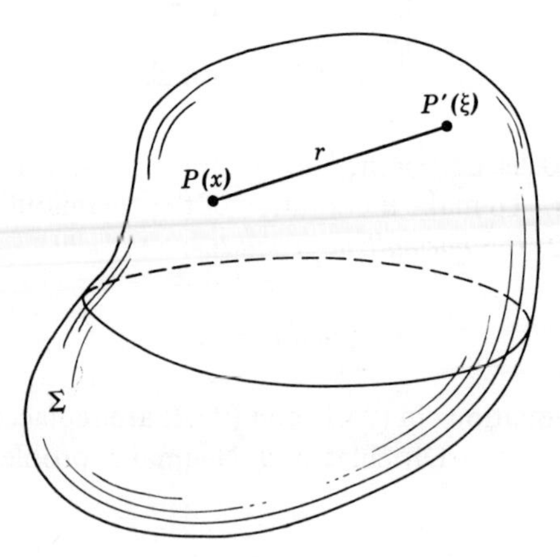

Fig. 45

The integral formula will include the solution of the boundary value problem 95.1 as a special case.

Green's symmetrical formula,

$$[92.6] \qquad \int_\tau (u\nabla^2 v - v\nabla^2 u)\, d\tau = \int_\Sigma \left(u\,\frac{\partial v}{\partial n} - v\,\frac{\partial u}{\partial n} \right) d\sigma,$$

cannot be applied to $u = G(P, P')$, $v = V$, since $G(P, P') \to \infty$ as $P' \to P$. If, however, we delete the point P by enclosing it in a sphere σ of radius δ, as in Fig. 44, the formula 92.6 is valid in the region $\tau - \epsilon$ bounded by Σ and σ. We can write

$$(95.5) \quad \int_{\tau-\epsilon} (G\nabla^2 V - V\nabla^2 G)\, d\tau = \int_\Sigma \left(G\frac{\partial V}{\partial n} - V\frac{\partial G}{\partial n} \right) d\sigma + \int_\sigma \left(G\frac{\partial V}{\partial n} - V\frac{\partial G}{\partial n} \right) d\sigma.$$

But in $\tau - \epsilon$, $G \equiv 1/r + w$ is harmonic, so that $\nabla^2 G = 0$ and $G = 0$ on Σ. Also $\nabla^2 V = -4\pi\rho$ by (95.4), so that (95.5) reduces to

$$(95.6) \quad -4\pi \int_{\tau-\epsilon} G\rho(\xi)\, d\tau = -\int_\Sigma V\frac{\partial G}{\partial n}\, d\sigma - \int_\sigma \left(\frac{1}{r} + w \right) \frac{\partial V}{\partial r}\, d\sigma + \int_\sigma V\frac{\partial(w + 1/r)}{\partial r}\, d\sigma.$$

In writing (95.6) we observed that, since n is an exterior normal, $\partial V/\partial n = -\partial V/\partial r$ and $\partial G/\partial n = -\partial G/\partial r$ on σ.

Since $\partial V/\partial r$ and w are continuous on σ and $d\sigma = r^2\, d\omega$, where $d\omega$ is an element of solid angle, it is obvious that the second integral on the right in (95.6) tends to zero as $\delta \to 0$. Similarly,

$$\int_\sigma V\frac{\partial w}{\partial r}\, d\sigma = 0 \qquad \text{as } \delta \to 0,$$

whereas

$$\int_\sigma \frac{V\partial(1/r)}{\partial r}\, r^2\, d\omega = -4\pi V(P) \qquad \text{as } \delta \to 0.$$

Accordingly, on letting $\delta \to 0$ in (95.6), we get

$$(95.7) \qquad 4\pi V(P) = \int_\tau 4\pi G\rho\, d\tau - \int_\Sigma V\frac{\partial G}{\partial n}\, d\sigma,$$

which is the desired solution of the problem 95.4. If we set $\rho = 0$, we get the solution of the corresponding problem in Laplace's equation,

$$V(P) = -\frac{1}{4\pi}\int_\Sigma V\frac{\partial G}{\partial n}\,d\sigma. \tag{95.8}$$

To apply this formula we must first obtain Green's function G for the region τ, that is, we must solve the special Dirichlet problem

$$\nabla^2 w = 0 \qquad \text{in } \tau,$$

$$w = -\frac{1}{r} \qquad \text{on } \Sigma.$$

Similar considerations are applicable to the Neumann problem 95.2. We introduce the *Neumann function*

$$N(P, P') = \frac{1}{r} + w(P'),$$

where $w(P')$ is harmonic in τ and satisfies on the boundary Σ of τ the condition[25]

$$\frac{\partial w}{\partial n} = -\frac{\partial}{\partial n}\left(\frac{1}{r}\right) + \text{constant}.$$

Computations entirely similar to those carried out previously for the Dirichlet problem yield for the boundary value problem 95.2, the formula

$$V(P) = \frac{1}{4\pi}\int_\Sigma gN\,d\sigma.$$

Physically, Green's function $G(P, P')$ can be interpreted as the electrostatic potential in the interior of a grounded conducting surface Σ produced by a unit charge at the point P. The potential produced by a unit charge alone is $1/r$, and $w(P')$ represents the potential produced by the induced surface charges on Σ. Since Σ is grounded, $G(P, P') = 1/r + w(P') = 0$ on Σ. The Neumann function can be interpreted as steady heat flow from a source of strength 4π placed at P, when the heat flows across the surface Σ at a uniform rate.

96. Green's Functions for Semi-infinite Space and Spherical Regions

A physical interpretation of Green's function given in Sec. 95 enables us to construct Green's functions for the half-space $z \geq 0$ and for the regions interior and exterior to the sphere.

[25] To make w unique we can normalize it by requiring $\int_\Sigma w\,d\sigma = 0$.

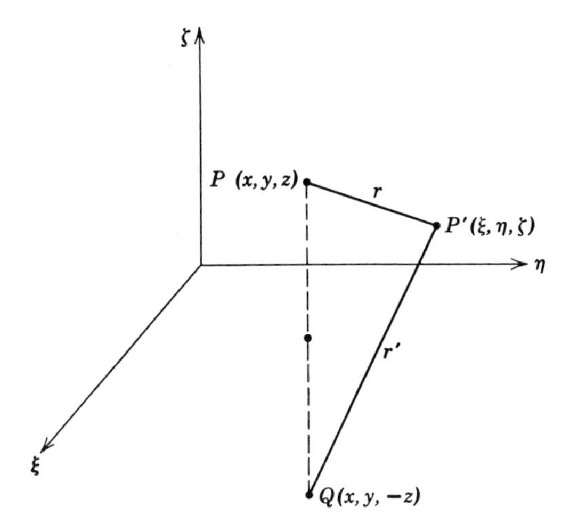

Fig. 46

When a positive unit charge is placed at $P(x, y, z)$ (Fig. 46) and a negative unit charge at the mirror image $Q(x, y, -z)$, the electrostatic potential G produced by these charges at $P'(\xi, \eta, \zeta)$ is

$$G = \frac{1}{r} - \frac{1}{r'}, \tag{96.1}$$

where $r = \overline{PP'} = \sqrt{(\xi - x)^2 + (\eta - y)^2 + (\zeta - z)^2}$

and

$$r' = \overline{QP'} = \sqrt{(\xi - x)^2 + (\eta - y)^2 + (\zeta + z)^2}.$$

Obviously, $G = 0$ on the plane $z = 0$, and since $w(P') = -1/r'$ is harmonic for $z > 0$, equation 96.1 gives the desired Green's function for the region $z \geq 0$.

On the plane $z = 0$,

$$\frac{\partial G}{\partial n} = -\left.\frac{\partial G}{\partial \zeta}\right|_{\zeta=0} = \left[\frac{1}{r^2}\frac{\partial r}{\partial \zeta} - \frac{1}{r'^2}\frac{\partial r'}{\partial \zeta}\right]_{\zeta=0},$$

and after performing simple calculations and substituting in formula 95.8, we obtain the solution of the Dirichlet problem 95.1 for the region $z > 0$ in the form of Poisson's integral

$$V(P) = \frac{z}{2\pi}\int_{-\infty}^{\infty}\int_{-\infty}^{\infty} \frac{f(\xi, \eta)\, d\xi\, d\eta}{[(\xi - x)^2 + (\eta - y)^2 + z^2]^{3/2}}. \tag{96.2}$$

The specified values $f(\xi, \eta)$ of V on $z = 0$, must, clearly, be such that (96.2) has a meaning.

A similar procedure enables us to construct Green's function for the spherical region $x^2 + y^2 + z^2 \leq R^2$ and obtain the solution of the Dirichlet problem 95.1 for the sphere.

We take $P(x, y, z)$ in the interior of the sphere S of radius R (Fig. 47) and construct the image Q of P with respect to S, so that $\overline{OP} \cdot \overline{OQ} = R^2$. Let P' be a variable point in S. When P' is on S, similar triangles $OP''P$ and $OP''Q$ yield the relation

$$\frac{P''Q}{P''P} = \frac{OP''}{OP}, \qquad \text{or} \qquad \frac{r'}{r} = \frac{R}{\rho},$$

where $\rho = \overline{OP}$. Thus

$$\frac{1}{r} = \frac{R}{\rho}\frac{1}{r'}, \qquad \text{on } S.$$

If, for any interior point P', we define

$$G(P, P') = \frac{1}{r} - \frac{R}{\rho}\frac{1}{r'}, \tag{96.3}$$

then (96.3) gives the desired Green's function, since $w = -(R/\rho)(1/r')$, is harmonic in the interior of S and $G(P, P') = 0$ on S.

A simple calculation of dG/dn from (96.3) gives

$$\frac{dG}{dn} = -\frac{R^2 - \rho^2}{Rr^3} \qquad \text{on } S,$$

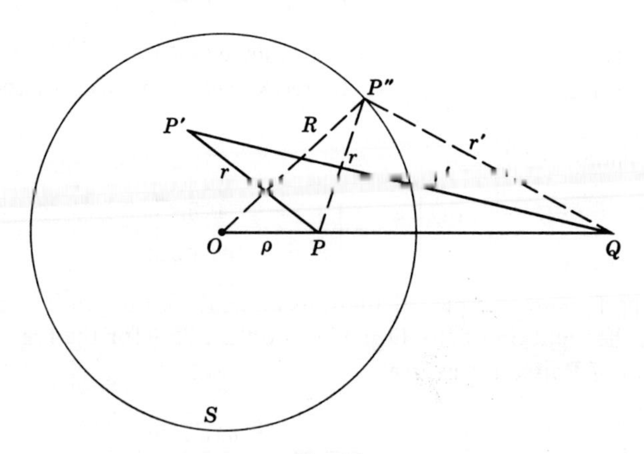

Fig. 47

and formula 95.8 yields the solution of the Dirichlet problem 95.1 for the sphere in the form of the Poisson integral

$$V(P) = \frac{1}{4\pi}\int_S f(P')\frac{R^2 - \rho^2}{Rr^3}\,d\sigma. \tag{96.4}$$

This integral is usually written in spherical coordinates (ρ, θ, ϕ) as

$$V(\rho, \theta, \phi) = \frac{1}{4\pi}\int_0^{\pi} R\sin\theta'\,d\theta' \int_0^{2\pi} \frac{(R^2 - \rho^2)f(\theta', \phi')\,d\phi'}{(R^2 - 2\rho R\cos\gamma + \rho^2)^{3/2}} \tag{96.5}$$

where $\cos\gamma = \cos\theta\cos\theta' + \sin\theta\sin\theta'\cos(\phi' - \phi)$, θ being the co-latitude and ϕ the longitude of P.

Green's function for the region $x^2 + y^2 + z^2 \geq R$ is obtained from (96.3) by interchanging the roles of P and Q.

Problems

1. Show by using (96.4) that for every position of P in the interior of the sphere

$$\frac{1}{4\pi R}\int_S \frac{R^2 - \rho^2}{r^3}\,d\sigma = 1.$$

Hint: Take the z-axis along $\overline{OP}$, so that $r^2 = R^2 - 2R\rho\cos\theta + \rho^2$. For a fixed position of P, ρ is fixed. Express $d\sigma = R^2\sin\theta\,d\theta\,d\phi$ in spherical coordinates and evaluate the integral.

2. Show that the solution of the exterior problem 95.1 for the sphere S is given by

$$V(P) = \frac{1}{4\pi R}\int_S f(P')\frac{\rho^2 - R^2}{r^3}\,d\delta,$$

where $\rho = \overline{OP} > R$.

3. Deduce (96.5) from (96.4). *Hint:* Let γ be the angle between $\overline{OP}$ and $\overline{OP'}$ when P' is on S.

97. The Problem of Two Bodies

The problem of two bodies can be stated as follows: *Given a system of two particles interacting in accordance with the law of universal gravitation, what is the trajectory of the system?* This problem was solved by Newton in the *Principia*, Book I, Sec. III. It lies at the basis of all considerations in astronomy.

Since there is no particular advantage in using general curvilinear coordinates in specific problems, we refer our system to a set of orthogonal cartesian axes. We denote the coordinates of mass points m_1, m_2 (at any given instant of time t) by $(x_1^{\,1}, x_1^{\,2}, x_1^{\,3})$ and $(x_2^{\,1}, x_2^{\,2}, x_2^{\,3})$ (Fig. 48). We also introduce another cartesian reference frame Y moving with the mass m_1 in such a way that m_1 is always at the origin O of the Y-system, and the axes Y^i always remain parallel to the axes X^i. The coordinates of the mass point m_2, relative to the Y-axes, are denoted by y^i, and we have the relations

$$y^i = x_2^{\,i} - x_1^{\,i}, \qquad (i = 1, 2, 3). \tag{97.1}$$

We choose the coordinates y^i of the mass m_2 as three of our generalized coordinates, and for the remaining three generalized coordinates we take those of the center of mass of the system. Thus

$$u^i = \frac{m_1 x_1^{\,i} + m_2 x_2^{\,i}}{m_1 + m_2}, \qquad (i = 1, 2, 3). \tag{97.2}$$

Clearly, the u^i lie on the line joining the points $(x_1^{\,i})$ and $(x_2^{\,i})$, and our choice of the generalized coordinates is then as follows:

$$q^1 = y^1, \quad q^2 = y^2, \quad q^3 = y^3, \quad q^4 = u^1, \quad q^5 = u^2, \quad q^6 = u^3.$$

If we solve equations 97.1 and 97.2 for the $x_1^{\,i}$ and $x_2^{\,i}$, we obtain

$$\begin{aligned} x_1^{\,i} &= u^i - \frac{m_2}{m_1 + m_2}\, y^i, \\ x_2^{\,i} &= u^i + \frac{m_1}{m_1 + m_2}\, y^i, \end{aligned} \tag{97.3}$$

and these equations enable us to determine the positional coordinates x^i in terms of the generalized coordinates q^i.

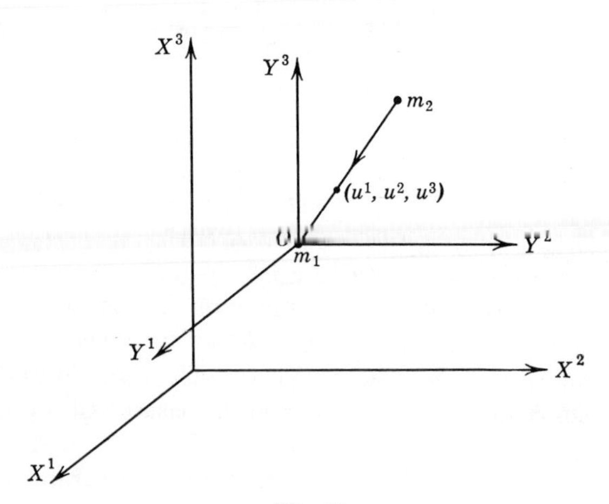

Fig. 48

This particular choice of generalized coordinates is made with a view toward obtaining a simple expression for the potential energy V of our system of particles. Indeed, since the magnitude of the force of attraction **F** is given by $F = m_1m_2/r^2$, where r is the distance between the particles, the potential energy V is

$$V = \frac{m_1m_2}{r} = \frac{m_1m_2}{[(x_1^{\ 1} - x_2^{\ 1})^2 + (x_1^{\ 2} - x_2^{\ 2})^2 + (x_1^{\ 3} - x_2^{\ 3})^2]^{1/2}}$$

and it follows from (97.1) that the coordinates u^i do not appear in V, so that V is a function of y^1, y^2, and y^3.

We recall the Lagrangean equations

$$\frac{d}{dt}\left(\frac{\partial T}{\partial \dot{q}^i}\right) - \frac{\partial T}{\partial q^i} = -\frac{\partial V}{\partial q^i}, \tag{86.11}$$

and compute

$$\begin{aligned} T &= \tfrac{1}{2}m_1\dot{x}_1^{\ i}\dot{x}_1^{\ i} + \tfrac{1}{2}m_2\dot{x}_2^{\ i}\dot{x}_2^{\ i} \\ &= \frac{1}{2}(m_1 + m_2)\dot{u}^i\dot{u}^i + \frac{1}{2}\frac{m_1m_2}{m_1 + m_2}\dot{y}^i\dot{y}^i. \end{aligned}$$

Since $\partial V/\partial q^i = 0$, for $i = 4, 5, 6$, an easy calculation makes equations 86.11 reduce to

$$\begin{aligned} \frac{m_1m_2}{m_1 + m_2}\ddot{y}^i &= -\frac{\partial V}{\partial y^i}, \qquad (i = 1, 2, 3), \\ \ddot{u}^i &= 0, \qquad (i = 1, 2, 3). \end{aligned} \tag{97.4}$$

Equations 97.4 are the differential equations characterizing the motion of our system. We note first that the motion of the mass m_2 relative to m_1 is the same as though the mass m_1 were fixed and m_2 attracted toward it with a force whose potential is $[(m_1 + m_2)/m_1]V$. This follows at once from the first three of equations 97.4 if we rewrite them in the form

$$m_2\ddot{y}^i = -\frac{m_1 + m_2}{m_1}\frac{\partial V}{\partial y^i}. \tag{97.5}$$

Thus our problem is reduced to a study of motion under the action of central forces. The second set of equations 97.4 states that the center of mass moves in a straight line with constant velocity.

We shall carry out the integration of equations 97.4 under the assumption that m_1 (the mass of the sun) is much larger than m_2 (the mass of the earth). If $m_1 \gg m_2$ the center of mass u^i will lie very close to the mass m_1 and hence the coordinates u^i will nearly coincide with those of the mass m_1. Thus $x_1^{\ i} \doteq u^i$, and from the second set of equations 97.4 we conclude

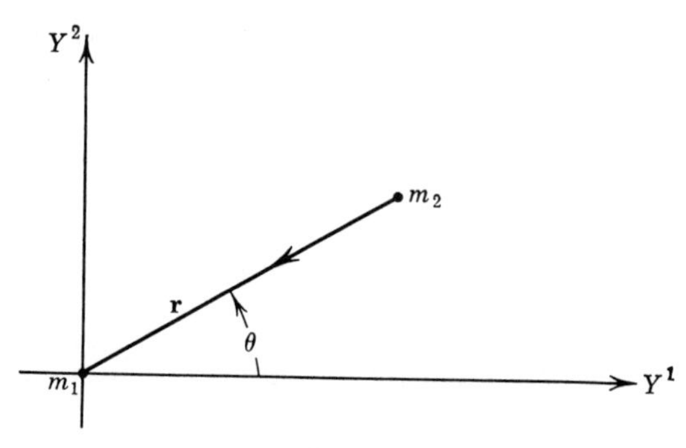

Fig. 49

that m_1 moves through space with constant velocity. Accordingly, we need to examine only the motion of mass m_2 relative to m_1.

If $m_1 \gg m_2$,

$$\frac{m_1 + m_2}{m_1} \doteq 1,$$

and equations 97.5 become

$$m_2 \ddot{y}^i = -\frac{\partial V}{\partial y^i} \quad \text{(approximately).}$$

Let us suppose that our coordinate axes are so oriented that the motion of the mass m_2 relative to m_1 initially is in the Y^1Y^2-plane. Then, since the force field is central, the motion will remain in this plane, for there is no component of force at right angles to the plane. Let r and θ (Fig. 49) be the polar coordinates of mass m_2, where

$$\begin{cases} y^1 = r\cos\theta, \\ y^2 = r\sin\theta; \end{cases}$$

then the kinetic energy of mass m_2 is

$$\begin{aligned} T &= \tfrac{1}{2}m_2[(\dot{y}_1)^2 + (\dot{y}_2)^2] \\ &= \tfrac{1}{2}m_2(\dot{r}^2 + r^2\dot{\theta}^2). \end{aligned}$$

Using this expression for T, and $V = -m_1m_2/r$, in the Lagrangean equations 86.11, with $q^1 = r$ and $q^2 = \theta$, we get[26]

$$\begin{cases} m_2\ddot{r} - m_2r\dot{\theta}^2 = -\dfrac{m_1m_2}{r^2}, \\ \dfrac{d}{dt}(m_2r^2\dot{\theta}) = 0, \end{cases}$$

[26] We consider the force directed *from* m_2 to m_1.

or

$$\begin{cases} \ddot{r} - r\dot{\theta}^2 + \dfrac{m_1}{r^2} = 0, \\ r^2\dot{\theta} = h, \end{cases} \tag{97.6}$$

where h is a constant of integration.

Equations 97.6 are simultaneous ordinary differential equations for the determination of the trajectory. The second of these states that the sectorial velocities are constant. This is one of the Kepler laws.[27] We can use the relation $r^2\dot{\theta} = h$ to determine the time required to describe the orbit.

If $h \neq 0$, so that the trajectory is not a straight line, we can eliminate the time parameter t by noting that $r^2\, d\theta = h\, dt$, or

$$t = \frac{1}{h}\int_0^\theta r^2\, d\theta.$$

Since $df/dt = df/d\theta \cdot d\theta/dt$, we have the relation $d/dt = (h/r^2)(d/d\theta)$, and, making use of this in the first equation in (97.6), we get

$$\frac{h}{r^2}\frac{d}{d\theta}\left(\frac{h}{r^2}\frac{dr}{d\theta}\right) - r\frac{h^2}{r^4} + \frac{m_1}{r^2} = 0,$$

or multiplying by r^2,

$$h\frac{d}{d\theta}\left(\frac{h}{r^2}\frac{dr}{d\theta}\right) - \frac{h^2}{r} + m_1 = 0. \tag{97.7}$$

If we further change the dependent variable r in (97.7) by setting $u = 1/r$, we get a simple second-order linear equation

$$\frac{d^2u}{d\theta^2} + u = \frac{m_1}{h^2},$$

whose solution is

$$u = \frac{1}{l}[1 - e\cos(\theta - \alpha)],$$

or

$$r = \frac{l}{1 - e\cos(\theta - \alpha)}, \tag{97.8}$$

where $l \equiv h^2/m_1$, and α and e are constants of integration.

[27] See also an illustrative example at the end of Sec. 90.

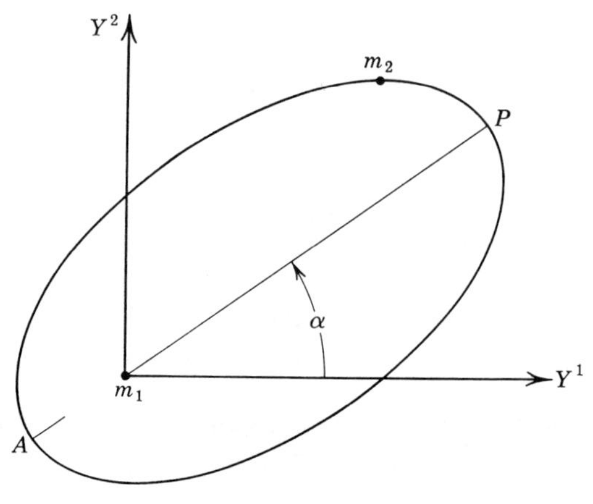

Fig. 50

We thus see that the orbit is a conic section (Fig. 50) whose eccentricity is e, with the position of the apse line determined by α. The constant α is known as the *perihelion constant.* We shall not go to the trouble of determining these constants[28] in terms of the initial position and velocity of mass m_2, since the main object of this section is to obtain formula 97.8 for the purpose of comparing it with the corresponding equation of the orbit in the relativistic dynamics.

[28] See P. Appell, *Mecanique rationelle*, vol. 1, Chapter 11, and J. L. Synge and B. A. Griffith, *Principles of Mechanics* (1959), pp. 160–169.

5

RELATIVISTIC MECHANICS

98. Invariance of Physical Laws

The formulation of the fundamental laws of classical mechanics in the preceding chapter is based on the hypothesis that physical phenomena take place in a three-dimensional Euclidean space. It is also assumed that these phenomena can be ordered in the one-dimensional continuum of the time variable t. The time variable t is regarded to be independent not only of the space variables x^i but also of the possible motion of the space reference systems. The mass m of a body is likewise supposed to be independent of the motion of reference systems, and, in particular, it is invariant with respect to a group of *Galilean transformations* of coordinates. By a Galilean transformation we mean a transformation that represents a translation of one coordinate system relative to another with constant velocity. Thus, if Y is a given cartesian frame, then a Galilean transformation of this frame has the form

$$\bar{y}^i = y^i + u^i t, \qquad (i = 1, 2, 3), \tag{98.1}$$

where u^i is a constant vector representing the velocity of the origin of the Y-system relative to the cartesian system $\bar{Y}$. It is supposed in (98.1) that the origins of the systems Y and $\bar{Y}$ coincide at the time $t = 0$.

From the linear character of (98.1), it is obvious that the accelerations d^2y^i/dt^2 and $d^2\bar{y}^i/dt^2$ of a particle referred to the frames Y and $\bar{Y}$, respectively, have the same value. It follows from this that the force $\mathbf{F}$ acting on a particle has the same value $\mathbf{F} = m\mathbf{a}$ in all reference systems moving relative to one another with constant velocity. In other words, Newton's second law of motion is formally invariant relative to a group of Galilean transformations 98.1.

Although the values of accelerations a^i are the same in all inertial systems,[1] the estimates of velocities differ in accordance with the formula

$$\bar{v}^i = v^i + u^i. \tag{98.2}$$

[1] See Sec. 76.

Hence a statement of any law that depends on the velocity relative to a primary inertial system will not be formally invariant when expressed in a secondary system. Consequently, the fundamental laws of electrodynamics and, particularly, of optics are not invariant with respect to a group of Galilean transformations 98.1, since these laws depend on the velocity of propagation of light. For this reason the primary inertial system has occupied a unique position in the theory of optics. In order to explain the observed fact of the independence of the velocity of light from the velocity of its source, and to imbed optics in the framework of analytical mechanics, physicists invented ether as a hypothetical carrier of light waves. This carrier was endowed with whatever physical properties were essential to ensure the same constant value for the velocity of propagation of light in all inertial systems, even when these properties did great violence to the established theories of elasticity and hydrodynamics. For instance, it was supposed that ether is an all-pervading, frictionless fluid that remains stationary relative to the primary inertial system, and that, when physical objects are forced to move through it, they suffer changes in shape, produced by elastic stresses that arise in a body moving in a quiescent fluid. It was then merely necessary to assume that the linear dimensions of measuring instruments suffer contractions depending on the velocity u^i, these contractions being of precisely the right amount to make the velocity of light come out to be independent of the velocity of its source.

A suitable formula expressing the dependence of the linear dimensions of a body on its velocity relative to a primary inertial system was developed by Lorentz, and a considerable body of the theory of relativity was phrased by him, in 1904, in terms of the quiescent ether. Lorentz's mathematics appeared to fit well the observed results in the domain of electrodynamics and provided a simple explanation of a puzzling behavior of the electrical field of a moving spherical charge, but the physics of the situation still remained in great doubt. However, all experimental attempts to detect the existence of ether have led to null results, and, in 1905, Albert Einstein achieved an explanation of the so-called Lorentz-Fitzgerald contraction by a sort of fiat which called for a profound revision in the prevailing notions of space and time.

99. Restricted, or Special, Theory of Relativity

In 1905, Einstein proposed two postulates, one of which relates to the formal invariance of physical laws, and the other epitomizes the results of certain remarkable experiments on the determination of the speed of light.[2]

[2] A. Einstein, *Annalen der Physik*, **18** (1905), p. 891.

These postulates can be stated as follows:

1. *Physical laws and principles have the same form in all Galilean systems; that is, reference systems that move relative to one another with uniform velocities.*

2. *The speed of light in free space has the same constant value in all inertial systems.*

In a sense there is nothing startling about these pronouncements since the ideas involved were in a state of ferment and discussion at the close of the nineteenth century and are quite explicit in the writings of Poincaré, Lorentz, Voigt, and others. But deductions to which Einstein was led from these postulates served to clarify and revise our concepts of space, time, and matter in a truly remarkable way. When viewed in the light of the fundamental laws of dynamics of a particle, the first postulate, as already remarked in Sec. 98, contains nothing novel. The laws of optics, on the other hand, are not invariant under the group of transformations 98.1, and one can set out to modify them so as to achieve the invariance of the fundamental laws of optics as well as mechanics. One way of accomplishing this is to abrogate the hypothesis that the estimates of time t are identical for observers located in two different Galilean reference systems. Mathematically this puts the time variable t on the same footing with the space variables y^i.

Thus, let us suppose that we have two cartesian reference frames Y and $\bar{Y}$, and an observer in the Y-frame recording the occurrence of some event at the point (y^i) at the time t, by means of four variables (y^1, y^2, y^3, t). The four-dimensional manifold S_4 of the variables (y^1, y^2, y^3, t) consists of E_3 and the range $-\infty < t < +\infty$. The same event is recorded by an observer in the $\bar{Y}$-frame as a point $(\bar{y}^1, \bar{y}^2, \bar{y}^3, \bar{t})$, in S_4, where $\bar{t}$ is the estimate of time based on the clock in the $\bar{Y}$-system of coordinates. As yet the variables (y^1, y^2, y^3, t) and $(\bar{y}^1, \bar{y}^2, \bar{y}^3, \bar{t})$ are unrelated, but, since we are in search of coordinate transformations which preserve the laws of dynamics of a particle, let the word "event" mean the track of a particle moving in the Y-frame under the action of a zero force. The trajectory of such a particle in the Y-frame is a straight line, and we shall suppose that the motion of the $\bar{Y}$-system relative to the Y-system is such that the trajectory in it also appears as a straight line.

This hypothesis implies the invariance of Newton's first law and requires that the variables (y^1, y^2, y^3, t) and $(\bar{y}^1, \bar{y}^2, \bar{y}^3, \bar{t})$ be related linearly. Thus

$$\begin{aligned} \bar{y}^i &= \alpha_j{}^i y^j + \alpha_4{}^i t, \qquad (i, j = 1, 2, 3), \\ \bar{t} &= \alpha_j{}^4 y^j + \alpha_4{}^4 t. \end{aligned} \tag{99.1}$$

It follows from these equations that the origin of the system $\bar{Y}$ moves relative to the system Y with constant velocity. To see this, note that the coordinates of the origin O of the system Y are (0, 0, 0), and hence the trajectory of the origin O relative to $\bar{Y}$ is given by (99.1) as

$$C: \quad \begin{cases} \bar{y}^i = \alpha_4{}^i t, \\ \bar{t} = \alpha_4{}^4 t. \end{cases}$$

Hence $d\bar{y}^i/d\bar{t} = \alpha_4{}^i/\alpha_4{}^4 =$ constant.

It can be shown in a similar way that the coordinate planes move with constant velocity, so that the reference frames Y and $\bar{Y}$ are Galilean.

Let us suppose next that a spherical pulse of light is sent out from the point $P(y^1, y^2, y^3)$ of the system Y at the time t. According to Einstein's second postulate, light travels with constant speed c in all directions; hence in dt seconds a photon starting from the point (y^i) will be at the point $(y^i + dy^i)$, and

$$dy^i\, dy^i = c^2\, dt^2. \tag{99.2}$$

Relative to an observer located in the $\bar{Y}$-system, the light pulse originates at the point $(\bar{y}^1, \bar{y}^2, \bar{y}^3)$, and his equation for the spherical wave front, $d\bar{t}$ seconds later, is

$$d\bar{y}^i\, d\bar{y}^i = c^2\, d\bar{t}^2. \tag{99.3}$$

Now if we substitute in (99.3) from (99.1) and compare the result with (99.2), we find that a particular set of equations

$$\begin{aligned} \bar{y}^1 &= k(y^1 - vt), \\ \bar{y}^2 &= y^2, \\ \bar{y}^3 &= y^3, \\ \bar{t} &= k\left(t - \frac{\beta}{c} y^1\right), \end{aligned} \tag{99.4}$$

where $k \equiv 1/\sqrt{1 - \beta^2}$, $\beta \equiv v/c$, leaves the quadratic form

$$d\sigma^2 = c^2\, dt^2 - dy^i\, dy^i \tag{99.5}$$

invariant. These equations correspond to the circumstance when the system $\bar{Y}$ moves relative to Y with the velocity v along the Y^1-axis.[3]

Equations 99.4 are known as the Lorentz-Einstein equations of transformation.[4] We shall not launch into extensive discussion of their

[3] We note that for the pulse of light $d\sigma = 0$

[4] These equations have been derived in many different ways. See, for example, J. Rice, *Relativity*, p. 89; R. Tolman, *Theory of Relativity of Motion;* A. Einstein, *Annalen der Physik*, **18** (1905); Frank, Ignatowsky, and Rothe, *Archiv für Mathematik und Physik*, **17** and **18**; J. L. Synge, *Relativity: The Special Theory* (1959), p. 69.

implications since most books on theoretical physics and special theory of relativity discuss them at great length, and there is no need to duplicate these considerations here. We shall mention only one example which has a direct bearing on the Lorentz-Fitzgerald contraction mentioned in Sec. 98.

Consider a rod moving with the system $\bar{Y}$. The end points of the rod have the coordinates $(\bar{y}_2{}^1, 0, 0)$, $(\bar{y}_1{}^1, 0, 0)$, so that its length, as measured by an observer in the $\bar{Y}$-system, is $\bar{L} = \bar{y}_2{}^1 - \bar{y}_1{}^1$. Since $\bar{y}_2{}^1 = k(y_2{}^1 - vt)$ and $\bar{y}_1{}^1 = k(y_1{}^1 - vt)$,

$$L = y_2{}^1 - y_1{}^1 = \sqrt{1 - \beta^2}(\bar{y}_2{}^1 - \bar{y}_1{}^1).$$

Accordingly, the estimate of the length L of the rod by an observer in Y-system is smaller than $\bar{L}$ in the ratio $\sqrt{1 - \beta^2}:1$. Thus the observer in the Y-system concludes that moving objects suffer a contraction in length. The magnitude of this contraction is the same as that deduced by Lorentz and Fitzgerald in connection with their study of the electrical field of a moving spherical charge. Whereas Lorentz and Fitzgerald thought of their contraction as a "real contraction" produced by the passage of objects through a quiescent ether, in the foregoing calculation it appears as a property of the space-time manifold subjected to a transformation 99.4, in which the space variables y^i are such that an element of arc ds is given by the formula $ds^2 = dy^i\, dy^i$.

If instead of cartesian variables y^i we had chosen curvilinear coordinates x^i, related to cartesian coordinates y^i by the formulas

$$y^i = y^i(x^1, x^2, x^3),$$

then the form 99.5 would have read

$$d\sigma^2 = c^2\, dt^2 - g_{ij}\, dx^i\, dx^j, \quad \left(g_{ij} = \frac{\partial y^k}{\partial x^i} \frac{\partial y^k}{\partial x^j}\right).$$

We note that the determinant of coefficients of this form has the value $-c^2 g$.

The foregoing formulas can be cast in a symmetric form by setting $t = x^4$; then

$$d\sigma^2 = a_{\alpha\beta}\, dx^\alpha\, dx^\beta, \qquad (\alpha, \beta = 1, 2, 3, 4), \tag{99.6}$$

where

$$a_{ij} = -g_{ij}, \qquad (i, j = 1, 2, 3),$$
$$a_{i4} = 0, \qquad a_{44} = c^2, \qquad \text{and} \qquad a = |a_{\alpha\beta}| = -c^2 g$$

If we now introduce a class of admissible functional transformations T in the four-dimensional manifold X,

$$T\colon\ \bar{x}^\alpha = \bar{x}^\alpha(x^1, x^2, x^3, x^4), \qquad (\alpha = 1, 2, 3, 4), \tag{99.7}$$

and require that the form 99.6 be invariant under the class of transformations 99.7, we can formulate the calculus of tensors as we did in Chapter 2.

Problems

1. Show, with the aid of equations 99.4, that events that are simultaneous from the point of view of an observer in the Y-system are not in general simultaneous in the $\bar{Y}$-system.

2. Discuss the slowing down of moving clocks.

3. Differentiate equations 99.4, and establish the relations between the components of velocity w^i of a moving point, as measured by an observer in the Y-system, with the corresponding quantities $\bar{w}^i$ measured in the $\bar{Y}$-system.

Ans.
$$\frac{dy^1}{dt} = \frac{\bar{w}^1 + v}{1 + (\beta/c)\bar{w}^1}, \qquad \frac{dy^\alpha}{dt} = \frac{\bar{w}^\alpha}{k(1 + (\beta/c)\bar{w}^1)}, \qquad (\alpha = 2, 3).$$

4. With the aid of the formulas given in Problem 3, show that, if $\bar{w}$ and v are both less than c, then $w/c < 1$. Thus, if $v = 0.9c$, $\bar{w} = 0.9c$, then $w = 0.994c$ instead of $1.8c$ given by the usual law of composition of velocities.

5. The expression arctanh w/c is sometimes called the *rapidity*. Show that the usual law of composition of velocities is obeyed by the rapidities. Thus

$$\operatorname{arctanh}\frac{w}{c} = \operatorname{arctanh}\frac{w}{c} - \operatorname{arctanh}\frac{v}{c}.$$

100. Proper or Local Coordinates

Consider a point P whose space coordinates relative to some reference frame X are (x^1, x^2, x^3). Let the velocity of P, relative to this frame at the instant t, be $\mathbf{v}$. We shall introduce a Galilean reference frame $\bar{X}$ moving with the point P so that, at each instant t, the point P is at rest relative to the system $\bar{X}$. We shall call the system $\bar{X}$ a *local* or *proper* coordinate system.

Obviously the choice of local coordinate systems is not unique, since the definition just laid down merely requires that the velocity of the local frame be the same as that of the particle. This implies that the estimates of time (measured by the clocks carried in two different local coordinate frames) are the same. Hence the transformation from one local system $\bar{X}$ to another $\bar{X}'$ has the form

$$\begin{cases} \bar{x}'^i = \bar{x}'^i(\bar{x}^1, \bar{x}^2, \bar{x}^3), \\ \bar{t}' = \bar{t}. \end{cases}$$

The *interval* $d\sigma$ is defined by the formula

$$(100.1) \qquad d\sigma^2 = a_{\alpha\beta}\, dx^\alpha\, dx^\beta = c^2\, dt^2 - g_{ij}\, dx^i\, dx^j,$$

so that

$$\left(\frac{d\sigma}{dt}\right)^2 = c^2 - g_{ij}\frac{dx^i}{dt}\frac{dx^j}{dt} = c^2 - v^2, \tag{100.2}$$

where v is the magnitude of the velocity $\mathbf{v}$ of the point P relative to the X-coordinate frame. If a local coordinate system $\bar{X}$ is introduced at P, then, relative to $\bar{X}$, $v = 0$ and equation 100.2 yields

$$\frac{d\sigma}{d\bar{t}} = c \tag{100.3}$$

in the local system. We define the *Minkowski velocity vector* u^α by the formula

$$u^\alpha = \frac{dx^\alpha}{d\sigma}, \qquad (\alpha = 1, 2, 3, 4), \tag*{[100.4]}$$

and observe that its components in a local system $\bar{X}$ are[5] $(0, 0, 0, 1/c)$.

Since $a = |a_{\alpha\beta}| = -c^2 g \neq 0$, we can construct the reciprocal tensor $a^{\alpha\beta}$, the Christoffel symbols

$$[\alpha\beta, \gamma] = \frac{1}{2}\left(\frac{\partial a_{\alpha\gamma}}{\partial x^\beta} + \frac{\partial a_{\beta\gamma}}{\partial x^\alpha} - \frac{\partial a_{\alpha\beta}}{\partial x^\gamma}\right),$$

$$\begin{Bmatrix}\gamma\\ \alpha\beta\end{Bmatrix} = a^{\gamma\delta}[\alpha\beta, \delta],$$

and define the operations of covariant and intrinsic differentiation as was done in Chapters 2 and 3. This permits us to define the *Minkowski acceleration vector* f^α by the formula

$$f^\alpha = \frac{\delta u^\alpha}{\delta\sigma} \equiv \frac{d^2x^\alpha}{d\sigma^2} + \begin{Bmatrix}\alpha\\ \beta\gamma\end{Bmatrix}\frac{dx^\beta}{d\sigma}\frac{dx^\gamma}{d\sigma}, \qquad (\alpha, \beta, \gamma = 1, \ldots, 4). \tag{100.5}$$

If our local reference frame $\bar{X}$ is cartesian so that $d\sigma^2 = c^2\,d\bar{t}^2 - d\bar{y}^i\,d\bar{y}^i$, the components $\bar{f}^\alpha$ of the Minkowski acceleration relative to it are

$$\bar{f}^\alpha = \frac{d^2\bar{y}^\alpha}{d\sigma^2} = \frac{d}{d\bar{t}}\left(\frac{d\bar{y}^\alpha}{d\sigma}\right)\frac{d\bar{t}}{d\sigma} = \frac{1}{c}\frac{d}{d\bar{t}}\left(\frac{d\bar{y}^\alpha}{d\sigma}\right) = \frac{1}{c^2}\frac{d^2\bar{y}^\alpha}{d\bar{t}^2},$$

[5] For, $u^\alpha = \dfrac{dx^\alpha}{d\sigma} = \dfrac{dx^\alpha}{(c^2\,dt^2 - g_{ij}\,dx^i\,dx^j)^{\frac{1}{2}}} = \dfrac{dx^\alpha}{dt}\cdot\dfrac{1}{\sqrt{c^2 - v^2}} = \dfrac{v^\alpha}{\sqrt{c^2 - v^2}}$, and $v^1 = v^2 = v^3 = 0$, $x^4 = t$ in a local system.

so that

$$\bar{f}^i = \frac{1}{c^2}\frac{d^2\bar{y}^i}{d\bar{t}^2}, \qquad (i = 1, 2, 3),$$

$$\bar{f}^4 = 0, \qquad \text{since } \bar{y}^4 = \bar{t}.$$

We shall show next how Newton's second law can be written in an invariant form relative to all Galilean reference frames. Consider the formula suggested by Newton's second law,

$$F^\alpha = \frac{\delta}{\delta\sigma}(m_0 u^\alpha), \qquad (\alpha = 1, 2, 3, 4),$$

where $u^\alpha = dx^\alpha/d\sigma$ is the Minkowski velocity and m_0 is a constant whose significance will appear presently. Now

$$\begin{aligned} F^\alpha &= \frac{\delta}{\delta t}(m_0 u^\alpha)\frac{dt}{d\sigma} \\ &= \frac{1}{\sqrt{c^2 - v^2}}\frac{\delta}{\delta t}\left(m_0 \frac{dx^\alpha}{d\sigma}\right) \\ &= \frac{1}{\sqrt{c^2 - v^2}}\frac{\delta}{\delta t}\left(\frac{m_0}{\sqrt{c^2 - v^2}}\frac{dx^\alpha}{dt}\right) \\ &= \frac{1}{c^2\sqrt{1 - \beta^2}}\frac{\delta}{\delta t}\left(\frac{m_0}{\sqrt{1 - \beta^2}}\frac{dx^\alpha}{dt}\right), \end{aligned}$$

where we made use of the relation 100.2, and set $\beta = v/c$. If we define

$$m \equiv \frac{m_0}{\sqrt{1 - \beta^2}},$$

the foregoing equation can be written in the form

$$\sqrt{1 - \beta^2}\, F^\alpha = \frac{1}{c^2}\frac{\delta}{\delta t}\left(m \frac{dx^\alpha}{dt}\right), \tag{100.6}$$

and since, in the local coordinate system $\bar{Y}$, $\beta = 0$ and $m = m_0$,

$$\begin{aligned} \bar{F}^\alpha &= \frac{m_0}{c^2}\frac{d^2\bar{y}^\alpha}{d\bar{t}^2} \\ &= m_0 \bar{f}^\alpha. \end{aligned} \tag{100.7}$$

This has the form of Newton's second law used in classical mechanics. We see that the invariant m_0 is the mass of the particle P referred to a local reference frame. It is called the *rest* (or *proper*) mass of the particle.

Since equation 100.7 is a tensor equation, we can write the force equation as

$$F^\alpha = m_0 f^\alpha,$$

which is valid in all Galilean reference frames.

We shall rewrite (100.6) in the form

$$\mathscr{F}^\alpha = \frac{\delta}{\delta t}\left(\frac{m_0 v^\alpha}{\sqrt{1-\beta^2}}\right), \tag{100.8}$$

where $v^\alpha = dx^\alpha/dt$, and $\mathscr{F}^\alpha \equiv c^2\sqrt{1-\beta^2}\,F^\alpha$, and shall take it as the equation of motion of a particle in the restricted theory of relativity.

101. Einstein's Energy Equation

We conclude our sketch of the rudiments of mechanics in the restricted theory of relativity by establishing an important connection between mass and energy.

For simplicity in writing we suppose that the coordinates x^i used in this section are *rectangular cartesian*; and we recall that the work done by the force F_i, $(i = 1, 2, 3)$, in producing a displacement dx^i is equal to the change in the kinetic energy. Indeed, the classical theory gives

$$\begin{aligned} T - T_0 &= \int_{v_0}^{v} mv\,dv = \int_{v_0}^{v} m\frac{dx^i}{dt}\,d\left(\frac{dx^i}{dt}\right) \\ &= \int_{t_0}^{t} m\frac{dx^i}{dt}\frac{d^2x^i}{dt^2}\,dt \\ &= \int_{P_0}^{P} m\frac{d^2x^i}{dt^2}\,dx^i \\ &= \int_{P_0}^{P} F_i\,dx^i. \end{aligned}$$

If we take as our definition of the kinetic energy in the restricted theory of relativity the expression

$$T = \int_{P_0}^{P} \mathscr{F}_i\,dx^i \tag{101.1}$$

and insert for the $\mathscr{F}_i$ from equation 100.8, we get[6]

$$\begin{aligned} T &= \int_{P_0}^{P} \mathscr{F}_i\,dx^i = \int_{P_0}^{P} \frac{d}{dt}\left(\frac{m_0 v^i}{\sqrt{1-\beta^2}}\right) dx^i \\ &= m_0 \int_{t_0}^{t} \left[\frac{d}{dt}\left(\frac{1}{\sqrt{1-\beta^2}}\right) v^i \frac{dx^i}{dt} + \frac{dv^i}{dt}\frac{1}{\sqrt{1-\beta^2}}\frac{dx^i}{dt}\right] dt. \end{aligned}$$

[6] Since the reference frame is cartesian, the intrinsic derivative reduces to the ordinary derivative.

But

$$\beta^2 = \frac{v^2}{c^2} = \frac{v^i v^i}{c^2}, \qquad v^i = \frac{dx^i}{dt},$$

hence $v^i(dx^i/dt) = \beta^2 c^2$, and $\beta\dot{\beta} = (v^i/c^2)(dv^i/dt)$. Substituting these expressions in the integral, we get

$$\begin{aligned} T &= m_0 \int_{t_0}^{t} \left[\frac{d}{dt}\left(\frac{1}{\sqrt{1-\beta^2}}\right)\beta^2 c^2 + c^2\beta\dot{\beta}\frac{1}{\sqrt{1-\beta^2}} \right] dt \\ &= m_0 \int_{t_0}^{t} \left[\beta^2 c^2 \frac{\beta\dot{\beta}}{(1-\beta^2)^{3/2}} + \frac{c^2\beta\dot{\beta}}{(1-\beta^2)^{1/2}} \right] dt \\ &= m_0 c^2 \int_{t_0}^{t} \frac{\beta\dot{\beta}}{(1-\beta^2)^{3/2}}\, dt \\ &= m_0 c^2 \int_{P_0}^{P} \frac{\beta\, d\beta}{(1-\beta^2)^{3/2}} \\ &= m_0 c^2 \int_{P_0}^{P} d\left[\frac{1}{(1-\beta^2)^{1/2}}\right]. \end{aligned}$$

Thus

$$T = \frac{m_0 c^2}{(1-\beta^2)^{1/2}} + \text{constant}.$$

If we wish to have $T = 0$ when $\beta = v/c = 0$, the constant of integration is $-m_0c^2$, so that

$$\begin{aligned} T &= \left[\frac{m_0}{(1-\beta^2)^{1/2}} - m_0\right] c^2 \\ &= (m - m_0)c^2. \end{aligned}$$

Thus

$$m = m_0 + \frac{T}{c^2} \tag{101.2}$$

We see that the mass m depends on the kinetic energy. If this result is assumed to hold in dissipative systems, then the decrease in mass m must be accounted for by the loss of energy by radiation.[7]

We see from the foregoing that the principles of conservation of energy and conservation of mass, which appeared to be quite distinct in the classical theory, can be united into one law in the restricted theory. We also see from equation 101.2 that, if a particle takes up an amount of energy ΔT, then its inertial mass m is increased by an amount $\Delta T/c^2$.

[7] In vol. 41 (1935) of the *Bulletin of the American Mathematical Society*, Einstein gave an elementary derivation of this mass-energy relation by basing his considerations on the principles of conservation of energy and momentum. For a definitive treatment see J. L. Synge, *Relativity: The Special Theory* (1956), Chapter VI.

Thus the inertial mass m can be considered a measure of the energy of the particle, and the law of conservation of mass holds if, and only if, the particle neither receives nor gives up its energy. Einstein associated with every mass m an amount of energy $E = mc^2$. Then equation 101.2 can be written in the form

$$E = m_0 c^2 + T,$$

in which $m_0 c^2$ appears as the intrinsic energy and T as the kinetic energy.

102. Restricted Theory. Retrospect and Prospect

In our development of mechanics in the manifold of the special theory of relativity we maintained the distinction between the space coordinates x^i, $(i = 1, 2, 3)$, of a particle and the time variable $t = x^4$. The metric of the space was assumed to be Euclidean. The novel features of the theory lie in the abandonment of the concept of universal time and in the demand that the mass of the particle change with velocity in a predetermined way, if the Newtonian law of motion is to be invariant with respect to a group of Lorentz-Einstein transformations.

The distinction between the space and time variables can be suppressed by introducing a single-valued reversible transformation of the S_4 manifold,

$$\bar{x}^\alpha = \bar{x}^\alpha(x^1, x^2, x^3, x^4), \qquad (\alpha = 1, 2, 3, 4),$$

where the coordinates $\bar{x}^\alpha$ are quite analogous to the generalized coordinates of analytical mechanics. We suppose that our space S_4 is so metrized that the quadratic form

$$d\sigma^2 = a_{\alpha\beta}\, dx^\alpha\, dx^\beta \tag{102.1}$$

reduces to

$$d\sigma^2 = c^2\, dt^2 - dy^i\, dy^i, \tag{102.2}$$

when the space coordinates x^i are orthogonal cartesian. Since the coefficients in the form 102.2 are constants, it follows that the Riemann curvature tensor $R_{\alpha\beta\gamma\delta}$ of the S_4 manifold vanishes, and hence the geodesics in S_4, determined by

$$\frac{d^2x^\alpha}{d\sigma^2} + \begin{Bmatrix} \alpha \\ \beta\gamma \end{Bmatrix} \frac{dx^\beta}{d\sigma}\frac{dx^\gamma}{d\sigma} = 0, \tag{102.3}$$

are straight lines.

We note, with reference to equations 100.5, that equations 102.3 characterize the motion of a particle in the absence of acceleration f^α. This suggests the possibility of interpreting the trajectories of particles, subjected to the action of nonvanishing forces, as geodesics in some manifold of the variables x for which the curvature tensor does not vanish.[8] Physically, this corresponds to the introduction of accelerated reference frames moving in such a way that the forces acting on the particles vanish. If this is done, the concept of force need not enter dynamics, and dynamical trajectories can then be viewed as geodesics determined by the metric properties of space.

In the remaining section of this chapter we discuss the problem of two bodies from a general relativistic point of view. This portion of the general theory of relativity was developed in the early 1920's, and its mathematical elegance and success in explaining the advance of the perihelion of Mercury gave hope that the time when all mathematical physics would be imbedded in the framework of the general theory of relativity was not too far away. However, the researches of the following two decades make it appear unlikely that general relativity will prove useful in the domain of microscopic physics, because of the failure of the theory to unify mechanics and electrodynamics. It is likely that the future usefulness of the theory will be in whatever stimulus it may provide to speculations in cosmology. These remarks do not detract from the profound effect which Einstein's paper,[9] setting forth the foundations of the general theory of relativity, had on the revision of the concepts of space, time, and matter.

103. Einstein's Gravitational Equations

In order to conform to the usual notation in books on general theory of relativity, we denote the metric coefficients of the four-dimensional relativity manifold by $g_{ij}(x^1, x^2, x^3, x^4)$, and write the fundamental quadratic form as

$$ds^2 = g_{ij}\, dx^i\, dx^j, \qquad (i, j = 1, 2, 3, 4). \tag{103.1}$$

In the special instance of the restricted theory the form 103.1 can be reduced by a suitable transformation to the canonical form

$$ds^2 = c^2(dt)^2 - dy^i\, dy^i. \tag{103.2}$$

[8] A similar situation arose in classical mechanics (Sec. 84), where we introduced a Riemannian manifold, with the arc element dS of the form

$$dS = \sqrt{2m(h - V)g_{ij}\, dx^i\, dx^j},$$

in which the trajectories are geodesics.

[9] A. Einstein, *Annalen der Physik*, **49** (1916), p. 769.

Our hypothesis is that the coefficients g_{ij}, which we will term *potential functions*,[10] can be so chosen that the trajectories of particles satisfy the equations of geodesics,

$$\frac{d^2x^i}{ds^2} + \begin{Bmatrix} i \\ jk \end{Bmatrix} \frac{dx^j}{ds}\frac{dx^k}{ds} = 0. \tag{103.3}$$

The Riemann curvature tensor R^i_{jkl}, associated with the manifold of restricted theory, vanishes, and the rectilinear geodesics of the manifold correspond to the trajectories of particles in the absence of a gravitational field. Consequently, if the manifold with the quadratic form 103.1 is to account for nonrectilinear trajectories, the Riemann curvature tensor must not vanish. We assume, with Einstein, that the field of a large gravitating mass (the sun) is such that the potential functions g_{ij} satisfy in vacuum the equations

$$G_j^{\ i} = R_j^{\ i} - \tfrac{1}{2}\delta_j^{\ i}R = 0,$$

where $G_j^{\ i}$ is the Einstein tensor defined in Sec. 38. If we contract $G_j^{\ i}$, we get the equation $R - \frac{1}{2}4R = 0$, so that $R = 0$. Accordingly,

$$R_{ij} \equiv R^\alpha_{ij\alpha} = 0, \tag{103.4}$$

where R_{ij} is the Ricci tensor. These equations include the flat manifold of restricted theory and admit the case for which the components of the curvature tensor do not vanish.

Equations 103.4 are analogous to Laplace's equation, $g^{ij}V_{i,j} = 0$, of Newtonian potential theory, which is valid at all points outside gravitating matter.[11]

We recall[12] that the Ricci tensor R_{ij} appearing in the left-hand member of equation 103.4 is given by

$$R_{ij} = \frac{\partial^2 \log\sqrt{|g|}}{\partial x^i\,\partial x^j} - \frac{\partial}{\partial x^\alpha}\begin{Bmatrix} \alpha \\ ij \end{Bmatrix} + \begin{Bmatrix} \alpha \\ \beta j \end{Bmatrix}\begin{Bmatrix} \beta \\ i\alpha \end{Bmatrix} - \begin{Bmatrix} \beta \\ ij \end{Bmatrix}\frac{\partial \log\sqrt{|g|}}{\partial x^\beta},$$

where we write $|g|$ since the determinant of the form 103.1 may be negative.

[10] This terminology can be justified by examining the form of the coefficients in equation 84.9 in a related problem in Newtonian mechanics.

[11] This equation is suggested by a chain of reasoning making use of equations of motion in the form $G_{,j}^{\ ij} = 0$, where $G^{ij} = -\rho u^i u^j$ with $u^i = dx^i/dt$. A delightful account of this approach is contained in G. Y. Rainich's, *Mathematics of Relativity* (1950). See also Problem 2, Sec. 38.

[12] These equations are not independent, and it can be shown that there are four relations connecting them. See, for example, A. S. Eddington, *The Mathematical Theory of Relativity*, 2d ed. (1924), p. 115. This fact, however, has no bearing on the calculations given below.

It is obvious from the foregoing that the system of ten nonlinear partial differential equations (see Sec. 38)

$$R_{ij} = 0$$

for the ten unknown functions g_{ij} is extremely complicated.[13] The general solution of this system is not known, and one is obliged to seek particular solutions, essentially by trial, and use Newtonian mechanics as a guide in selecting sensible forms for the coefficients g_{ij}. Once a set of g_{ij}'s satisfying equations 103.4 is found, we can form the equations of geodesics 103.3, and if the solution of equations 103.3 agrees to the first order of small quantities with the corresponding situations in Newtonian theory, all is well.

We shall illustrate this procedure in Sec. 104, where we will obtain the Schwarzschild[14] solution of the gravitational equations 103.4.

Before we proceed to that topic we note that equations 103.3 can be written in a neat form,

$$\dot{x}^i{}_{,j}\dot{x}^j = 0, \tag{103.5}$$

where $\dot{x}^i = dx^i/ds$. If we regard the vector $dx^i/ds = \lambda^i$ as the tangent vector, then equations 103.5, or $\lambda^i{}_{,j}\lambda^j = 0$, are precisely the equations for the parallel displacement of the tangent vector λ^i along a geodesic. Our problem has thus been reduced to the solution of a deceptively simple-looking system

$$\begin{cases} R_{ij} = 0, \\ \dot{x}^i{}_{,j}\dot{x}^j = 0, \end{cases}$$

with which we will occupy ourselves in Secs. 104 and 105.

104. Spherically Symmetric Static Field

We proceed to deduce a solution of Einstein's equations

$$R_{ij} = 0, \tag{104.1}$$

[13] It is interesting to note that as an argument for adopting this system of equations as the law of gravitation it is frequently stated that the law 103.4 represents a simple relation involving the curvature tensor R^i_{jkl}, and hence a desirable one. A skeptic might feel that the Creator was not greatly concerned with the simplicity of mathematical physics.

[14] K. Schwarzschild, *Berlin Sitzungsberichte* (1916), p. 189. See also some important special solutions in G. D. Birkhoff's *Relativity and Modern Physics*, pp. 219–227. There is also the solution of H. Weyl and T. Levi-Civita, corresponding to rotational symmetry. See P. G. Bergmann, *Introduction to the Theory of Relativity* (1942), pp. 206–210. For a comprehensive discussion of spherically symmetric fields see a treatise by J. L. Synge, *Relativity: The General Theory* (1960), Chapter VII.

for the gravitational field produced by a spherically symmetric mass particle, which will be shown to correspond to the gravitational field of the sun fixed at the origin of our reference frame. In obtaining this solution we will be guided by the properties of the Newtonian gravitational field and by the form of the corresponding solution in classical mechanics.

The discussion of the two-body problem in Sec. 97 suggests that we adopt as our reference frame a system of coordinates which at great distance from the gravitating mass specializes to the ordinary spherical coordinate system. Moreover, since the field is spherically symmetric, and since the metric of the manifold is determined by the field, the metric tensor g_{ij} must be spherically symmetric. Thus we shall select the coordinates in such a way that, at great distance from the center of attraction (the origin),

$$x^1 = r, \qquad x^2 = \theta, \qquad x^3 = \phi, \qquad x^4 = t,$$

where r, θ, ϕ are the usual spherical coordinates.

The trajectories of particles far away from gravitating matter should be straight lines, so that $R^i_{jkl} = 0$. We write the limiting form for the space-time interval as

$$ds^2 = (dt)^2 - (dr)^2 - r^2(d\theta)^2 - r^2 \sin^2 \theta (d\phi)^2, \tag{104.2}$$

where we have adopted a new unit for the velocity of light c so that it is 1. This leads us to assume that, in the presence of a spherically symmetric static gravitational field,

$$ds^2 = f_1(r)(dt)^2 - f_2(r)(dr)^2 - r^2(d\theta)^2 - r^2 \sin^2 \theta(d\phi)^2, \tag{104.3}$$

where f_1 and f_2 are unknown functions of r, each reducing to unity when r is increased indefinitely.

The cross-product terms $dr\, d\theta$, $d\phi\, d\theta$, etc., are omitted in the form 104.3 since ds^2 must be independent of the signs of $d\theta$ and $d\phi$ because of the spherical symmetry. Likewise, we reject the cross-product terms involving dt, since we assume that the field is static and reversible in time, and hence must be independent of the sign of dt. Our procedure in determining the functions f_1 and f_2 will be to insert the expressions for metric coefficients g_{ij} from (104.3) in the gravitational equations 104.1, and use equation 104.2 as a boundary condition at infinity.

For the purpose of calculating f_1 and f_2 it is convenient to set

$$f_1 = e^{\mu}, \qquad f_2 = e^{\lambda},$$

where λ and μ are functions of r. Since effects of the gravitational field diminish as $r \to \infty$, the functions λ and μ must tend to zero when r increases indefinitely.

We can write the form 104.3 in the new notation as

$$(104.4)\qquad ds^2 = -e^{\lambda}(dr)^2 - r^2(d\theta)^2 - r^2\sin^2\theta(d\phi)^2 + e^{\mu}(dt)^2,$$

so that the metric coefficients g_{ij} are

$$g_{11} = -e^{\lambda}, \qquad g_{22} = -r^2, \qquad g_{33} = -r^2\sin^2\theta, \qquad g_{44} = e^{\mu}$$
$$g_{ij} = 0, \qquad i \neq j.$$

The determinant g of the quadratic form 104.4 is

$$g = g_{11}g_{22}g_{33}g_{44} = -e^{\lambda+\mu}r^4\sin^2\theta,$$

and the contravariant tensor g^{ij} is given by the matrix

$$(g^{ij}) = \begin{bmatrix} -e^{-\lambda} & 0 & 0 & 0 \\ 0 & -\dfrac{1}{r^2} & 0 & 0 \\ 0 & 0 & -\dfrac{1}{r^2\sin^2\theta} & 0 \\ 0 & 0 & 0 & e^{-\mu} \end{bmatrix}.$$

In order to form equations 104.1, we construct the Christoffel symbols $\left\{ {k \atop ij} \right\}$, and, since $g_{ij} = 0$ when $i \neq j$, we have

$$\left\{ {k \atop ij} \right\} = \frac{1}{2} g^{kk}\left(\frac{\partial g_{ik}}{\partial x^j} + \frac{\partial g_{jk}}{\partial x^i} - \frac{\partial g_{ij}}{\partial x^k}\right), \qquad \text{(no sum on } k\text{)}.$$

It is easy to verify that distinct, nonvanishing Christoffel's symbols are

$$\left\{ {1 \atop 11} \right\} = \tfrac{1}{2}\lambda', \qquad \left\{ {2 \atop 12} \right\} = \frac{1}{r}, \qquad \left\{ {3 \atop 13} \right\} = \frac{1}{r},$$
$$\left\{ {4 \atop 14} \right\} = \tfrac{1}{2}\mu', \qquad \left\{ {1 \atop 22} \right\} = -re^{-\lambda}, \qquad \left\{ {3 \atop 23} \right\} = \cot\theta,$$
$$\left\{ {1 \atop 33} \right\} = -r\sin^2\theta e^{-\lambda}, \qquad \left\{ {2 \atop 33} \right\} = -\sin\theta\cos\theta, \qquad \left\{ {1 \atop 44} \right\} = \tfrac{1}{2}e^{\mu-\lambda}\mu',$$

where primes denote the derivatives with respect to r.

We can now insert these symbols in the formula

$$R_{ij} = \frac{\partial^2 \log\sqrt{|g|}}{\partial x^i\,\partial x^j} - \frac{\partial}{\partial x^\alpha}\left\{ {\alpha \atop ij} \right\} + \left\{ {\alpha \atop \beta j} \right\}\left\{ {\beta \atop i\alpha} \right\} - \left\{ {\beta \atop ij} \right\}\frac{\partial \log\sqrt{|g|}}{\partial x^\beta},$$

and obtain after tedious but simple calculations the following set of differential equations:

$$(104.5)\qquad R_{11} = \tfrac{1}{2}\mu'' - \tfrac{1}{4}\lambda'\mu' + \tfrac{1}{4}(\mu')^2 - \frac{\lambda'}{r} = 0,$$

$$(104.6)\qquad R_{22} = e^{-\lambda}[1 + \tfrac{1}{2}r(\mu' - \lambda')] - 1 = 0,$$

$$(104.7)\qquad R_{33} = \sin^2\theta\{e^{-\lambda}[1 + \tfrac{1}{2}r(\mu' - \lambda')] - 1\} = 0,$$

$$(104.8)\qquad R_{44} = e^{\mu-\lambda}\left[-\tfrac{1}{2}\mu'' + \tfrac{1}{4}\lambda'\mu' - \tfrac{1}{4}(\mu')^2 - \frac{\mu'}{r}\right] = 0,$$

$$R_{ij} = 0, \qquad \text{if } i \neq j.$$

Equation 104.7 in this set is a mere repetition of equation 104.6. We thus have only three equations on λ and μ to consider.

From equations 104.5 and 104.8 we deduce that

$$\lambda' = -\mu',$$

so that

$$\lambda = -\mu + \text{constant}.$$

However, as $r \to \infty$, λ and μ tend to zero; hence,

$$\lambda(r) = -\mu(r).$$

Epuation 104.6 thus becomes

$$(104.9)\qquad e^{\mu}(1 + r\mu') = 1.$$

We set

$$e^{\mu} = \gamma,$$

and equation 104.9 becomes

$$\gamma + r\gamma' = 1.$$

Integrating this first-order linear equation, we get

$$(104.10)\qquad \gamma = 1 - \frac{2m}{r} \equiv e^{\mu},$$

where $2m$ is a constant of integration. We shall identify m, in Sec. 105, with the mass of the sun.

It is easily checked that the solution just obtained satisfies all equations in our system. Inserting $e^{-\lambda} = e^{\mu} = \gamma$ in equation 104.4, we get the desired quadratic form

$$(104.11)\qquad ds^2 = -\gamma^{-1}(dr)^2 - r^2(d\theta)^2 - r^2\sin^2\theta(d\phi)^2 + \gamma(dt)^2,$$

where $\gamma = 1 - 2m/r$. If the constant of integration $2m$ vanishes, $\gamma = 1$, and the resulting manifold is the flat manifold of restricted theory. For $m \neq 0$, the manifold is curved.

The reader may feel uneasy about the Schwarzschild solution of Einstein's gravitational equations, since it was obtained on the basis of several fortuitous guesses with one eye cocked on results of the classical theory. He may feel that a different mode of attack might yield a different solution. That this is not so was shown[15] by G. D. Birkhoff, who demonstrated that all spherically symmetric static solutions of the gravitational equations $R_{ij} = 0$, which yield a flat metric at infinity (that is, the one characterized by equation 104.2), are equivalent to the Schwarzschild solution. Thus the solution obtained previously is of interest because it is the only static solution of our equations satisfying specified boundary conditions at infinity.

105. Planetary Orbits

We are in a position now to determine the trajectory of a particle moving in a spherically symmetric static field determined by the quadratic form 104.11. The trajectory of the particle is a geodesic, so that we have to solve the set of equations[16]

$$\frac{d^2x^i}{ds^2} + \begin{Bmatrix} i \\ \alpha\beta \end{Bmatrix} \frac{dx^\alpha}{ds}\frac{dx^\beta}{ds} = 0,$$

where $x^1 = r$, $x^2 = \theta$, $x^3 = \phi$, $x^4 = t$.

Making use of the table of values of Christoffel's symbols given in Sec. 104, we find that for $i = 2$, for example, we have the equation

$$\frac{d^2x^2}{ds^2} + \begin{Bmatrix} 2 \\ 12 \end{Bmatrix} \frac{dx^1}{ds}\frac{dx^2}{ds} + \begin{Bmatrix} 2 \\ 21 \end{Bmatrix} \frac{dx^2}{ds}\frac{dx^1}{ds} + \begin{Bmatrix} 2 \\ 33 \end{Bmatrix} \frac{dx^3}{ds}\frac{dx^3}{ds} = 0,$$

or

$$\frac{d^2\theta}{ds^2} + \frac{2}{r}\frac{dr}{ds}\frac{d\theta}{ds} - \cos\theta\sin\theta\left(\frac{d\phi}{ds}\right)^2 = 0. \tag{105.1}$$

In a similar way we form equations for $i = 1, 3, 4$. The results are

$$\frac{d^2r}{ds^2} - \frac{1}{2\gamma}\frac{d\gamma}{dr}\left(\frac{dr}{ds}\right)^2 - \gamma r\left(\frac{d\theta}{ds}\right)^2 - \gamma r\sin^2\theta\left(\frac{d\phi}{ds}\right)^2 + \frac{\gamma}{2}\frac{d\gamma}{dr}\left(\frac{dt}{ds}\right)^2 = 0, \tag{105.2}$$

$$\frac{d^2\phi}{ds^2} + \frac{2}{r}\frac{dr}{ds}\frac{d\phi}{ds} + 2\cot\theta\frac{d\theta}{ds}\frac{d\phi}{ds} = 0, \tag{105.3}$$

$$\frac{d^2t}{ds^2} + \frac{1}{\gamma}\frac{d\gamma}{dr}\frac{dt}{ds}\frac{dr}{ds} = 0. \tag{105.4}$$

[15] G. D. Birkhoff, *Relativity and Modern Physics*, p. 253.

[16] For an elegant treatment of planetary orbits by means of Lagrangean equations see J. L. Synge, *Relativity: The General Theory* (1960), pp. 289–298.

The last of these equations can be written

$$\frac{d^2t}{ds^2} + \frac{1}{\gamma}\frac{d\gamma}{ds}\frac{dt}{ds} = 0,$$

or

$$\frac{d}{ds}\left(\gamma \frac{dt}{ds}\right) = 0. \tag{105.5}$$

We will prove that the analytic solution of equation 105.1, satisfying the initial condition $d\theta/ds = 0$, when $\theta = \pi/2$, is $\theta(s) \equiv \pi/2$. Since $d\theta/ds = (d\theta/dt)(dt/ds)$, and $dt/ds \neq 0$, this is equivalent to showing that the trajectory of the particle lies in the plane $\theta = \pi/2$, provided that the initial component $d\theta/dt$ of the velocity, in the direction of increasing θ, vanishes. We thus assume that the solution $\theta(s)$ can be represented by the series

$$\theta(s) = (\theta)_0 + \left(\frac{d\theta}{ds}\right)_0 s + \left(\frac{d^2\theta}{ds^2}\right)_0 \frac{s^2}{2!} + \cdots \tag{105.6}$$

Since $d\theta/ds = 0$, when $\theta = \pi/2$, equation 105.1 for $\theta = \pi/2$ gives $(d^2\theta/ds^2)_0 = 0$.

To obtain $(d^3\theta/ds^3)_0$ we differentiate equation 105.1, and insert in the result the values $\theta = \pi/2, d\theta/ds = 0$, and $d^2\theta/ds^2 = 0$. We find $d^3\theta/ds^3 = 0$. In this manner we can show that $\theta(s)$ in (105.6) is $\theta(s) = (\theta)_0 = \pi/2$.

The corresponding result in the Newtonian case is obvious since, under the assumption of the central field of force, there can be no component of force at right angles to the plane of motion. Thus, if the motion had once started in the plane $\theta = \pi/2$, it would continue in that plane. If we insert the solution $\theta = \pi/2$ of equation 105.1 in equation 105.3, we get

$$\frac{d^2\phi}{ds^2} + \frac{2}{r}\frac{dr}{ds}\frac{d\phi}{ds} = 0; \tag{105.7}$$

and integrating equations 105.5 and 105.7 we obtain

$$r^2 \frac{d\phi}{ds} = h, \tag{105.8}$$

$$\frac{dt}{ds} = \frac{a}{\gamma}, \tag{105.9}$$

where a and h are arbitrary constants.

Substituting in equation 105.2 from 105.8 and 105.9, and using the previously found solution $\theta = \pi/2$, we have

$$\frac{d^2r}{ds^2} - \frac{1}{2\gamma}\frac{d\gamma}{dr}\left(\frac{dr}{ds}\right)^2 - \gamma r\left(\frac{h}{r^2}\right)^2 + \frac{\gamma}{2}\frac{d\gamma}{dr}\left(\frac{a}{\gamma}\right)^2 = 0. \tag{105.10}$$

The expression for $(dr/ds)^2$, appearing in equation 105.10, can be obtained from formula 104.11 by using equations 105.8 and 105.9 and $\theta = \pi/2$. We have

$$\left(\frac{dr}{ds}\right)^2 = a^2 - \frac{h^2\gamma}{r^2} - \gamma,$$

which, upon insertion in (105.10), gives

$$\frac{d^2r}{ds^2} + \frac{m}{r^2} = \frac{h^2}{r^3}\left(1 - \frac{3m}{r}\right), \tag{105.11}$$

since $\gamma = 1 - 2m/r$. But

$$\frac{dr}{ds} = \frac{dr}{d\phi}\frac{d\phi}{ds}, \qquad \frac{d^2r}{ds^2} = \frac{d^2r}{d\phi^2}\left(\frac{d\phi}{ds}\right)^2 + \frac{d^2\phi}{ds^2}\frac{dr}{d\phi}$$
$$= \frac{d^2r}{d\phi^2}\frac{h^2}{r^4} - \frac{2h^2}{r^5}\left(\frac{dr}{d\phi}\right)^2,$$

where we made use of equation 105.8.

Thus equation 105.11 can be written in the form

$$\frac{h^2}{r^4}\frac{d^2r}{d\phi^2} - \frac{2h^2}{r^5}\left(\frac{dr}{d\phi}\right)^2 + \frac{m}{r^2} = \frac{h^2}{r^3}\left(1 - \frac{3m}{r}\right). \tag{105.12}$$

If we introduce a new dependent variable $u = 1/r$,

$$\frac{dr}{d\phi} = -\frac{1}{u^2}\frac{du}{d\phi}, \qquad \frac{d^2r}{d\phi^2} = \frac{2}{u^3}\left(\frac{du}{d\phi}\right)^2 - \frac{1}{u^2}\frac{d^2u}{d\phi^2},$$

and equation 105.12 reduces to

$$\frac{d^2u}{d\phi^2} + u = \frac{m}{h^2} + 3mu^2. \tag{105.13}$$

Equation 105.13 together with equation 105.8, which we write as

$$\frac{d\phi}{ds} = \frac{h}{r^2}, \tag{105.14}$$

suffices to determine the trajectory.

It is interesting to write down here the corresponding equations of the classical theory obtained in Sec. 97:

$$\frac{d^2u}{d\phi^2} + u = \frac{km_1}{h^2}, \qquad \frac{d\phi}{dt} = \frac{h}{r^2}, \tag{105.15}$$

where we write ϕ for the angular variable θ used in that section and introduce the gravitational constant $k = 6.7 \times 10^{-8}$ and

$$m_1 = 1.98 \times 10^{33} \text{ gr}$$

is the mass of the sun. Because of our choice of units for the velocity of light, we note that far away from gravitating matter

$$ds^2 = (dt)^2 - dy^i\, dy^i,$$

so that

$$\left(\frac{ds}{dt}\right)^2 = 1 - \frac{dy^i}{dt}\frac{dy^i}{dt} = 1 - v^2.$$

For planetary velocities, v is very small compared with the velocity of light, which we took to be 1, so that to a high degree of approximation $ds = dt$. Thus, in both classical and relativistic sets of equations, h can be interpreted as the sectorial velocity. The constant of integration m corresponds to km_1, so that the relativistic equation 105.13 differs from the corresponding classical equation only in the appearance of the term $3mu^2$.

Now, the ratio of $3mu^2$ to m/h^2 is $3h^2u^2$, or using equation 105.14 it is $3(r\, d\phi/ds)^2$. For ordinary planetary speeds this ratio is small. For example, the average radius of the earth's orbit is $r = 1.5 \times 10^{13}$ cm, the angular velocity $d\phi/dt = 2 \cdot 10^{-7}$ rad/sec, and, if we take as a first approximation $dt/ds = 1/c$, we find the value of $3r^2(d\phi/ds)^2$ to be of the order 10^{-8}.

Consequently, in ordinary planetary motion "the correction term" in the relativistic equation 105.13 is negligible, as far as the shape of the orbit is concerned, but the influence of this term on the behavior of the perihelion, as will be seen in Sec. 106, is significant.

It will be shown in the next section that the perihelion rotates through an angle $6m^2\pi/h^2$ rad during each revolution. This value proves to be too small for all planets in the solar system with the exception of Mercury, for which it corresponds to nearly 42″ of arc per century. This advance of the perihelion of Mercury has found no satisfactory explanation on the basis of the Newtonian theory, and we will see that the calculations based on the relativistic equation 105.13 give results which agree extraordinarily well with observed values.

We conclude this section by remarking that, if the foregoing calculations were performed with the quadratic form

$$ds^2 = c^2\gamma(dt)^2 - \frac{(dr)^2}{\gamma} - r^2[(d\theta)^2 + \sin^2\theta(d\phi)^2]$$

as a basis, we would have arrived at the equation[17]

$$\frac{d^2u}{d\phi^2} + u = \frac{km_1}{h^2} + \frac{3km_1u^2}{c^2},$$

where $m_1 = 1.98 \times 10^{33}$ gr (mass of the sun), $k = 6.7 \times 10^{-8}\ \text{gr}^{-1}\ \text{cm}^3/\text{sec}^2$, $c = 3 \cdot 10^{10}$ cm/sec.

For the motion of Mercury the term km_1/h^2 is of the order 10^{-12}, whereas $3km_1u^2/c^2$ is of the order 10^{-21}. These estimates justify us in attempting to solve equation 105.13 by a method of successive approximations sketched in the following section.

106. The Advance of Perihelion

A comparison of analytical results of this section with observed astronomical data provides us with the best available evidence in support of the general theory of relativity. In Sec. 107 we mention the deflection of the light beam by the sun and the shift of the Fraunhofer lines toward the red end of the spectrum, but the quantitative agreement for these phenomena between observations and theoretical predictions is still in some doubt.

The relativistic equation for the orbit of a planet

$$\frac{d^2u}{d\phi^2} + u = \frac{m}{h^2}(1 + 3h^2u^2), \tag{106.1}$$

deduced in Sec. 105, can be integrated in closed form with the aid of elliptic functions, but the solution obtained in this way does not lend itself to a convenient comparison with the corresponding result obtained in Sec. 97 on the basis of the Newtonian theory.

We noted in Sec. 105 that the magnitude of the term $3h^2u^2$, appearing in the right-hand member of equation 106.1, is small compared with unity, and this justifies us in attempting to obtain a solution of this equation by the method of perturbations. Accordingly, we neglect the small term $3mu^2$ and obtain for our first approximation u_1 the Newtonian equation

$$\frac{d^2u_1}{d\phi^2} + u_1 = \frac{m}{h^2},$$

the solution of which is

$$u_1 = \frac{m}{h^2}[1 + e\cos(\phi - \omega)], \tag{106.2}$$

[17] In this equation the sectorial velocity h is the sectorial velocity of the classical theory.

where e is the eccentricity of the orbit and ω is the longitude of the perihelion. Inserting from equation 106.2 in the right-hand member of equation 106.1 yields

$$\frac{d^2u}{d\phi^2} + u = \frac{m}{h^2}(1 + 3h^2u_1^2)$$
$$= \frac{m}{h^2} + \frac{6m^3}{h^4} e \cos(\phi - \omega)$$
$$+ \frac{3m^3}{2h^4} e^2[1 + \cos 2(\phi - \omega)] + \frac{3m^3}{h^4}. \tag{106.3}$$

Since planetary orbits are nearly circular (for Mercury, $e^2 = 0.04$), the contribution of the perturbation term containing e^2 will be negligible. Also the term $3m^3/h^4$ will not have a significant effect on the shape of the orbit, but the second term, containing $\cos(\phi - \omega)$, may have a pronounced cumulative effect on the displacement of the perihelion. Accordingly, we simplify equation 106.3 to read

$$\frac{d^2u}{d\phi^2} + u = \frac{m}{h^2} + \frac{6m^3}{h^4} e \cos(\phi - \omega).$$

The solution of this linear equation is clearly made up of the solution u_1 and the solution of

$$\frac{d^2u}{d\phi^2} + u = \frac{6m^3}{h^4} e \cos(\phi - \omega).$$

The result of easy calculations gives us the second approximation u_2 in the form

$$u_2 = \frac{m}{h^2}\left[1 + e\cos(\phi - \omega) + \frac{3m^2}{h^2} e\phi \sin(\phi - \omega)\right]. \tag{106.4}$$

It will suffice for our purposes to terminate the sequence of steps in the scheme of successive approximations at this stage and to regard u_2 as representing the solution of equation 106.1 to a sufficiently high degree of accuracy. If we set

$$\delta\omega \equiv \frac{3m^2}{h^2}\phi \tag{106.5}$$

and note that

$$\cos(\phi - \omega) + \delta\omega \sin(\phi - \omega) = \sqrt{1 + (\delta\omega)^2}\cos(\phi - \omega - \alpha),$$

where $\alpha = \tan^{-1}\delta\omega \doteq \delta\omega$, we can write (106.4) as

$$u_2 \doteq u = \frac{m}{h^2}[1 + e\cos(\phi - \omega - \delta\omega)], \tag{106.6}$$

if we neglect in comparison with unity terms of the order $(\delta\omega)^2$. It is clear from equations 106.5 and 106.6 that when a planet moves through one revolution, the perihelion advances through an angle

$$\epsilon = \frac{3m^2}{h^2} 2\pi \text{ rad.} \tag{106.7}$$

Equation 106.6 represents a closed orbit, only approximately elliptical in shape, because $\delta\omega$ is a function of ϕ. Since $u = 1/r$, we have

$$r = \frac{h^2/m}{1 + e\cos(\phi - \omega - \delta\omega)},$$

so that the "semilatus rectum" $l = h^2/m$.

Recalling from the geometry of conics that $l = a(1 - e^2)$, where a is the major axis of the conic, we get

$$h^2 = ml = ma(1 - e^2).$$

Inserting this result in equation 106.7 we have[18]

$$\epsilon = \frac{6\pi m^2}{am(1 - e^2)} = \frac{6\pi m}{a(1 - e^2)}.$$

In this expression m is the mass of the sun.

For Mercury the quantity ϵ works out to be 4.90×10^{-7} rad. This angle is very small, but the observational data on the location of Mercury during the last century are available, and since this planet has a period of 88 days, it completes 415 revolutions per century. Thus the cumulative advance of the perihelion in 100 years should amount to $415\epsilon = 2.04 \times 10^{-4}$ rad $= 42''$ of arc. For planets other than Mercury the corresponding advance is too small for accurate experimental determination. Thus for Venus it is only $9''$, for Earth $4''$, and for Mars $1''$.

The actual path of Mercury about the sun is not an ellipse, of course, because of the perturbing effects of other planets. We are not in reality dealing with a two-body problem. However, perturbations due to other planets can be taken into account and the deviations from an elliptical path calculated. Such calculations have been performed with great care, and it has been found that the advance of Mercury's perihelion should amount to about $42''$ of arc per century. The Newtonian theory is unable to account for the advance of this amount, and the remarkably close

[18] For a different way of deducing the value of ϵ see J. L. Synge, *Relativity: The General Theory* (1960), pp. 294–296, and G. Y. Rainich, *Mathematics of Relativity* (1950), p. 162.

agreement between the relativistic calculations and the best observed value can hardly be viewed as fortuitous.[19]

It is worth noting that the calculations based on the restricted theory of relativity also give a precessional effect when one assumes that a particle moves in a field of force with potential $V = km/r$. However, the precession based on such calculations yields results that are not as close to the observed value as those furnished by the general theory.

107. Concluding Remarks

We conclude this chapter with a mention of the relativistic prediction of deflection of light rays by the sun and of the shift toward the red end of the spectrum of spectral lines of light originating in dense stars.[20]

Since light is material in nature it must be affected by the gravitational field of the sun, and the deviation from the rectilinear path of the light ray from a distant star, as it grazes the sun, can be readily calculated.

The deflection of light rays passing near a large mass can be observed during eclipses of the sun when fixed stars in the apparent neighborhood of the sun become visible. However, because of the uncertainty about the magnitude of experimental errors arising from the difficulty of obtaining sharp photographic images, it is generally conceded that these results neither prove nor disprove the general theory. It may be remarked that the calculations based on Newtonian theory of gravitation can be made to account for about one-half of the observed values.

Among other experimental evidence cited in favor of the general theory is the observed displacement of spectral lines of light emitted from the stars toward the red end of the spectrum. Elementary considerations indicate that the frequency of vibration of the emitted light from a distant star is less than the corresponding frequency on the surface of the earth.[21] If this frequency is associated with the emitted light from the sun, the lines of the solar spectrum should be shifted slightly toward the long-wave end of the spectrum as contrasted with the corresponding lines of terrestrial

[19] G. M. Clemence gives $42''.56 \pm 0.94$ in *Reviews of Modern Physics*, vol. 19 (1947), p. 361, See also G. C. McVittie, *General Relativity and Cosmology* (1956). These authors make incisive comments on the difficulty of performing meaningful astronomical observations.

[20] See J. L. Synge, *Relativity: The General Theory* (1960), pp. 298–308, and Secs. 36 and 37 of G. Y. Rainich's *Mathematics of Relativity* (1950). See, also, P. G. Bergmann, *Introduction to the Theory of Relativity* (1942), Chapter XIV, and A. S. Eddington, *Mathematical Theory of Relativity* (1924), pp. 90–93. A critical survey of the validity of predictions of Einstein's theory is provided by G. C. McVittie, *General Relativity and Cosmology* (1956).

[21] See references given in the preceding footnote.

spectra. The expected shift for the light emitted by the sun is very small, but for the companions of Sirius it is estimated to be about thirty times as great as for vibrating solar particles and should be observed with a reasonable accuracy. In 1925, Adams measured the "red shift" for the companion of Sirius[22] and found it to be $\Delta\lambda = 0.27$ for the line of wavelength $\lambda = 4000$ A. From this determination the diameter of the star can be estimated, and it is found to be of the right order of magnitude. The evidence here is not conclusive, but it is generally regarded as favorable.

The law of gravitation $R_{ij} = 0$ was generalized by Einstein to the form $R_{ij} = \lambda g_{ij}$, where λ is a small "universal constant." Solutions of the generalized equation have led to various cosmological theories and have given rise to speculations about the expanding universe. We refer the reader for detailed accounts to specialized treatises on this subject.[23]

[22] The shift of the corresponding line in the sun's spectrum is calculated to be $\Delta\lambda = 0.008$.

[23] A. Eddington, *Mathematical Theory of Relativity* (1924).
R. C. Tolman, *Relativity, Thermodynamics and Cosmology* (1934).
P. Bergmann, *Introduction to the Theory of Relativity* (1942).
G. Y. Rainich, *Mathematics of Relativity* (1950).
L. Landau and E. Lifshitz, *The Classical Theory of Fields* (1951).
J. L. Synge, *Relativity: The Special Theory* (1956).
J. L. Synge, *Relativity: The General Theory* (1960).

6

MECHANICS OF CONTINUOUS MEDIA

108. Introductory Remarks

This chapter contains a general formulation of the basic concepts of mechanics of continua and a derivation of the fundamental equations governing the behavior of continuous media. The treatment contained here forms a substantial introduction to nonlinear mechanics of fluids and elastic solids. The linearized equations of classical theory appear as special cases of nonlinear equations, and throughout the chapter emphasis is placed on the unified formulation of equations of mechanics of continua in the most general tensor form.

A systematic development of tensor calculus, with an eye to applications to mechanics of continuous media, is contained in P. Appell's definitive, *Traité de mécanique rationnelle*, vol. 5 (1926), and in A. J. McConnell's *Applications of the Absolute Differential Calculus* (1931). These are largely concerned with the linearized cases. The landmarks in the domain of nonlinear theory of elasticity are papers by Leon Brillouin, "Les lois de l'élasticité sous forme tensorielle valable pour des coordonnées quelconques," *Annales de physique*, **3** (1925), pp. 251–298, and F. D. Murnaghan,[1] "Finite Deformations of an Elastic Solid," *American Journal of Mathematics*, **59** (1937), pp. 235–260. The essence of Brillouin's contributions appears also in his book *Les tenseurs en mécanique et en élasticité*, first published by Masson et Cie in 1938, and reprinted by the Dover Press in 1946.

Among more recent contributions to nonlinear theory of elasticity are the books of V. V. Novozhilov, *Foundations of Non-Linear Theory of Elasticity*, Moscow (1947), A. E. Green and W. Zerna, *Theoretical Elasticity*, Oxford (1954), and A. Signorini, *Questioni di Elasticita non Linearezzata*, Rome (1960). An exhaustive critical survey of the foundations of elasticity and fluid mechanics is contained in two extensive

[1] A brief exposition of the central ideas of Murnaghan's contributions will be found in Chapters 14 and 15 of A. D. Michal's *Matrix and Tensor Calculus* (1947).

memoirs by C. Truesdell in *Journal of Rational Mechanics and Analysis*, vol. 1 (1952) and vol. 2 (1953).

A development of the foundations of continuum mechanics, primarily within the framework of linear theories (including applications to mechanics of fluids, elasticity, and plasticity) is contained in W. Prager's *Introduction to Mechanics of Continua*, Boston (1961). A general unified development of geometrically and dynamically nonlinear mechanics of continuous media will be found in an excellent monograph by L. I. Sedov, *Introduction to Mechanics of a Continuous Medium*, Moscow (1962). Sedov's monograph to a large extent is based on the close union of classical mechanics and macroscopic thermodynamics. This unification permits one to construct the general models of gases, liquids, elastic and thermoelastic solids, and of several types of plastic media from a single point of view.

109. Deformation of a Continuous Medium

We consider a continuum of identifiable material points which at a given time $t = t_0$ fill a certain region of space τ_0. We shall refer to t_0 as the *initial time* and shall call τ_0 the *initial region*. With the passage of time the points P of τ_0 undergo displacements and at some time t fill a certain region τ. In the course of displacement, the initial region τ_0 is usually deformed, and we suppose that the deformation of τ_0 into τ is fully determined when the motion of every point P is known. To describe the motion of points P we introduce a coordinate system X which moves with the medium in such a way that the coordinates (x^1, x^2, x^3) of any given point P initially in τ_0 do not change with t. In addition to the system X we consider a fixed reference frame Y, relative to which the coordinates of the point $P(x^1, x^2, x^3)$ are given by

$$y^i = y^i(x^1, x^2, x^3, t). \tag{109.1}$$

The functional form of relations (109.1) clearly depends on the nature of deformation of τ_0 into τ. We shall assume that the functions $y^i(x, t)$ in (109.1) are single-valued, piecewise smooth, and possess for each value of time t a single-valued, piecewise smooth inverse

$$x^i = x^i(y^1, y^2, y^3, t). \tag{109.2}$$

The fixed coordinate system Y, without loss of generality, can be assumed to be orthogonal cartesian.

A material point P in τ_0, relative to an orthogonal cartesian frame Y, is determined by the position vector (Fig. 51)

$$\mathbf{r}_0 = \mathbf{c}_i y_0{}^i \equiv \mathbf{c}_i y^i(x^1, x^2, x^3, t_0), \tag{109.3}$$

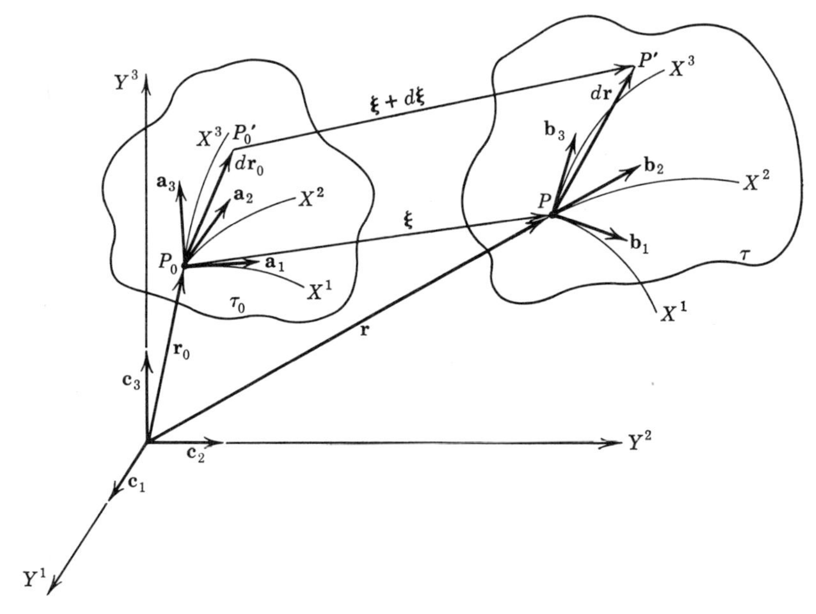

Fig. 51

where the $\mathbf{c}_i$ are the orthonormal base vectors associated with the frame Y. The location of the same point P in the region τ is determined by the vector

$$\mathbf{r} = \mathbf{c}_i y^i(x^1, x^2, x^3, t). \tag{109.4}$$

The base vectors $\mathbf{b}_j$ in the moving frame X are given by

$$\mathbf{b}_j = \frac{\partial \mathbf{r}}{\partial x^j} = \mathbf{c}_i \frac{\partial y^i(x, t)}{\partial x^j}, \tag{109.5}$$

and these vectors obviously depend not only on the coordinates x^i of P, but also on t. When $P(x^1, x^2, x^3)$ is in τ_0, we denote the base vectors $\mathbf{b}_j$ by $\mathbf{a}_j$ so that

$$\mathbf{a}_j = \frac{\partial \mathbf{r}_0}{\partial x^j} \equiv \mathbf{c}_i \frac{\partial y^i(x, t_0)}{\partial x^j}. \tag{109.6}$$

Thus, in analyzing the deformation of a continuous medium, we can speak of three reference frames: a fixed reference system Y determined by the basis $\mathbf{c}_i$, a moving reference frame X with the basis $\mathbf{b}_i$, and a fixed reference frame X with the basis $\mathbf{a}_i$. We emphasize that the labels (x^1, x^2, x^3) of a given material point P in both curvilinear coordinate systems X have the same values, but to avoid circumlocution we shall denote the point $P(x^1, x^2, x^3)$ when it is located in the initial region τ_0 by P_0.

Let P_0' be a point in the neighborhood of $P_0(x^1, x^2, x^3)$. The vector $\overrightarrow{P_0P_0'} = d\mathbf{r}_0$ can be represented in the form

(109.7) $$d\mathbf{r}_0 = \mathbf{a}_i \, dx^i$$

and the square of the arc element ds_0 in τ_0 is

$$(ds_0)^2 = d\mathbf{r}_0 \cdot d\mathbf{r}_0 = \mathbf{a}_i \cdot \mathbf{a}_j \, dx^i \, dx^j,$$

or

(109.8) $$(ds_0)^2 = h_{ij} \, dx^i \, dx^j,$$

where $h_{ij} = \mathbf{a}_i \cdot \mathbf{a}_j$ are metric coefficients in τ_0. Similarly, the square of the element of arc ds determined by the corresponding vector $\overrightarrow{PP'} = d\mathbf{r} = \mathbf{b}_i \, dx^i$ in τ is

$$ds^2 = \mathbf{b}_i \cdot \mathbf{b}_j \, dx^i \, dx^j,$$

or

(109.9) $$ds^2 = g_{ij} \, dx^i \, dx^j,$$

where the $g_{ij} = \mathbf{b}_i \cdot \mathbf{b}_j$ are metric coefficients in τ. Ordinarily the lengths and the orientations of vectors $d\mathbf{r}_0$ and $d\mathbf{r}$ will be different, and we shall say that the medium occupying τ is strained whenever $ds_0 \neq ds$. We can take as our measure of strain the difference

(109.10) $$(ds)^2 - (ds_0)^2 = (g_{ij} - h_{ij}) \, dx^i \, dx^j,$$

and, if we set

(109.11) $$g_{ij} - h_{ij} = 2\epsilon_{ij},$$

we can write (109.10) as

(109.12) $$(ds)^2 - (ds_0)^2 = 2\epsilon_{ij} \, dx^i \, dx^j.$$

Since (109.12) is an invariant and $\epsilon_{ij} = \epsilon_{ji}$, we conclude that the set of functions $\epsilon_{ij}(x, t)$ represents a tensor E_0 with respect to a class of admissible transformations of coordinates X, with the basis $\mathbf{a}_i$, covering the region τ_0. The same set of functions $\epsilon_{ij}(x, t)$ also determines a tensor E with respect to a set of transformations of coordinates determined by the basis $\mathbf{b}_i$ of the final state τ. In the notation of the concluding paragraph of Sec. 45, the tensor E_0 is specified by the multilinear form $E_0 = \epsilon_{ij}\mathbf{a}^i\mathbf{a}^j$, whereas the tensor E is determined by $E = \epsilon_{ij}\mathbf{b}^i\mathbf{b}^j$. Thus the operations of covariant differentiation and those of raising and lowering indices on the components of E_0 involve the metric tensor h_{ij}, whereas the corresponding operations on E make use of the tensor g_{ij}. Accordingly,

$$h^{ij}\epsilon_{ik} = \epsilon_k^{\ j} \quad \text{and} \quad g^{ij}\epsilon_{ik} = \epsilon_k^{\ j}.$$

However, the two sets of functions $\epsilon_k^{\ j}$ so computed in general are distinct, and to indicate the origin of the set $\epsilon_k^{\ j}$ obtained with the use of the tensor h_{ij}, we shall write

$$h^{ij}\epsilon_{ik} = \epsilon_{0k}^{\ \ j}.$$

It will be shown in the following section that either of the tensors E_0 or E can serve to characterize the state of deformation of the neighborhood of P_0. Tensors E_0 and E are sometimes called, respectively, the Lagrangean and Eulerian strain tensors in accordance with the two viewpoints of hydrodynamics, associated with the choices of coordinates of the initial or final states as independent variables in the formulation of hydrodynamical equations.

110. Geometric Interpretation of Strain Tensors E_0 and E

In the preceding section we defined the set of functions ϵ_{ij} by the formula

$$(ds)^2 - (ds_0)^2 = 2\epsilon_{ij}\,dx^i\,dx^j,$$

where

$$2\epsilon_{ij} = g_{ij} - h_{ij}. \tag{110.1}$$

Since $g_{ij} = \mathbf{b}_i \cdot \mathbf{b}_j$ and $h_{ij} = \mathbf{a}_i \cdot \mathbf{a}_j$, we can write (110.1) as

$$\begin{aligned} 2\epsilon_{ij} &= \mathbf{b}_i \cdot \mathbf{b}_j - \mathbf{a}_i \cdot \mathbf{a}_j \\ &= |\mathbf{b}_i| \cdot |\mathbf{b}_j| \cos\theta_{ij} - |\mathbf{a}_i| \cdot |\mathbf{a}_j| \cos\theta_{ij}^0, \end{aligned} \tag{110.2}$$

where θ_{ij} is the angle between the base vectors $\mathbf{b}_i$ and $\mathbf{b}_j$, and $\theta_{ij}{}^0$ is the angle between $\mathbf{a}_i$ and $\mathbf{a}_j$. If we denote by e the change in length per unit length of the vector $d\mathbf{r}_0 = \overrightarrow{P_0P_0'}$ in Fig. 51, so that

$$e = \frac{|d\mathbf{r}| - |d\mathbf{r}_0|}{|d\mathbf{r}_0|} = \frac{ds - ds_0}{ds_0},$$

we have

$$|d\mathbf{r}| = (1 + e)\,|d\mathbf{r}_0|. \tag{110.3}$$

We call e the *elongation* of $d\mathbf{r}_0$ and we see from (110.3) that the elongations e_i in the directions of base vectors $\mathbf{a}_i$ are given by

$$|\mathbf{b}_i| = (1 + e_i)\,|\mathbf{a}_i|. \tag{110.4}$$

However, $|\mathbf{b}_i| = \sqrt{g_{ii}}$ and $|\mathbf{a}_i| = \sqrt{h_{ii}}$, so that

(110.5) $$\sqrt{g_{ii}} = (1 + e_i)\sqrt{h_{ii}}, \qquad \text{(no sum on } i\text{)},$$

and hence formula 110.2 can be written with the aid of (110.4) and (110.5) as

(110.6) $$\frac{2\epsilon_{ij}}{\sqrt{h_{ii}}\sqrt{h_{jj}}} = (1 + e_i)(1 + e_j)\cos\theta_{ij} - \cos\theta_{ij}^0.$$

Since $\theta_{ij}{}^0 = \theta_{ij} = 0$ for $i = j$, equation 110.6 yields

$$\frac{2\epsilon_{ii}}{h_{ii}} = (1 + e_i)^2 - 1,$$

or

(110.7) $$e_i = \sqrt{1 + \frac{2\epsilon_{ii}}{h_{ii}}} - 1, \qquad \text{(no sum)}.$$

When the coordinates of the initial state are rectangular cartesians, $h_{ii} = 1$, and we see from (110.7) that for $2\epsilon_{ii}/h_{ii} \ll 1, e_i \doteq \epsilon_{ii}$. Accordingly, the functions $\epsilon_{11}, \epsilon_{22}, \epsilon_{33}$ are related to the elongations of arc elements directed along the base vectors $\mathbf{a}_1, \mathbf{a}_2, \mathbf{a}_3$.

The significance of the ϵ_{ij} for $i \neq j$ follows from (110.6) on noting that when $\mathbf{a}_i$ and $\mathbf{a}_j$ are orthogonal unit vectors, $\theta_{ij}{}^0 = \pi/2$. If we set $\theta_{ij} = \pi/2 - \alpha_{ij}$, so that α_{ij} represents the change in the initially right angle between the pair of arc elements directed along $\mathbf{a}_i$ and $\mathbf{a}_j$, formula 110.6 gives

$$2\epsilon_{ij} = (1 + e_i)(1 + e_j)\sin\alpha_{ij},$$

or

(110.8) $$\sin\alpha_{ij} = \frac{2\epsilon_{ij}}{\sqrt{1 + 2\epsilon_{ii}}\sqrt{1 + 2\epsilon_{jj}}},$$

where we recalled (110.7). If $2\epsilon_{ii} \ll 1$ and the angle α_{ij} is small, we have an approximate equality, $\alpha_{ij} \doteq 2\epsilon_{ij}$. Thus the functions ϵ_{ij} for $i \neq j$ provide a measure of the decrease in the initially right angle between the arc elements parallel to the vectors $\mathbf{a}_i$ and $\mathbf{a}_j$. The components ϵ_{ij} for $i \neq j$ are called *shearing components* of the strain tensor E_0, and the components ϵ_{ij} for $i = j$ are the *normal components* of E_0.

Quite analogous interpretations can be provided for the functions ϵ_{ij} when these are viewed as components of the tensor $E = \epsilon_{ij}\mathbf{b}^i\mathbf{b}^j$. Indeed, if we now define the elongation e as the change in length per unit final length $|d\mathbf{r}|$ of the arc element so that

$$e = \frac{ds - ds_0}{ds},$$

the calculations similar to those that have led to formulas 110.7 and 110.8 now yield

$$e_i = 1 - \sqrt{1 - \frac{2\epsilon_{ii}}{g_{ii}}} \tag{110.9}$$

and

$$\sin \beta_{ij} = \frac{2\epsilon_{ij}}{\sqrt{1 - 2\epsilon_{ii}}\sqrt{1 - 2\epsilon_{jj}}}, \qquad \text{(no sums)}, \tag{110.10}$$

where $\beta_{ij} = \theta_{ij}{}^0 - \pi/2$.

We conclude, as before, that the components ϵ_{ii} in (110.9) are associated with elongations of the arc elements originally parallel to the base vectors $\mathbf{b}_i$, whereas the components ϵ_{ij} for $i \neq j$ measure the corresponding shearing deformations.

111. Strain Quadric. Principal Strains

The defining formula 109.12 for components ϵ_{ij} of the strain tensor $E = \epsilon_{ij}\mathbf{b}^i\mathbf{b}^j$ can be written as

$$\frac{(ds)^2 - (ds_0)^2}{2(ds)^2} = \epsilon_{ij}\frac{dx^i}{ds}\frac{dx^j}{ds}, \tag{111.1}$$

where $dx^i/ds = \lambda^i$ is the unit vector determining the direction of the vector $d\mathbf{r}$ in the final state. We seek to determine those directions λ^i for which (111.1) takes on extreme values. Accordingly, we set

$$Q(\lambda) = \epsilon_{ij}\lambda^i\lambda^j \tag{111.2}$$

and maximize the quadratic form $Q(\lambda)$ subject to constraining relation

$$\phi(\lambda) = g_{ij}\lambda^i\lambda^j - 1 = 0,$$

requiring that λ^i be a unit vector.

The familiar procedure for determining the extreme values of (111.2) by the method of Lagrange multipliers leads to the system of equations

$$\frac{\partial Q}{\partial \lambda^i} - \epsilon\frac{\partial \phi}{\partial \lambda^i} = 0,$$

or

$$(\epsilon_{ij} - \epsilon g_{ij})\lambda^j = 0, \tag{111.3}$$

where ϵ is the Lagrange multiplier.

This system possesses nontrivial solutions for λ^i if, and only if,

$$|\epsilon_{ij}(x) - \epsilon g_{ij}(x)| = 0$$

at each point $P(x)$ of the region τ. In order to reduce this system 111.3 to the form 13.10 considered in Sec. 13, we multiply (111.3) by g^{ik}, sum on i, and obtain

$$(\epsilon_j^{\ k} - \epsilon\delta_j^{\ k})\lambda^j = 0, \tag{111.4}$$

where

$$\epsilon_j^{\ k} = g^{ik}\epsilon_{ij}. \tag{111.5}$$

The system 111.4 has three nontrivial solutions $\lambda^i_{(1)}, \lambda^i_{(2)}, \lambda^i_{(3)}$ $(i = 1, 2, 3)$, corresponding to the roots ϵ_i of the cubic

$$|\epsilon_j^{\ i} - \epsilon\delta_j^{\ i}| \equiv -\epsilon^3 + \vartheta_1\epsilon^2 - \vartheta_2\epsilon + \vartheta_3 = 0. \tag{111.6}$$

The coefficients ϑ_i in this cubic are the invariants

$$\begin{cases} \vartheta_1 = \epsilon_1 + \epsilon_2 + \epsilon_3, \\ \vartheta_2 = \epsilon_2\epsilon_3 + \epsilon_3\epsilon_1 + \epsilon_1\epsilon_2, \\ \vartheta_3 = \epsilon_1\epsilon_2\epsilon_3. \end{cases} \tag{111.7}$$

It was shown in Secs. 13–15 that the roots ϵ_i are necessarily real and the directions $\lambda^i_{(1)}, \lambda^i_{(2)}, \lambda^i_{(3)}$ associated with them are orthogonal.

The quadratic form 111.2, where we regard the λ^i as the running coordinates, reduces to the canonical form

$$Q(y) = \epsilon_1(y^1)^2 + \epsilon_2(y^2)^2 + \epsilon_3(y^3)^2, \tag{111.8}$$

provided that the *principal directions* $\lambda^i_{(1)}, \lambda^i_{(2)}, \lambda^i_{(3)}$ are chosen as the base vectors of a suitable orthogonal cartesian reference system Y in τ.

We can interpret these results geometrically by introducing a *strain quadric*

$$\epsilon_{ij}(x)\lambda^i\lambda^j = \text{constant}, \tag{111.9}$$

which, at each point $P(x)$, represents a quadric surface with the λ^i as the running coordinates. The principal directions $\lambda^i_{(j)}$ coincide with the axes of the quadric 111.9, and it follows from (111.8) that the strain tensor ϵ_{ij}, when referred to the frame Y, has the form

$$\begin{bmatrix} \epsilon_1 & 0 & 0 \\ 0 & \epsilon_2 & 0 \\ 0 & 0 & \epsilon_3 \end{bmatrix}.$$

From the geometrical significance of components ϵ_{ij}, $i \neq j$ (see equation 110.10), it follows that the *principal directions are those orthogonal directions in the undeformed state which remain orthogonal after deformation.*

The strains $\epsilon_1, \epsilon_2, \epsilon_3$ are termed the *principal strains.*

The invariants ϑ_1, ϑ_2, ϑ_3 defined by (111.7) play an important role in the construction of models of continuous media. If we expand the determinant in (111.6) and equate the coefficients of like powers of ϵ in the result, we find

$$
\begin{aligned}
\vartheta_1 &= \epsilon_1^{\ 1} + \epsilon_2^{\ 2} + \epsilon_3^{\ 3} \equiv \epsilon_i^{\ i}, \\
\vartheta_2 &= \begin{vmatrix} \epsilon_2^{\ 2} & \epsilon_3^{\ 2} \\ \epsilon_2^{\ 3} & \epsilon_3^{\ 3} \end{vmatrix} + \begin{vmatrix} \epsilon_3^{\ 3} & \epsilon_1^{\ 3} \\ \epsilon_3^{\ 1} & \epsilon_1^{\ 1} \end{vmatrix} + \begin{vmatrix} \epsilon_1^{\ 1} & \epsilon_2^{\ 1} \\ \epsilon_1^{\ 2} & \epsilon_2^{\ 2} \end{vmatrix} \equiv \frac{1}{2!}\,\delta^{ij}_{\alpha\beta}\epsilon_i^{\ \alpha}\epsilon_j^{\ \beta}, \\
\vartheta_3 &= \begin{vmatrix} \epsilon_1^{\ 1} & \epsilon_2^{\ 1} & \epsilon_3^{\ 1} \\ \epsilon_1^{\ 2} & \epsilon_2^{\ 2} & \epsilon_3^{\ 2} \\ \epsilon_1^{\ 3} & \epsilon_2^{\ 3} & \epsilon_3^{\ 3} \end{vmatrix} \equiv \frac{1}{3!}\,\delta^{ijk}_{\alpha\beta\gamma}\epsilon_i^{\ \alpha}\epsilon_j^{\ \beta}\epsilon_k^{\ \gamma}.
\end{aligned}
\tag{111.10}
$$

We will see in the following section how these invariants enter in the expression for the ratio of the volume elements $d\tau_0$ and $d\tau$ of the initial and deformed states.

We could have equally well considered the quadratic form

$$Q_0 = \epsilon_{ij}\lambda_0^{\ i}\lambda_0^{\ j}, \tag{111.11}$$

with $\lambda_0^{\ i} = dx^i/ds_0$ specifying the direction of the vector $d\mathbf{r}_0$ of the initial state, and with ϵ_{ij}'s regarded as components of the strain tensor $E_0 = \epsilon_{ij}\mathbf{a}^i\mathbf{a}^j$.

For the determination of principal directions we now have the set of equations of the type 111.4 in which

$$\epsilon_j^{\ k} = h^{ik}\epsilon_{ij} \tag{111.12}$$

and the values of ϵ are the roots of the characteristic equation

$$|\epsilon_j^{\ k} - \epsilon\delta_j^{\ k}| = 0, \tag{111.13}$$

in which the $\epsilon_j^{\ k}$ are given by (111.12). The quadratic form 111.11 can thus be reduced to a canonical form

$$Q_0 = \epsilon_1^{\ 0}(y_0^{\ 1})^2 + \epsilon_2^{\ 0}(y_0^{\ 2})^2 + \epsilon_0^{\ 3}(y_0^{\ 3})^2$$

when the principal directions $\lambda^i_{0(1)}$, $\lambda^i_{0(2)}$, $\lambda^i_{0(3)}$ are taken as the basis of a suitable orthogonal frame Y_0 in τ_0.

It follows from formulas 110.7 that the elongations $e_i^{\ 0}$ along the principal directions are

$$e_i^{\ 0} = \frac{ds^i - ds_0^{\ i}}{ds_0^{\ i}} = \sqrt{1 + 2\epsilon_i^{\ 0}} - 1, \tag{111.14}$$

whereas the elongations e_i, reckoned per unit length in the final state (cf. equation 110.9), are

$$e_i = \frac{ds^i - ds_0{}^i}{ds^i} = 1 - \sqrt{1 - 2\epsilon_i}. \tag{111.15}$$

We conclude from (111.14) and (111.15) that

$$\epsilon_i{}^0 = \frac{\epsilon_i}{1 - 2\epsilon_i} \quad \text{and} \quad \epsilon_i = \frac{\epsilon_i{}^0}{1 + 2\epsilon_i{}^0}. \tag{111.16}$$

Formulas 111.16 permit us to express the invariants $\vartheta_i{}^0$ of the cubic (111.13) in terms of the invariants ϑ_i given in (111.7), and the invariants ϑ_i of the cubic (111.6) in terms of the $\vartheta_i{}^0$.

Problem

Show that

$$\vartheta_1 = \frac{\vartheta_1{}^0 + 4\vartheta_2{}^0 + 12\vartheta_3{}^0}{1 + 2\vartheta_1{}^0 + 4\vartheta_2{}^0 + 8\vartheta_3{}^0},$$

$$\vartheta_2 = \frac{\vartheta_2{}^0 + 6\vartheta_3{}^0}{1 + 2\vartheta_1{}^0 + 4\vartheta_2{}^0 + 8\vartheta_3{}^0},$$

$$\vartheta_3 = \frac{\vartheta_0{}^3}{1 + 2\vartheta_1{}^0 + 4\vartheta_2{}^0 + 8\vartheta_3{}^0}.$$

112. Distortion of Volume Elements

We investigate next the change in volume elements $d\tau_0$ and $d\tau$ of the initial and deformed states and indicate its connection with the invariants ϑ_i introduced in Sec. 111.

It follows from the definition of the volume element in Sec. 44 that

$$d\tau_0 = \sqrt{h}\, dx^1\, dx^2\, dx^3 \quad \text{and} \quad d\tau = \sqrt{g}\, dx^1\, dx^2\, dx^3,$$

where $h = |h_{ij}|$ and $g = |g_{ij}|$ are the determinants of the quadratic forms

$$ds_0{}^2 = h_{ij}\, dx^i\, dx^j \quad \text{and} \quad ds^2 = g_{ij}\, dx^i\, dx^j.$$

Thus

$$\frac{d\tau_0}{d\tau} = \sqrt{h/g}, \tag{112.1}$$

The set of functions $h_{ij}(x)$ can be regarded as components of the tensor $H = h_{ij}\mathbf{b}^i\mathbf{b}^j$ defined in the space of the variables x^i in the final state, so that

$$g^{ik}h_{ij} = h_j{}^k$$

and

$$g_{ik}h_j{}^k = h_{ij}.$$

We conclude that

$$|g_{ik}h_j^{\ k}| = |h_{ij}|,$$

so that

$$g\,|h_j^{\ i}| = h.$$

Consequently the ratio 112.1 assumes the form

$$\frac{d\tau_0}{d\tau} = \sqrt{|h_j^{\ i}|}. \tag{112.2}$$

But from definition 110.1 we have

$$h_{ij}(x) = g_{ij}(x) - 2\epsilon_{ij}(x),$$

which, upon raising the indices, reads

$$h_j^{\ i} = \delta_j^{\ i} - 2\epsilon_j^{\ i}.$$

We can therefore write formula 112.2 as

$$\frac{d\tau_0}{d\tau} = \sqrt{|\delta_j^{\ i} - 2\epsilon_j^{\ i}|}. \tag{112.3}$$

If we expand the determinant appearing under the radical sign, we find

$$|\delta_j^{\ i} - 2\epsilon_j^{\ i}| = 1 - 2\vartheta_1 + 4\vartheta_2 - 8\vartheta_3, \tag{112.4}$$

where the ϑ_i are the invariants 111.10.

In the linear theory of deformation the products of strains $\epsilon_j^{\ i}$ can be disregarded, so that an approximate expression for the ratio 112.3 is

$$\frac{d\tau_0}{d\tau} \doteq \sqrt{1 - 2\vartheta_1}$$
$$\doteq 1 - \vartheta_1.$$

Thus approximately

$$\frac{d\tau - d\tau_0}{d\tau} = \vartheta_1. \tag{112.5}$$

This represents the change in volume per unit volume, and, for this reason, ϑ_1 is called the *dilatation*. It figures prominently in linear elasticity and hydrodynamics.

Formula 112.3 can be cast in the form

$$\frac{d\tau_0}{d\tau} = \frac{1}{\sqrt{|\delta_j^{\ i} + 2\epsilon_i^{\ 0}|}} = \frac{1}{\sqrt{1 + 2\vartheta_1^{\ 0} + 4\vartheta_2^{\ 0} + 8\vartheta_3^{\ 0}}} \tag{112.6}$$

by expressing the invariants ϑ_i in terms of the ϑ_i^0 as in the Problem of Sec. 111. When the deformations are small, it follows from (112.6) that

$$\frac{d\tau - d\tau_0}{d\tau_0} \doteq \vartheta_1^0 \tag{112.7}$$

and, since for small deformations $\epsilon_i^0 \doteq \epsilon_i$, $\vartheta_1^0 \doteq \vartheta_1$, both formulas 112.5 and 112.7 give the same value for the dilatation.

Problem

Obtain formulas 112.3 and 112.6 directly from (111.14) and (111.15).

113. Displacements in Continuous Media

We define the displacement vector $\boldsymbol{\xi}$ of the point P_0 (Fig. 51) by

$$\boldsymbol{\xi} = \mathbf{r} - \mathbf{r}_0 \tag{113.1}$$

and denote the components of $\boldsymbol{\xi}$ relative to the basis $\mathbf{a}_i$ by u^i and its components relative to the basis $\mathbf{b}_i$ by w^i. Thus

$$\boldsymbol{\xi} = u^i\mathbf{a}_i, \qquad \boldsymbol{\xi} = w^i\mathbf{b}_i. \tag{113.2}$$

From (113.1) we have

$$\frac{\partial \boldsymbol{\xi}}{\partial x^i} = \frac{\partial \mathbf{r}}{\partial x^i} - \frac{\partial \mathbf{r}_0}{\partial x^i} = \mathbf{b}_i - \mathbf{a}_i,$$

so that

$$\mathbf{b}_i = \mathbf{a}_i + \frac{\partial \boldsymbol{\xi}}{\partial x^i}. \tag{113.3}$$

On computing $g_{ij} = \mathbf{b}_i \cdot \mathbf{b}_j$ with the aid of (113.3) and subtracting $h_{ij} = \mathbf{a}_i \cdot \mathbf{a}_j$ from the result, we find

$$\begin{aligned} g_{ij} - h_{ij} &= \frac{\partial \boldsymbol{\xi}}{\partial x^i} \cdot \frac{\partial \boldsymbol{\xi}}{\partial x^j} + \mathbf{a}_i \cdot \frac{\partial \boldsymbol{\xi}}{\partial x^j} + \mathbf{a}_j \cdot \frac{\partial \boldsymbol{\xi}}{\partial x^i} \\ &= 2\epsilon_{ij}, \quad \text{by (110.1)}. \end{aligned} \tag{113.4}$$

Equations 113.4 can be regarded as a set of differential equations for the components of $\boldsymbol{\xi}$ when the functions ϵ_{ij} are specified. This set of equations assumes quite simple form when the displacement vector $\boldsymbol{\xi}$ is expressed in terms of its covariant components u_j or w_j, so that

$$\boldsymbol{\xi} = u_j\mathbf{a}^j, \qquad \boldsymbol{\xi} = w_j\mathbf{b}^j, \tag{113.5}$$

the $\mathbf{a}^j$ and $\mathbf{b}^j$ being the reciprocal base vectors introduced in Sec. 45.

On differentiating (113.5) with respect to x^i, we get (cf. Sec. 45)

$$\frac{\partial \boldsymbol{\xi}}{\partial x^i} = u_{j|i}\mathbf{a}^j, \quad \frac{\partial \boldsymbol{\xi}}{\partial x^i} = w_{j,i}\mathbf{b}^j, \tag{113.6}$$

where

$$u_{j|i} = \frac{\partial u_j}{\partial x^i} - {}_h\begin{Bmatrix} k \\ ji \end{Bmatrix} u_k \tag{113.7}$$

is the covariant derivative of u_j with respect to the metric h_{ij} of the initial state and

$$w_{j,i} = \frac{\partial w_j}{\partial x^i} - {}_g\begin{Bmatrix} k \\ ji \end{Bmatrix} w_k \tag{113.8}$$

is the covariant derivative of w_j with respect to the metric g_{ij} of the final state. The prescripts on the Christoffel symbols in (113.7) and (113.8) indicate that these symbols in (113.7) are constructed from the tensor h_{ij}, whereas those in (113.8) from the g_{ij}'s.

If we insert from the first of formulas 113.6 in 113.4, we get

$$\begin{aligned} 2\epsilon_{ij} &= (u_{l|i}\mathbf{a}^l \cdot u_{k|j}\mathbf{a}^k) + (\mathbf{a}_i \cdot \mathbf{a}^k u_{k|j} + \mathbf{a}_j \cdot \mathbf{a}^k u_{k|i}) \\ &= u_{l|i}u_{k|j}\mathbf{a}^l \cdot \mathbf{a}^k + \delta_i^k u_{k|j} + \delta_j^k u_{k|i} \\ &= u^k_{|i}u_{k|j} + u_{i|j} + u_{j|i}, \end{aligned}$$

since $\quad \mathbf{a}^l \cdot \mathbf{a}^k = h^{lk}.$

Thus

$$2\epsilon_{ij} = u_{i|j} + u_{j|i} + u^k_{|i}u_{k|j}. \tag{113.9}$$

On the other hand, when $\boldsymbol{\xi}$ is represented in the form $\boldsymbol{\xi} = w_j\mathbf{b}^j$, we recall (113.3) and write

$$h_{ij} = \mathbf{a}_i \cdot \mathbf{a}_j = \left(\mathbf{b}_i - \frac{\partial \boldsymbol{\xi}}{\partial x^i}\right) \cdot \left(\mathbf{b}_j - \frac{\partial \boldsymbol{\xi}}{\partial x^j}\right).$$

The substitution into this expression from the second of the formulas in (113.6) yields

$$2\epsilon_{ij} = w_{i,j} + w_{j,i} - w^k_{,i}w_{k,j}. \tag{113.10}$$

Formulas 113.9 enable us to compute the strain components ϵ_{ij} from components u_i of the vector $\boldsymbol{\xi}$ referred to the basis $\mathbf{a}_i$ of the initial state. Formulas 113.10, on the other hand, involve the components of $\boldsymbol{\xi}$ relative to the basis $\mathbf{b}_i$ of the final state. Alternatively, when the functions ϵ_{ij} are specified, equations 113.9 and 113.10 are differential equations serving to determine the components of the displacement vector $\boldsymbol{\xi}$.

When the reference system X is orthogonal cartesian, we set $y^i = x^i$ and obtain from (113.9) and (113.10)

$$2\epsilon_{ij} = \frac{\partial u_i}{\partial y_0{}^j} + \frac{\partial u_j}{\partial y_0{}^i} + \frac{\partial u_k}{\partial y_0{}^i}\frac{\partial u_k}{\partial y_0{}^j}, \tag{113.11}$$

$$2\epsilon_{ij} = \frac{\partial w_i}{\partial y^j} + \frac{\partial w_j}{\partial y^i} - \frac{\partial w_k}{\partial y^i}\frac{\partial w_k}{\partial y^j}, \tag{113.12}$$

where the labels $y_0{}^j$ refer to the cartesian coordinates in the initial state.

In special problems the derivatives of the displacement components may be sufficiently small to justify one in neglecting products of these derivatives in comparison with the first-order terms in these derivatives. In this event equations 113.11 and 113.12 become linear and the theory of deformation based on a study of resulting linear differential equations is called the *linear theory*. In the linear theory it is usually assumed that the displacement vector $\boldsymbol{\xi}$ is so small that one is justified in identifying the coördinates $y_0{}^j$ and y^j of the initial and final states. The resulting theory is called the *infinitesimal* theory of deformation. In the infinitesimal theory, formulas 113.11 and 113.12 coalesce and we write

$$2e_{ij} = u_{i,j} + u_{j,i} \tag{113.13}$$

where the e_{ij} are the *infinitesimal components* of the strain tensor ϵ_{ij}. In classical theory of elasticity, the strain tensor ϵ_{ij} is taken in the form (113.13). The strain invariant $\vartheta_1 = e_{11} + e_{22} + e_{33}$, as follows from (113.13), is then equal to divergence of the displacement vector u_i, and hence the dilation $\vartheta_1 = (d\tau - d\tau_0)/d\tau = u^i_{,i}$.

114. Equations of Compatibility

Equations 113.10 or, in cartesian form, 113.12, can be viewed as a system of six simultaneous partial differential equations for the determination of three components of displacement from prescribed values of the strain tensor. Clearly, if a solution of this system is to exist, components of the strain tensor cannot be specified arbitrarily. To ensure the integrability of the system it is necessary to impose certain restrictions on the choice of functions ϵ_{ij}. Such conditions were deduced and the proof of their necessity,[2] for the linearized case typified by equation 113.13, was given by B. Saint Venant in 1860. We indicate here how these *integrability*, or *compatibility*, conditions can be deduced in the general case.

[2] For a proof of necessity and sufficiency of Saint Venant's conditions see I. S. Sokolnikoff, *Mathematical Theory of Elasticity* (1946), pp. 24–25.

We recall that the space in which the deformations take place is Euclidean, and hence the Riemann tensor, associated with the metric of Euclidean space specified by $ds_0{}^2 = h_{ij}\,dx^i\,dx^j$, vanishes (see Sec. 39). Thus

$$(114.1)\qquad R^0_{ijkl} = \frac{\partial}{\partial x^k}[jl, i] - \frac{\partial}{\partial x^l}[jk, i] + \begin{Bmatrix} \alpha \\ jk \end{Bmatrix}[il, \alpha] - \begin{Bmatrix} \alpha \\ jl \end{Bmatrix}[ik, \alpha] = 0,$$

where the Riemann tensor R^0_{ijk} is formed from the metric coefficients h_{ij}. If we recall that (see 110.1)

$$h_{ij} = g_{ij} - 2\epsilon_{ij},$$

compute the Christoffel symbols needed in (114.1) in terms of the g_{ij} and ϵ_{ij}, and make use of the fact that the Riemann tensor R_{ijkl} based on the g_{ij}'s also vanishes, we get the condition

$$(114.2)\qquad \epsilon_{ijkl} + \bar{h}^{\alpha\beta}(\epsilon_{jk\beta}\epsilon_{il\alpha} - \epsilon_{jl\beta}\epsilon_{ik\alpha}) = 0,$$

where

$$\epsilon_{ijkl} \equiv \epsilon_{jl,ik} + \epsilon_{ik,jl} - \epsilon_{ij,kl} - \epsilon_{kl,ij},$$

$$\epsilon_{ijk} \equiv \epsilon_{ik,j} + \epsilon_{kj,i} - \epsilon_{ij,k},$$

and

$$\bar{h}^{\alpha\beta} = \frac{H^{\alpha\beta}}{h},$$

$H^{\alpha\beta}$ being the cofactor[3] of $h_{\alpha\beta}$ in $|h_{ij}|$.

If we linearize (114.2) by dropping terms involving the products of the ϵ_{ijk}, we get Saint Venant's compatibility equations

$$(114.3)\qquad e_{ij,kl} + e_{kl,ij} - e_{ik,jl} - e_{jl,ik} = 0,$$

familiar in the linear theory of strain.[4]

From the fact that in a three-dimensional space the Riemann tensor has six independent nonzero components, if follows that there are six independent equations in (114.2) and (114.3).

115. Analysis of Stressed State

In analyzing the state of stress in a deformed body, it is natural to use the variables x^i of the final state as the independent variables. We will demonstrate that the state of stress at a point $P(x)$ of a body, in equilibrium

[3] Note that the contravariant tensor h^{ij} is the associated tensor of h_{ij} *with respect to the metric tensor* g_{ij}. See Sec. 30.

[4] See in this connection a paper by W. R. Seugling, *American Mathematical Monthly*, vol. 57 (1950), pp. 679–681; also F. D. Murnaghan, *Finite Deformation of an Elastic Solid* (1951) and L. I. Sedov, *Introduction to Mechanics of a Continuous Medium* (1962), pp. 128–130.

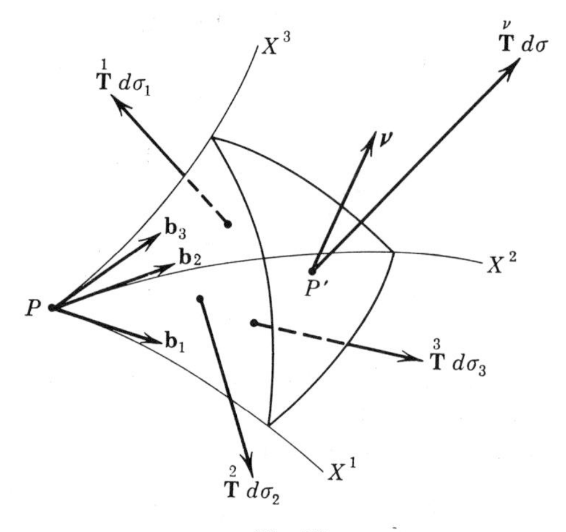

Fig. 52

under prescribed surface and body forces, is characterized by a symmetric tensor, the *stress tensor.*

Let a body τ be referred to a curvilinear coordinate system X, and consider an element of surface area at some point P' of the body. Let a small tetrahedral volume element $d\tau$ be formed by the coordinate surfaces at a nearby point P and by the surface element $d\sigma$ (Fig. 52). If $\boldsymbol{\nu}$ is the unit normal to $d\sigma$ then the elements of area $d\sigma_i$ lying in the coordinate surfaces are given by the formulas

$$d\sigma_i = \nu_i\, d\sigma, \tag{115.1}$$

where the ν_i are the covariant components of $\boldsymbol{\nu}$.

We denote the *stress vector* (force per unit area) acting on $d\sigma$ by $\overset{\nu}{\mathbf{T}}$ where the superscript ν brings into evidence the dependence of the stress vector on the orientation of the element $d\sigma$. The stress vectors acting on the surface elements $d\sigma_i$ are denoted by $\overset{i}{\mathbf{T}}$, and we take as their positive directions the directions of the exterior normals to the volume element. We can write

$$\overset{i}{\mathbf{T}} = -\tau^{ij}\mathbf{b}_j, \tag{115.2}$$

where the $\mathbf{b}_j$ are base vectors directed along the coordinate lines and the τ^{ij} are the contravariant components of $\overset{i}{\mathbf{T}}$.

Now, if $\mathbf{F} = F^i\mathbf{b}_i$ denotes the force per unit volume acting on the mass contained in $d\tau$, the first condition of equilibrium requires that

$$\mathbf{F}\, d\tau + \overset{\nu}{\mathbf{T}}\, d\sigma + \overset{i}{\mathbf{T}}\, d\sigma_i = \mathbf{0}. \tag{115.3}$$

If we note the definitions 115.1 and 115.2 and observe that $d\tau = l\,d\sigma$, where l is the appropriate factor depending on the linear dimension of the volume element, the equilibrium condition 115.3 becomes

$$F^i \mathbf{b}_i l\,d\sigma + \overset{\nu}{T}{}^j \mathbf{b}_j\,d\sigma - \tau^{ij}\nu_i\,d\sigma \mathbf{b}_j = \mathbf{0},$$

where $\overset{\nu}{T}{}^j \mathbf{b}_j \equiv \overset{\nu}{\mathbf{T}}$.

If the point P' is now made to approach P so that the direction of ν remains fixed, $l \rightarrow 0$, and the first term in the above relation will surely vanish whenever the body force $\mathbf{F}$ is bounded. This leads to the result that the components T^j of the stress $\overset{\nu}{\mathbf{T}}$, acting on a surface element with the orientation $\boldsymbol{\nu}$, are given by the formula

$$T^j = \tau^{ij}\nu_i. \tag{115.4}$$

Since T^i is a vector and ν_i is an arbitrary covariant vector, we conclude that the τ^{ij} are the contravariant components of a tensor, the *stress tensor*. Formula 115.4 permits us to calculate the stress vector acting on a surface element with the specified orientation whenever a set of nine functions τ^{ij} is known. We will see in Sec. 116 that the application of the remaining condition of equilibrium leads to the conclusion that the stress tensor is symmetric.

We can obviously write (115.4) in the form

$$T_j = \tau_{ij}\nu^i. \tag{115.5}$$

The component N of the vector $\overset{\nu}{\mathbf{T}}$ in the direction of the normal $\boldsymbol{\nu}$ is $\overset{\nu}{\mathbf{T}} \cdot \boldsymbol{\nu} = T_j\nu^j$, so that, using (115.5),

$$N = \tau_{ij}\nu^i\nu^j. \tag{115.6}$$

In regard to the quadratic form 115.6, we can raise the question of determining directions $\boldsymbol{\nu}^i$ such that N takes on the extreme values. As in Sec. 111, this leads to the consideration of the characteristic equation

$$|\tau_j{}^i - \tau\delta_j{}^i| \equiv -\tau^3 + \Theta_1\tau^2 - \Theta_2\tau + \Theta_3 = 0, \tag{115.7}$$

where

$$\Theta_1 = \tau_1 + \tau_2 + \tau_3,$$
$$\Theta_2 = \tau_2\tau_3 + \tau_3\tau_1 + \tau_1\tau_2,$$
$$\Theta_3 = \tau_1\tau_2\tau_3,$$

the τ_i being the roots of the cubic (115.7). The orthogonal directions ν^i corresponding to the *principal* stresses τ_i are determined from the set of linear equations (cf. equation 111.4)

$$(\tau_j{}^k - \tau\delta_j{}^k)\nu^j = 0, \tag{115.8}$$

and are called the *principal directions of stress*. If we choose an orthogonal cartesian frame Y whose axes coincide with the principal directions at P, the quadric surface

$$\tau_{ij}\nu^i\nu^j = \text{constant} \tag{115.9}$$

assumes the form

$$\tau_1(y^1)^2 + \tau_2(y^2)^2 + \tau_3(y^3)^2 = \text{constant}. \tag{115.10}$$

The quadric surface 115.9 was introduced by Cauchy, and it is called the *stress quadric*.

It is obvious from (115.10) that the components τ_{ij}, for $i \neq j$, vanish when a suitable reference frame is chosen at P. The components τ_{11}, τ_{22}, τ_{33} are called the *normal components* of stress, and the remaining ones are *shears*.

By analogy with formulas 111.10, we can write down the expressions for the *stress invariants* Θ_i. They are

$$\Theta_1 = \tau_i^{\ i}, \quad \Theta_2 = \frac{1}{2!}\,\delta^{ij}_{\alpha\beta}\tau_i^{\ \alpha}\tau_j^{\ \beta}, \quad \Theta_3 = \frac{1}{3!}\,\delta^{ijk}_{\alpha\beta\gamma}\tau_i^{\ \alpha}\tau_j^{\ \beta}\tau_k^{\ \gamma}. \tag{115.11}$$

116. Differential Equations of Equilibrium

Let a body τ be in a state of equilibrium under the action of prescribed body and surface forces. Since every portion of the body is in equilibrium, the resultant of all forces and the resultant moment of these forces acting on every subregion V of τ must vanish. The condition that the resultant force in every direction vanishes yields the equation

$$\int_V F^i\lambda_i\,d\tau + \int_S T^i\lambda_i\,d\sigma = 0, \tag{116.1}$$

where λ_i is the unit vector in an arbitrarily fixed direction.

We assume that the components of body force $F^i(x)$ are continuous functions and that the components T^i of the stress vector are of class C^1. The substitution for T^i from (115.4) and the application of divergence theorem 92.3 to the surface integral in (116.1) yields the equation

$$\int_V [F^i\lambda_i + (\tau^{ji}\lambda_i)_{,j}]\,d\tau = 0.$$

Since λ_i is a parallel vector field, $\lambda_{i,j} = 0$, so that the preceding equation can be written

$$\int_V (F^i + \tau^{ji}_{\ ,j})\lambda_i\,d\tau = 0. \tag{116.2}$$

Since the integrand in (116.2) is continuous and the direction of λ_i is arbitrary, we conclude that, at every point P of τ,

$$\tau^{ji}_{,j} + F^i = 0. \tag{116.3}$$

We apply next the condition that the resultant moment of the body and surface forces vanishes. If $\mathbf{r} = l^i\mathbf{b}_i$ is the position vector of the point $P'(x)$ relative to some point P, the component of the moment $(\mathbf{F} \times \mathbf{r})\, d\tau$ in the direction of the unit vector $\boldsymbol{\lambda}$ is $\mathbf{F} \times \mathbf{r} \cdot \boldsymbol{\lambda}\, d\tau$. The component of the moment due to the surface forces $\overset{\nu}{\mathbf{T}}$ is $\overset{\nu}{\mathbf{T}} \times \mathbf{r} \cdot \boldsymbol{\lambda}\, d\sigma$. Recalling (Sec. 49) the expression for the triple scalar product

$$\mathbf{A} \times \mathbf{B} \cdot \mathbf{C} = \epsilon_{ijk} A^i B^j C^k$$

enables us to write

$$\int_V \epsilon_{ijk} F^i l^j \lambda^k \, d\tau + \int_S \epsilon_{ijk} T^i l^j \lambda^k \, d\sigma = 0.$$

The substitution in the surface integral from (115.4) and the application of the divergence theorem yields

$$\int_V \epsilon_{ijk} \lambda^k [F^i l^j + (\tau^{mi} l^j)_{,m}] \, d\tau = 0,$$

since $\epsilon_{ijk,m} = 0$.

If we carry out the indicated covariant differentiation and make use of equations 116.3, we get

$$\int_V \epsilon_{ijk} \tau^{mi} l^j_{,m} \lambda^k \, d\tau = 0,$$

and, since[5] $l^j_{,m} = \delta^j_m$, and V is arbitrary, we conclude that

$$\epsilon_{ijk} \tau^{ji} \lambda^k = 0. \tag{116.4}$$

Noting that $\epsilon_{ijk} = -\epsilon_{jik}$ enables us to write this in the form

$$\tfrac{1}{2}\epsilon_{ijk}(\tau^{ji} - \tau^{ij})\lambda^k = 0. \tag{116.5}$$

Since $\epsilon_{ijk} = \sqrt{g}\, e_{ijk}$, and $\sqrt{g} \neq 0$, we have, upon expanding (116.5),

$$(\tau^{23} - \tau^{32})\lambda^1 + (\tau^{31} - \tau^{13})\lambda^2 + (\tau^{12} - \tau^{21})\lambda^3 = 0.$$

Inasmuch as the direction of $\boldsymbol{\lambda}$ is arbitrary, we conclude that

$$\tau^{ij} = \tau^{ji}. \tag{116.6}$$

Thus the stress tensor is symmetric.

[5] For: $\mathbf{b}_j = \dfrac{\partial \mathbf{r}}{\partial x^j} = l^i_{,j}\mathbf{b}_i$ by (46.6). Hence $l^i_{,j} = \delta^i_j$.

We summarize these results in a

THEOREM. *If a body is in equilibrium, under the action of prescribed body and surface forces, then the components of the stress tensor τ^{ij} at each point of the body satisfy the system of partial differential equations*

$$\tau^{ij}_{,j} + F^i = 0,$$

where $\tau^{ij} = \tau^{ji}$. On the surface Σ of the body where stress vectors T^i are assigned,

$$\tau^{ij}\nu_j = T^i,$$

ν_j being the exterior unit normal ot Σ.

We can write down at once the equations of motion by invoking the principle of D'Alembert. We merely have to add to the body force F^i the inertial force $-\rho a^i$, where ρ is the density and a^i is the acceleration. Thus the equations of motion are

$$\tau^{ij}_{,j} + F^i = \rho a^i, \tag{116.7}$$

where F^i is the body force per unit volume. If F^i represents the force *per unit mass*, the equations of motion read

$$\tau^{ij}_{,j} = \rho(a^i - F^i). \tag{116.8}$$

Since all equations in Secs. 115 and 116 appear in tensor form, they are valid in all admissible reference frames. In particular, in the reference frame X of the initial state τ_0, the covariant derivatives in (116.8) are taken with respect to the metric coefficients h_{ij} and the a^i and F^i are components of the acceleration and force vectors relative to the basis of the initial state.

117. Virtual Work

Let a continuous medium be maintained in the state of equilibrium by the body forces F^i and surface forces T^i. If $\overrightarrow{P_0P} = \boldsymbol{\xi}(x, t)$ is the displacement vector of the point P in (Fig. 51), we can consider a point P' in the neighborhood of P, and denote the vector $\overrightarrow{P_0P'}$ (not shown in Fig. 51) by $\boldsymbol{\xi}'(x, t)$, Thus

$$\boldsymbol{\xi}'(x, t) = \boldsymbol{\xi}(x, t) + \delta\boldsymbol{\xi}(x, t),$$

where the *variation*

$$\delta\boldsymbol{\xi} = \boldsymbol{\xi}' - \boldsymbol{\xi}, \tag{117.1}$$

or the *virtual displacement* of P, is an arbitrary vector $\overrightarrow{PP'}$ in the neighborhood of P. We consider the variation of vectors only in the final state τ

and we shall say that the variations of vectors and tensors associated with the points P_0 of the initial state τ_0 is zero.

We suppose that $\boldsymbol{\xi}$ is of class C^2 and define the variation of $\partial\boldsymbol{\xi}/\partial x^i$ by

$$\delta\left(\frac{\partial\boldsymbol{\xi}}{\partial x^i}\right) = \frac{\partial\boldsymbol{\xi}'}{\partial x^i} - \frac{\partial\boldsymbol{\xi}}{\partial x^i} = \frac{\partial(\boldsymbol{\xi}' - \boldsymbol{\xi})}{\partial x^i} = \frac{\partial(\delta\boldsymbol{\xi})}{\partial x^i}, \tag{117.2}$$

so that the variation of the derivative $\partial\boldsymbol{\xi}/\partial x^i$ is equal to the derivative of the variation (cf. Sec. 81). Since $\boldsymbol{\xi} = \mathbf{r} - \mathbf{r}_0$, and

$$\frac{\partial\boldsymbol{\xi}}{\partial x^i} = \frac{\partial\mathbf{r}}{\partial x^i} - \frac{\partial\mathbf{r}_0}{\partial x^i} = \mathbf{b}_i - \mathbf{a}_i,$$

we obtain, on utilizing the distributive property of the symbol δ,

$$\delta\left(\frac{\partial\boldsymbol{\xi}}{\partial x^i}\right) = \delta(\mathbf{b}_i - \mathbf{a}_i) = \delta\mathbf{b}_i,$$

for $\delta\mathbf{a}_i = 0$, since points in the initial state are not varied. Thus

$$\delta\mathbf{b}_i = \delta\left(\frac{\partial\boldsymbol{\xi}}{\partial x^i}\right) = \frac{\partial(\delta\boldsymbol{\xi})}{\partial x^i}. \tag{117.3}$$

The metric coefficients of the final state are given by $g_{ij} = \mathbf{b}_i \cdot \mathbf{b}_j$ and we find, as in Sec. 81,

$$\delta g_{ij} = \delta(\mathbf{b}_i \cdot \mathbf{b}_j) = \mathbf{b}_i \cdot \delta\mathbf{b}_j + \mathbf{b}_j \cdot \delta\mathbf{b}_i,$$

so that

$$\delta g_{ij} = \mathbf{b}_i \cdot \frac{\partial(\delta\boldsymbol{\xi})}{\partial x^j} + \mathbf{b}_j \frac{\partial(\delta\boldsymbol{\xi})}{\partial x^i}, \qquad \text{by (117.3).} \tag{117.4}$$

The strain tensor ϵ_{ij} was defined by

$$2\epsilon_{ij} = g_{ij} - h_{ij} \tag*{[109.11]}$$

and hence

$$2\delta\epsilon_{ij} = \delta g_{ij}, \tag{117.5}$$

since $\delta(h_{ij}) = \delta(\mathbf{a}_i \cdot \mathbf{a}_j) = 0$.

On substituting from (117.4) in (117.5), we get

$$2\delta\epsilon_{ij} = \mathbf{b}_i \cdot \frac{\partial(\delta\boldsymbol{\xi})}{\partial x^j} + \mathbf{b}_j \cdot \frac{\partial(\delta\boldsymbol{\xi})}{\partial x^i}, \tag{117.6}$$

But $\delta\boldsymbol{\xi} = (\delta\boldsymbol{\xi})_i\mathbf{b}^i$, where the $(\delta\boldsymbol{\xi})_i$ are the covariant components of the vector $\delta\boldsymbol{\xi}$, and, since

$$\frac{\partial(\delta\boldsymbol{\xi})}{\partial x^i} = (\delta\boldsymbol{\xi})_{j,i}\mathbf{b}^j$$

by (46.8), we conclude that (117.6) can be written as

$$2\delta\epsilon_{ij} = (\delta\boldsymbol{\xi})_{i,j} + (\delta\boldsymbol{\xi})_{j,i}. \tag{117.7}$$

If we form the inner product of the vector $(\delta\boldsymbol{\xi})_i$ with both members of the equilibrium equation

$$\tau^{ij}_{,j} = -\rho F^i, \tag{117.8}$$

where F^i is the body force per unit mass (cf. 116.3), and integrate over the body τ, we get

$$\int_\tau \tau^{ij}_{,j}(\delta\boldsymbol{\xi})_i \, d\tau = -\int_\tau \rho F^i(\delta\boldsymbol{\xi})_i \, d\tau, \tag{117.9}$$

But $\tau^{ij}_{,j}(\delta\boldsymbol{\xi})_i = [\tau^{ij}(\delta\boldsymbol{\xi})_i]_{,j} - \tau^{ij}(\delta\boldsymbol{\xi})_{i,j}$, so that (117.9) can be written as

$$\int_\tau [\tau^{ij}(\delta\boldsymbol{\xi})_i]_{,j} \, d\tau - \int_\tau \tau^{ij}(\delta\boldsymbol{\xi})_{i,j} \, d\tau = -\int_\tau \rho F^i(\delta\boldsymbol{\xi})_i \, d\tau,$$

or

$$\int_\Sigma \tau^{ij}(\delta\boldsymbol{\xi})_i \nu_j \, d\sigma - \int_\tau \tau^{ij}(\delta\boldsymbol{\xi})_{i,j} \, d\tau = -\int_\tau \rho F^i(\delta\boldsymbol{\xi})_i \, d\tau,$$

where we transformed the volume integral over τ into the surface integral over surface Σ bounding τ.

Since $\tau^{ij}\nu_j = T^i$ by (115.1) and

$$\tau^{ij}(\delta\boldsymbol{\xi})_{i,j} = \tfrac{1}{2}\tau^{ij}[(\delta\boldsymbol{\xi})_{i,j} + (\delta\boldsymbol{\xi})_{j,i}] = \tau^{ij}\,\delta\epsilon_{ij}$$

by (117.7), we have finally

$$\int_\tau \tau^{ij}\delta\epsilon_{ij} \, d\tau = \int_\Sigma T^i(\delta\boldsymbol{\xi})_i \, d\sigma + \int_\tau \rho F^i(\delta\boldsymbol{\xi})_i \, d\tau. \tag{117.10}$$

By definition, the surface integral in (117.10) represents the *virtual work* performed by the external surface forces T^i in a virtual displacement $(\delta\boldsymbol{\xi})_i$. The volume integral in the right-hand member of (117.10), on the other hand, represents the *virtual work done* by the body forces F^i. If we denote the virtual work done by body and surface forces by

$$\delta W = \int_\Sigma T^i\,(\delta\boldsymbol{\xi})_i \, d\sigma + \int_\tau \rho F^i(\delta\boldsymbol{\xi})_i \, d\tau, \tag{117.11}$$

we can write (117.10) as

$$\delta W = \int \tau^{ij}\,\delta\epsilon_{ij} \, d\tau. \tag{117.12}$$

If in the foregoing calculation instead of the equilibrium equations (117.8) we considered the dynamical equations,

$$\tau^{ij}_{,j} = \rho(a^i - F^i), \tag*{[116.8]}$$

we would have obtained in the left-hand member of (117.10) the additional term

$$\delta K \equiv \int_\tau \rho a^i (\delta \xi)_i \, d\tau. \tag{117.13}$$

This term has a simple mechanical interpretation when the virtual displacements $(\delta \xi)_i$ are the actual displacements $(d\xi)_i$ that take place in a body whose motion is governed by equations 116.8. In this event we write (117.3) as

$$dK = \int_\tau \rho a^i (d\xi)_i \, d\tau. \tag{117.14}$$

But the velocity of a point P in τ is

$$v_i = \frac{(d\xi)_i}{dt},$$

and we can thus write (117.14)

$$dK = \int_\tau \rho a^i v_i \, d\tau \, dt.$$

Now, in the orthogonal cartesian coordinates,

$$a^i v_i = \frac{1}{2} \frac{d}{dt} (v^i v_i) = \frac{1}{2} \frac{d(v)^2}{dt},$$

and hence

$$dK = \int_\tau \tfrac{1}{2} d(v)^2 (\rho \, d\tau).$$

The integrand in this integral represents an increment in kinetic energy of the element of mass $dm = \rho \, d\tau$ acquired by it in the interval of time $(t, t + dt)$. Thus dK represents an increment of kinetic energy $K = \int_\tau \frac{1}{2} \rho v^2 \, d\tau$. Accordingly, for the motion of a body τ governed by equations 116.8, we have an important result:

$$dK + dA = dW. \tag{117.15}$$

where

(117.16)

$$dA = \int_\tau \tau^{ij} \, d\epsilon_{ij} \, d\tau \quad \text{and} \quad dW = \int_\Sigma T^i (d\xi)_i \, d\sigma + \int_\tau \rho F^i (d\xi)_i \, d\tau.$$

In the static case $dK = 0$ and $dA = dW$.

The results of the section, coupled with some thermodynamic considerations, form the basis for constructing the theoretical models of elastic bodies, viscous fluids, and so on.

118. Laws of Thermodynamics

The construction of mathematical models of different types of continuous media hinges on the use of certain energy concepts that enter in the structures of mechanics and thermodynamics. We borrow from mechanics the notions of potential and kinetic energy and from thermodynamics the somewhat less sharply defined concepts of chemical energy, heat energy, electrical energy, and so on. We shall suppose that functions defining various kinds of energies depend on a number of parameters, some of which are variables (positional coordinates, temperature, densities, strain tensors, and so on), whereas others are physical or universal constants. The totality of constant parameters c_j and variable parameters q^i chosen to describe a given function need not be unique. But whatever particular choices of a set of parameters is made, we shall assume that the q^i $(i = 1, \ldots, n)$ are independent.

In some special situations the q^i may be determined as functions of a scalar t (usually time) so that one can regard them as defining a curve

$$C: \quad q^i = q^i(t),$$

characterizing a certain process.

In the preceding section we introduced the notion of work or *mechanical energy* by considering linear forms of the type

$$\delta W = Q_i(q^1, \ldots, q^n; \; c_1, \ldots, c_m)\, \delta q^i. \tag{118.1}$$

The line integral $\int_C Q_i \, dq^i$ then represents the work done along the path C by the generalized forces Q_i. Ordinarily, such integrals depend on the path C associated with a given process.

We shall suppose that a particle of mass $dm = \rho \, d\tau$, where ρ is the density and $d\tau$ is the volume element, may acquire energies other than mechanical, and we shall represent such accretions of energy in the form

$$\delta E = F_i(q^1, \ldots, q^n; \; c_1, \ldots, c_m)\, \delta q^i. \tag{118.2}$$

If δE includes *all* energies other than mechanical, the total amount of energy acquired by the particle is determined by the integral

$$\int_C (\delta W + \delta E) = \int_C (F_i + Q_i)\, dq^i. \tag{118.3}$$

From the principle of conservation of total energy we conclude that the integral 118.3 must vanish for an arbitrary closed path C, and hence the integrand $(F_i + Q_i)\, dq^i$ is an exact differential of some function $\bar{U}(q^1, \ldots, q^n; \; c_1, \ldots, c_m)$ determined to within a constant of integration. We shall call $\bar{U}$ the *total energy per unit mass* and define the *internal energy* U per unit mass by the formula

$$U = \bar{U} - \tfrac{1}{2}v^2, \tag{118.4}$$

where v is the velocity of the element of mass dm. The amount $K \equiv \frac{1}{2}v^2$ represents the kinetic energy per unit mass, so that the total energy

$$\bar{U} = U + K.$$

We can thus formulate the basic law of conservation of total energy in the form

$$\delta K + \delta U = \delta W + \delta E, \tag{118.5}$$

where the left-hand member in (118.5) is the sum of the increments of kinetic energy K and internal energy U acquired by the unit mass.

When δE consists only of the heat energy δQ, we have the statement of the First Law of Thermodynamics:

$$\delta K + \delta U = \delta W + \delta Q. \tag{118.6}$$

The heat energy δQ, as shown in works on thermodynamics, can be determined by specifying the temperature T. Experiments show that heat invariably passes from bodies with higher temperature to those with lower temperature and that the transfer of heat from one body to another is wholly determined by T and, of course, by certain physical parameters depending on constitutive properties of the bodies. Experiments further show that it is impossible to construct a machine which transforms the heat energy into mechanical energy from a body with the least temperature. It is a consequence of this Second Law of Thermodynamics that for every reversible thermodynamic process there exists a function S, called *entropy*, such that

$$\delta S\, dm = \frac{\delta Q}{T}, \tag{118.7}$$

T being the absolute temperature and δQ an increment of heat acquired by the element of mass dm.

When the medium is in the state of mechanical equilibrium, the kinetic energy K vanishes, and the law (118.6) assumes the form

$$\delta U = \delta W + \delta Q. \tag{118.8}$$

We shall make use of the laws 118.7 and 118.8 in Sec. 119 to construct a mathematical model of an elastic body.

119. Elastic Media

Some bodies possess the property of recovering their original size and shape when the impressed forces producing deformations are removed. The media of which such bodies are composed are called *elastic*. In constructing a model of an elastic body we shall suppose that all processes taking place in such a body are reversible, but we do not assume that the body is necessarily in the state of thermal equilibrium. Thus our thermoelastic model will take account of the effects of temperature on deformations.

As our points of departure we take the First Law of Thermodynamics in the form [cf. (118.8)]

$$\delta U = \delta Q + \delta W \tag{119.1}$$

in which

$$\delta W = \int_\tau \tau^{ij}\, \delta\epsilon_{ij}\, d\tau \tag{119.2}$$

is given by (117.12).

We also write the relation 118.7 in the form

$$\delta Q = \int_m T\,\delta S\, dm$$

or

$$\delta Q = \int_\tau T\,\delta S \rho\, d\tau. \tag{119.3}$$

If u denotes the internal energy U per unit mass of the body, then

$$\delta U = \int_m \delta u\, dm = \int_\tau \rho\, \delta u\, d\tau, \tag{119.4}$$

where δU stands for the increment of internal energy acquired by τ.

The substitution in (119.1) from (119.2), (119.3), and (119.4) gives

$$\int_\tau \rho\, \delta u\, d\tau = \int_\tau \rho T\,\delta S\, d\tau + \int_\tau \tau^{ij}\, \delta\epsilon_{ij}\, d\tau. \tag{119.5}$$

We suppose that the integrands in (119.5) are continuous functions and, since the equality 119.5 holds in an arbitrary subregion of τ, we conclude that

$$\delta u = T\,\delta S + \frac{1}{\rho}\tau^{ij}\,\delta\epsilon_{ij} \tag{119.6}$$

at all points of τ.

The formula 119.6 suggests that we regard u as a function of the independent variable S and of the nine independent parameters ϵ_{ij}. Since the components ϵ_{ij} of the stress tensor $E_0 = \epsilon_{ij}\mathbf{a}^i\mathbf{a}^j$ usually depend on the choices of a coordinate system X, the function u may also contain explicitly the metric tensor h_{ij} and the coordinates x^i. And, of course, u must depend on an assortment of parameters $\{c\}$ associated with the physical properties of the medium. Thus we are led to consider u in the form

$$u = u(h_{ij}, \epsilon_{ij}, S, \{c\}, x^i), \tag{119.7}$$

where the arguments of u are deemed independent. The relation (119.6) then permits us to assert that

$$\frac{\partial u}{\partial \epsilon_{ij}} = \frac{1}{\rho}\tau^{ij} \quad \text{and} \quad \frac{\partial u}{\partial S} = T.$$

The first of these relations

$$\tau^{ij} = \rho \frac{\partial u}{\partial \epsilon_{ij}} \tag{119.8}$$

connects components τ^{ij} of the stress tensor with components ϵ_{ij} of the strain tensor. It thus yields a set of *stress-strain relations*, in which the internal energy density u serves as a potential function.

A different potential function can be constructed by defining a function ϕ, known as the *free energy*, by

$$\phi = u - TS. \tag{119.9}$$

From (119.9), the increment $\delta\phi$ of ϕ is

$$\delta\phi = \delta u - T\,\delta S - S\,\delta T$$

and the substitution in this expression for δu from (119.6) gives

$$\delta\phi = \frac{1}{\rho}\tau^{ij}\,\delta\epsilon_{ij} - S\,\delta T. \tag{119.10}$$

Because of the appearance of $\delta\epsilon_{ij}$ and δT in the right-hand member of (119.10), we are now led to regard T and ϵ_{ij} as independent variables and consider ϕ in the form [cf. (119.7)]

$$\phi = (h_{ij}, \epsilon_{ij}T, \{c\}, x^i). \tag{119.11}$$

We conclude, then, from (119.10) that

$$\frac{\partial \phi}{\partial \epsilon_{ij}} = \frac{1}{\rho}\tau^{ij}, \quad \frac{\partial \phi}{\partial T} = -S,$$

so that the stress-strain relation now has the form

$$\tau^{ij} = \rho \frac{\partial \phi}{\partial \epsilon_{ij}}. \tag{119.12}$$

Thus either u or ϕ (when they exist) can be used to deduce the stress-strain relations. When the process is adiabatic, $S =$ constant and hence $\delta Q = 0$ by (119.3). It is then more convenient to use u as a stress-potential. In the isothermal case, $T =$ constant and ϕ appears to be more suitable.

We say that an elastic medium is *homogeneous* whenever the coordinates x^i do not appear explicitly in (119.7) or (119.11). The medium is *isotropic* when all parameters in the set $\{c\}$ are scalars, so that the values in $\{c\}$ are independent of the choices of the reference frames X. When the medium is both homogeneous and isotropic, the parameters $\{c\}$ have constant values throughout the medium.

If we consider a homogeneous elastic medium and suppose that $\phi(\epsilon_{ij}, T)$ is an analytic function of the ϵ_{ij} and of $\Delta T \equiv T - T_0$, where T_0 is the temperature of the initial state, we can expand ϕ in powers of ϵ_{ij} and ΔT. When the initial state of the body is that corresponding to $\epsilon_{ij} = 0$ and $\tau_{ij} = 0$, the expansions will begin with the second-order terms, so that

$$\phi = c^{ijkl}\epsilon_{ij}\epsilon_{kl} + k^{ij}\epsilon_{ij}\Delta T + m(\Delta T)^2 + \cdots.$$

For small deformations, the terms of order higher than two can be neglected, and we obtain with the aid of (119.12) a linear stress-strain relation that includes the effects of temperature on the stress tensor τ^{ij}. It is

$$\tau^{ij} = \rho_0[c^{ijkl}e_{kl} + k^{ij}(T - T_0)], \tag{119.13}$$

where we replaced ρ by ρ_0—the density of the initial state—and wrote e_{kl} for the linearized components ϵ_{kl}. The tensor c^{ijkl} characterizes the elastic properties of the medium and the k^{ij} are related to the coefficients of thermal expansion. For a given medium the tensors c^{ijkl} and k^{ij} must be determined from experiments. When $T = T_0$, the relation 119.13 reduces to the familiar generalized Hooke's law of linear elasticity.[6]

$$\tau^{ij} = c^{ijkl}e_{kl}. \tag{119.14}$$

In the next section we deduce a special form of stress-strain relations for large deformations for a homogeneous isotropic elastic medium and get from it the familiar Hooke's law of linear theory of elasticity.

[6] See I. S. Sokolnikoff, *Mathematical Theory of Elasticity* (1956), pp. 58–67, where it is shown that the number of independent *elastic coefficients* c^{ijkl} in the most general anisotropic case is 21.

120. Stress-strain Relations in Isotropic Elastic Media

When the orientation of coordinate axes is immaterial, the arguments of the potential ϕ in (119.11) are scalars or tensors that depend only on the metric tensor h_{ij}. In this event the scalar invariants of tensors h_{ij} and ϵ_{ij} can be considered as functions of the invariants ϑ_i, defined in Sec. 111, and can be taken in the form

$$\phi = \phi(\vartheta_1, \vartheta_2, \vartheta_3, T, \{c\}, x^i).$$

If the medium is both homogeneous and isotropic, ϕ assumes the form

$$\phi = \phi(\vartheta_1, \vartheta_2, \vartheta_3, T, \{c\}), \tag{120.1}$$

in which all parameters in $\{c\}$ are constants. The formula

$$\tau^{ij} = \rho \frac{\partial \phi}{\partial \epsilon_{ij}} \tag{[119.12]}$$

with ϕ specified by (120.1) can be written as[7]

$$\tau_j^{\ i} = \rho(\delta_k^{\ i} - 2\epsilon_k^{\ i}) \frac{\partial \phi}{\partial \epsilon_k^{\ j}} \tag{120.2}$$

where $\tau_j^{\ i} = g_{\alpha j}\tau^{i\alpha}$ and $\epsilon_{ij} = g_{i\alpha}\epsilon_j^{\ \alpha}$.

If we now suppose that ϕ in (120.1) with $T =$ constant can be expanded in a power series in the ϑ_i and consider the case when there is no initial stress, so that $\tau_j^{\ i} = 0$ when $\epsilon_j^{\ i} = 0$, the expansion takes the form

$$\rho_0\phi = c_1\vartheta_1^{\ 2} + c_2\vartheta_2 + c_3\vartheta_1^{\ 3} + c_4\vartheta_1\vartheta_2 + c_5\vartheta_3 + \cdots. \tag{120.3}$$

If in this expression we retain only the terms of third order in the $\epsilon_j^{\ i}$, we see from (120.2) that the expression for the stresses $\tau_j^{\ i}$ in terms of the strains $\epsilon_j^{\ i}$ will contain five elastic coefficients c_i. From the mass-conservation principle it follows that

$$\rho_0\, d\tau_0 = \rho\, d\tau,$$

[7] Note that

$$\frac{\partial \phi}{\partial \epsilon_{ij}} = \frac{\partial \phi}{\partial \epsilon_\beta^{\ \alpha}} \cdot \frac{\partial \epsilon_\beta^{\ \alpha}}{\partial \epsilon_{ij}} = \frac{1}{\rho}\tau^{ij}.$$

Since

$$g_{\alpha\beta} = 2\epsilon_{\alpha\beta} - h_{\alpha\beta}, \qquad \frac{\partial g_{\alpha\beta}}{\partial \epsilon_{ij}} = \frac{2\partial \epsilon_{\alpha\beta}}{\partial \epsilon_{ij}} = 2\delta_\alpha^{\ i}\delta_\beta^{\ j}.$$

Compute $\partial\epsilon_{\alpha\beta}/\partial\epsilon_{ij}$ from $\epsilon_{\alpha\beta} = g_{\alpha\gamma}\epsilon_\beta^{\ \gamma}$, use the above result, and conclude that

$$\delta_k^{\ i} - 2\epsilon_k^{\ i} = g_{j\alpha}\frac{\partial \epsilon_k^{\ \alpha}}{\partial \epsilon_{ij}}.$$

Formula 120.2 then follows on substituting this result in (a).

and formulas 112.3 and 112.4 yield the result

$$\rho = \rho_0\sqrt{1 - 2\vartheta_1 + 4\vartheta_2 - 8\vartheta_3}$$
$$\doteq \rho_0(1 - \vartheta_1 - \tfrac{1}{2}\vartheta_1^{\,2} + 2\vartheta_2),$$

if we discard the third-order terms in the $\epsilon_j^{\,i}$. The substitution from this formula and (120.3) in (120.2) gives the following expression for the stress-strain relation, where we retain only the second-order terms in the strains

$$\text{(120.4)}\quad \tau_j^{\,i} = [2c_1\vartheta_1 + (3c_3 - 2c_1)\vartheta_1^{\,2} + c_4\vartheta_2]\,\delta_j^{\,i} + [c_2 + (c_4 - c_2)\vartheta_1]\,\delta_{j\beta}^{i\alpha}\epsilon_\alpha^{\,\beta} - 4c_1\vartheta_1\epsilon_j^{\,i} + \tfrac{1}{2}c_5\,\delta_{j\alpha\delta}^{i\beta\gamma}\epsilon_\beta^{\,\alpha}\epsilon_\gamma^{\,\delta} - 2c_2\,\delta_{j\gamma}^{\beta\alpha}\epsilon_\alpha^{\,\gamma}\epsilon_\beta^{\,i}.$$

These involve five elastic constants. If, however, we retain in (120.4) only the first-degree terms in the $\epsilon_j^{\,i}$, we get the linear law

$$\text{(120.5)}\qquad \tau_j^{\,i} = (2c_1 + c_2)\vartheta_1\,\delta_j^{\,i} - c_2\epsilon_j^{\,i}.$$

We identify this result with the generalized Hooke's law for isotropic media

$$\text{(120.6)}\qquad \tau_j^{\,i} = \lambda\vartheta_1\,\delta_j^{\,i} + 2\mu e_j^{\,i}, \qquad \vartheta_1 = e_i^{\,i},$$

where λ and μ are *Lamé's constants*, related to *Young's modulus* E and *Poisson's ratio* σ by

$$\lambda = \frac{E\sigma}{(1 + \sigma)(1 - 2\sigma)}, \qquad \mu = \frac{E}{2(1 + \sigma)}.$$

We see that

$$c_1 = \tfrac{1}{2}(\lambda + 2\mu), \qquad c_2 = -2\mu.$$

If we replace c_1 and c_2 in (120.4) by these values and set $c_3 = l$, $c_4 = m$, $c_5 = n$, we can write it in the form

$$\text{(120.7)}\quad \tau_j^{\,i} = [\lambda\vartheta_1 + (3l + m - \lambda)\vartheta_1^{\,2} + m\vartheta_2]\,\delta_j^{\,i} + [2\mu - (m + 2\lambda + 2\mu)\vartheta_1]\epsilon_j^{\,i} - 4\mu\epsilon_\alpha^{\,i}\epsilon_j^{\,\alpha} + n\vartheta_3\phi_j^{\,i},$$

where $\phi_j^{\,i}$ is defined by the formula

$$\phi_j^{\,i} \equiv \frac{1}{2\vartheta_3}\,\delta_{j\alpha\delta}^{i\beta\gamma}\epsilon_\beta^{\,\alpha}\epsilon_\gamma^{\,\delta}.$$

The new elastic constants l, m, and n appearing in (120.7) are subject to experimental determination, just as Lamé's constants λ and μ are.[8]

[8] Assumptions, of varying degrees of plausibility, about the possible relations that might exist between the new constants (l, m, n) and the old ones (λ, μ) have been made by several authors. Murnaghan obtained a good agreement with experimental results (for solids subjected to high hydrostatic pressures) by setting $l = m = n = 0$ in formulas 120. A discussion of this appears in a paper by F. D. Murnaghan, "The Compressibility of Solids under Extreme Pressures," *Th. v. Kármán Anniversary Volume* (1941), pp. 112–136. See also P. Riz, *Comptes rendus* (*Doklady*) *Acad. Sci, U.R.S.S.*, **20** (1938), and P. M. Riz and N. V. Zvolinsky, *Journal of Applied Mathematics and Mechanics, Acad. Sci. U.S.S.R.*, **2** (1939).

An excellent discussion of a model of a thermoelastic isotropic medium is contained on pages 234–241 of L. I. Sedov's *Introduction to Mechanics of a Continuous Medium*, Moscow (1962).

121. Equations of Elasticity

If we write the stress-strain relations 120.6 in the form

$$\tau_{ij} = \lambda g_{ij}\vartheta + 2\mu e_{ij}, \tag{121.1}$$

where $\vartheta = g^{ij}e_{ij} \equiv e_i^{\ i}$, and use the equilibrium equations 116.3 in the form

$$g^{jk}\tau_{ij,k} + F_i = 0. \tag{121.2}$$

we can write down the linearized differential equations of equilibrium, in terms of the displacement vector u^i, by recalling that (equation 113.13)

$$e_{ij} = \tfrac{1}{2}(u_{i,j} + u_{j,i}). \tag{121.3}$$

The computation proceeds as follows. The substitution from (121.1) into (121.2) yields

$$g^{jk}\left(\lambda g_{ij}\frac{\partial \vartheta}{\partial x^k} + 2\mu e_{ij,k}\right) + F_i = 0,$$

or

$$\lambda\frac{\partial \vartheta}{\partial x^i} + 2\mu g^{jk}e_{ij,k} + F_i = 0. \tag{121.4}$$

But from (121.3)

$$\begin{aligned} g^{jk}e_{ij,k} &= \tfrac{1}{2}g^{jk}(u_{i,jk} + u_{j,ik}) \\ &= \frac{1}{2}g^{jk}u_{i,jk} + \frac{1}{2}\frac{\partial \vartheta}{\partial x^i}, \end{aligned}$$

since $g^{jk}u_{j,ik} = u^k_{\ ,ki}$ and $u^k_{\ ,k} = \vartheta$. Thus (121.4) becomes

$$(\lambda + \mu)\frac{\partial \vartheta}{\partial x^i} + \mu g^{jk}u_{i,jk} + F_i = 0. \tag{121.5}$$

If we recall the notation 92.7,

$$g^{jk}u_{i,jk} = \nabla^2 u_i,$$

we get

$$(\lambda + \mu)\frac{\partial \vartheta}{\partial x^i} + \mu\nabla^2 u_i + F_i = 0. \tag{121.6}$$

These are the celebrated *Navier equations* in the classical theory of elasticity.

The equations of motion,

$$(121.7) \qquad (\lambda + \mu)\frac{\partial \vartheta}{\partial x^i} + \mu \nabla^2 u_i + F_i = \rho a_i,$$

follow at once from (121.6) upon application of the D'Alembert principle.

The differential equations 121.6 and 121.7 for the displacement vector u_i can be shown to yield unique solutions when suitable boundary and initial conditions are specified. We refer interested readers to treatises on the mathematical theory of elasticity where such boundary value problems are discussed in detail.[9]

122. Fluid Mechanics. Equations of Continuity

We now turn to the formulation of equations governing the flow of liquids and gases. From the point of view of mechanics, fluids are continuous distributions of matter which cannot support shearing stresses when at rest. If follows from this definition that the stress vector T^i on a surface element $d\sigma$ of a fluid at rest is normal to the element. In symbols,

$$T^i = -p\nu^i,$$

where ν^i is the unit normal to the surface element and $p(x^1, x^2, x^3, t)$ is the invariant called the *hydrostatic* or *fluid pressure*. In general the pressure p is a function of the time t as well as of the coordinates x^i.

Since the vector T^i is expressible in terms of the stress tensor τ^{ij}, and $\nu^i = g^{ij}\nu_j$, we see that

$$T^i = \tau^{ij}\nu_j = -pg^{ij}\nu_j.$$

Hence

$$(122.1) \qquad \tau^{ij} = -pg^{ij}.$$

It follows from (122.1) that the hydrostatic pressure p is related to the stress invariant $\Theta = g_{ij}\tau^{ij}$ (see equation 115.11) by the formula

$$(122.2) \qquad p = -\tfrac{1}{3}g_{ij}\tau^{ij}.$$

When the fluid is not in motion, however, in addition to the normal stresses, new oblique stresses, produced by the interaction of moving particles, arise. For instance, if a fluid at rest is placed between two large parallel plates and one of the plates is caused to move parallel to the other plate (Fig. 53), the fluid particles adhering to the moving plate transmit

[9] See, for example, I. S. Sokolnikoff, *Mathematical Theory of Elasticity*, New York (1956).

A. E. H. Love, *A Treatise on the Mathematical Theory of Elasticity*, Cambridge (1927).

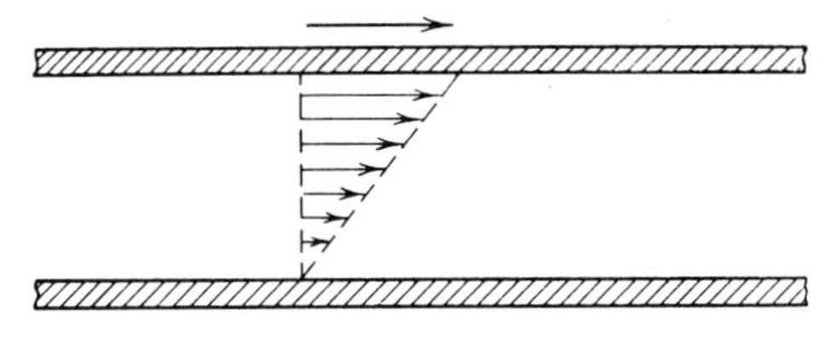

Fig. 53

their momentum to the particles in the interior. In this way the fluid between the plates is set in motion, and experiments show that the retarding force per unit area of the plate, exerted on the plate by the fluid, is proportional to its velocity and inversely proportional to the distance between the plates. The proportionality constant in this relation is the measure of *viscosity* of the fluid.

We shall say that the fluid is *viscous* if the stress tensor for a fluid in motion has the form

$$\tau^{ij} = -pg^{ij} + t^{ij}, \tag{122.3}$$

where the nonvanishing tensor t^{ij} is the *tensor of viscous stresses.* The fluid is called *ideal* if $t^{ij} \equiv 0$.

The mass-conservation principle of mechanics requires that

$$dm = \rho_0\, d\tau_0 = \rho\, d\tau,$$

where $\rho_0(x, t_0)$ is the density of matter in the volume element $d\tau_0 = \sqrt{h}\, dx^1\, dx^2\, dx^3$ of the initial state and $\rho(x, t)$ is the density in the volume element $d\tau = \sqrt{g}\, dx^1\, dx^2\, dx^3$ at time t. Thus

$$\frac{\rho_0}{\rho} = \frac{d\tau}{d\tau_0}, \tag{122.4}$$

which we can also write as

$$-\frac{\rho(x, t) - \rho(x, t_0)}{\rho(x, t)} = \frac{d\tau - d\tau_0}{d\tau_0}. \tag{122.5}$$

The numerator in the left-hand member of (122.5), $\Delta\rho \equiv \rho(x, t) - \rho(x, t_0)$, represents the change in density in the small interval of time $\Delta t = t - t_0$, whereas the right-hand member

$$\frac{d\tau - d\tau_0}{d\tau_0} \doteq \vartheta_1{}^0, \qquad \text{by (112.7)}$$

is the corresponding small change in volume per unit volume. Since $\vartheta_1{}^0 \doteq \operatorname{div} \boldsymbol{\xi} = u^i_{,i}$ (see Sec. 113), we can write (122.5) as

$$-\frac{\Delta\rho}{\Delta t}\cdot\frac{1}{\rho} = \frac{u^i_{,i}}{\Delta t}, \tag{122.6}$$

and since $u^i \doteq v^i \Delta t$, where the v^i denote the components of velocity $d\boldsymbol{\xi}/dt$, we conclude from (122.6) that $-(1/\rho)(d\rho/dt) = v^i_{,i}$. We thus get the *continuity equation* in the form

$$\frac{1}{\rho}\frac{d\rho}{dt} + v^i_{,i} = 0. \tag{122.7}$$

We recall that $\rho(x, t)$ is a function of the coordinates x^i in the reference frame X in which the x^i are independent of t. If Y is a *fixed* reference frame (cf. Sec. 109) in which the coordinates y^i of the particle are given by

$$y^i = y^i(x^1, x^2, x^3, t),$$

then the chain rule of differentiation gives for $\rho(y, t)$

$$\frac{d\rho}{dt} = \left(\frac{\partial\rho}{\partial t}\right)_{y^i\text{ fixed}} + \frac{\partial\rho}{\partial y^i}v^i.$$

Equation 122.7 then assumes the form

$$\frac{\partial\rho}{\partial t} + \frac{\partial\rho}{\partial y^i}v^i + \rho v^i_{,i} = 0,$$

or

$$\frac{\partial\rho}{\partial t} + (\rho v^i)_{,i} = 0. \tag{122.8}$$

In this formula, the covariant differentiation is performed with respect to the metric tensor in the frame Y and $v^i = dy^i/dt$. Formula 122.8 specializes to (122.7) when the system Y moves with the particle.

123. Ideal Fluids. Euler's Equations

In this section we deduce a set of equations governing the behavior of ideal fluids. We recall from Sec. 122 that in an ideal fluid the stress tensor has a simple form

$$\tau^{ij} = -pg^{ij}, \tag{123.1}$$

in which the scalar p is the pressure.

On substituting from (123.1) in the general dynamical equations

$$\rho(a^i - F^i) = \tau^{ij}_{,j}, \tag*{[116.8]}$$

we get three *Euler's equations*

$$\rho(a^i - F^i) = -g^{ij}p_{,j}, \tag{123.2}$$

or in vector form,

$$\rho(\mathbf{a} - \mathbf{F}) = -\mathbf{grad}\, p.$$

Equations (123.2) involve five unknowns: the density $\rho(x, t)$, the pressure $p(x, t)$, and three components of velocity $v^i(x, t)$, since $a^i = \delta v^i/\delta t$. The system of three equations (123.2) for the determination of the five unknowns, thus, is not complete, and we need two additional independent equations to complete the system. One such equation is the continuity equation

$$\frac{1}{\rho}\frac{d\rho}{dt} + v^i_{,i} = 0, \tag{123.3}$$

deduced in Sec. 122. The remaining equation, known as the *equation of state*, is furnished by the thermodynamical equations 118.7 and 118.8, which, for reversible processes in the fluid, we write in the form

$$\frac{dQ}{T} = dS\, dm, \tag{123.4}$$

$$dU = dW + dQ. \tag{123.5}$$

The work

$$dW = \frac{\tau^{ij}\, d\epsilon_{ij}}{\rho}\, dm, \tag{123.6}$$

performed by internal stresses τ^{ij} on an element of mass $dm = \rho\, d\tau$ can be written as [cf. (119.2)]

$$dW = \frac{-pg^{ij}}{\rho}\frac{d\epsilon_{ij}}{dt}\, dt\, dm, \tag{123.7}$$

when we note equations 123.1.

We shall suppose that the components ϵ_{ij} of the deformation tensor in a fluid are so small that they are represented with sufficiently high accuracy by linearized formulas 113.9 or 113.10, which coalesce in the infinitesimal theory. Thus we write

$$2e_{ij} = w_{i,j} + w_{j,i}$$

and

$$2\frac{de_{ij}}{dt} = \frac{d}{dt}(w_{i,j} + w_{j,i}),$$

where the e_{ij} are the linearized components ϵ_{ij}. Since the velocity components $v_i = dw_i/dt$, we conclude from the equation just written that

$$\frac{de_{ij}}{dt} = \tfrac{1}{2}(v_{i,j} + v_{j,i}). \tag{123.8}$$

The substitution for de_{ij}/dt from (123.8) in (123.7) then gives

$$dW = -\frac{p}{\rho} v^i_{,i}\, dt\, dm,$$

and, since $v^i_{,i} = -(1/\rho)(d\rho/dt)$ by (123.3), we have

$$dW = \frac{p}{\rho^2}\frac{d\rho}{dt}\, dt\, dm = -p d\left(\frac{1}{\rho}\right) dm.$$

On making use of this result in (123.5), we get for the amount of heat dQ acquired by an element of mass dm,

$$dQ = dU + pd\left(\frac{1}{\rho}\right) dm. \tag{123.9}$$

On the other hand, formula 123.4 states that

$$dQ = T dS\, dm. \tag{123.10}$$

If we let dq stand for the change in heat per unit mass, so that $dq = dQ/dm$, and denote the change in internal energy U per unit mass by du, we can write (123.9) and (123.10) as

$$\begin{cases} dq = du + pd\left(\dfrac{1}{\rho}\right), \\ dq = T\, dS. \end{cases} \tag{123.11}$$

In a variety of problems the absolute temperature T, the internal energy density u, and entropy S appear to depend only on the pressure p and density ρ, so that[10]

$$T = T(p, \rho), \qquad S = S(p, \rho), \qquad u = u(p, \rho). \tag{123.12}$$

It follows then from (123.11) that T, S, and u are not independent, since equations 123.11 require that

$$T dS = du + pd\left(\frac{1}{\rho}\right). \tag{123.13}$$

When T, S, and u in (123.12) are determined (either experimentally or from theoretical considerations), the differential equation 123.13, if integrable, specifies p as some function of ρ:

$$p = f(\rho). \tag{123.14}$$

Equation 123.14 is the desired *equation of state* needed to complete the system of four equations 123.2, 123.4 for the determination of the five unknown functions v^1, v^2, v^3, p, and ρ.

[10] These functions may (and usually do) depend on physical or chemical constants which characterize properties of a specific fluid.

124. Viscous Fluids. Navier's Equations

When viscous fluid is in motion, the components τ_{ij} of the stress tensor have the form

$$\tau^{ij} = -pg^{ij} + t^{ij}, \tag{124.1}$$

where the t^{ij}, as noted in Sec. 122, are associated with viscous stresses. As in Sec. 123, we limit ourselves to the consideration of small displacements and write formula 123.8 as

$$\dot{e}_{ij} = \tfrac{1}{2}(v_{i,j} + v_{j,i}), \tag{124.2}$$

where $\dot{e}_{ij} = de_{ij}/dt$ are components of the *strain velocity tensor.*

The construction of models of viscous fluids and the formulation of complete systems of equations now call for the introduction of additional assumptions about the nature of viscous stresses. The latter must obviously depend on the strain velocities $\dot{e}_{ij}$ and, to a first-order approximation, it is natural to suppose that

$$t^{ij} = c^{ijkl}\dot{e}_{kl}. \tag{124.3}$$

The coefficients c^{ijkl} are the *coefficients of viscosity*, which depend on the properties of a specific fluid under consideration. The linear law (124.3) is quite analogous to the generalized Hooke's law (119.14).

If the fluid is both homogeneous and isotropic, the number of independent viscosity coefficients reduces to two, and the relation 124.3 assumes the form [cf. (121.1)]

$$t^{ij} = \lambda v^k_{,k} g^{ij} + 2\mu\dot{e}^{ij}, \tag{124.4}$$

where λ and μ are constants and $v^k_{,k}$ is the divergence of the velocity field. Accordingly, the complete stress tensor which includes the effects of viscosity and hydrostatic pressure can be written in the form

$$\tau_{ij} = -pg_{ij} + \lambda\dot{\vartheta}g_{ij} + 2\mu\dot{e}_{ij}, \tag{124.5}$$

where $\dot{\vartheta} = v^i_{,i} = g^{ij}\dot{e}_{ij}$.

We recall next the equations of motion 116.8, and write them in the covariant form

$$g^{jk}\tau_{ij,k} = \rho(a_i - F_i), \tag{124.6}$$

and substitute in (124.6) for the τ_{ij} from (124.5). The result[11] is *Navier's equations of fluid motion*

$$(\lambda + \mu)\dot{\vartheta}_{,i} + \mu g^{jk}v_{i,jk} - p_{,i} = \rho(a_i - F_i), \tag{124.7}$$

[11] Note that $a^i = \dfrac{\delta v^i}{\delta t} = \dfrac{\partial v^i}{\partial t} + v^i_{,j}v^j$, so that $a_i = \dfrac{\partial v_i}{\partial t} + v_{i,j}v^j$.

or, in vector form,

$$(\lambda + \mu)\nabla\dot{\vartheta} + \mu\nabla^2\mathbf{v} - \nabla p = \rho(\mathbf{a} - \mathbf{F}).$$

The set of three Navier's equations 124.7 involves five unknowns: $v^i(x, t)$, $(i = 1, 2, 3)$, $p(x, t)$, and $\rho(x, t)$. To complete the system, we adjoin (as in the case of ideal fluids) the equation of state and the continuity equation 123.3. For incompressible fluids $d\rho/dt = 0$, and hence $v^i_{,i} = \dot{\vartheta} = 0$ by (123.3). Accordingly, for incompressible fluids, equation 124.7 yields

$$\mu g^{jk} v_{i,jk} - p_{,i} = \rho(a_i - F_i). \tag{124.8}$$

Furthermore if the fluid is ideal, $\mu = 0$ and equations 124.8 reduce to Euler's equations (123.2).

Stokes simplified equations 124.7 by introducing a hypothesis to the effect that the mean pressure p in a viscous fluid is given by the same formula 122.2 as in the case of fluids at rest. This assumption leads to the conclusion that the constants λ and μ are not independent. Indeed, from (124.1)

$$t_{ij} = \tau_{ij} + p g_{ij};$$

hence

$$\begin{aligned} g^{ij} t_{ij} &= g^{ij}\tau_{ij} + p g^{ij} g_{ij} \\ &= -3p + 3p = 0, \end{aligned}$$

if we use formula 122.2. But since t_{ij} is given by (124.4), we have upon multiplying those equations by g^{ij},

$$\lambda g^{ij} g_{ij} \dot{\vartheta} + \mu 2 g^{ij} \dot{e}_{ij} = 0,$$

or

$$(3\lambda + 2\mu)\dot{\vartheta} = 0.$$

Thus

$$3\lambda + 2\mu = 0. \tag{124.9}$$

As a consequence of this relation, equations 124.7 depend only on one viscosity coefficient μ, and the substitution from (124.9) in (124.7) yields the set of *Navier-Stokes's hydrodynamical equations*

$$\mu g^{jk} v_{i,jk} + \frac{\mu}{3}\frac{\partial \dot{\vartheta}}{\partial x^i} - \frac{\partial p}{\partial x^i} = \rho(a_i - F_i). \tag{124.10}$$

If the fluid is ideal we get, on setting $\mu = 0$ and

$$a_i = \frac{\partial v_i}{\partial t} + v_{i,j} v^j,$$

the *Eulerian hydrodynamical equations*

$$\frac{\partial v_i}{\partial t} = F_i - \frac{1}{\rho}\frac{\partial p}{\partial x^i} - v_{i,j} v^j, \tag{124.11}$$

for ideal compressible fluids.

If the motion is slow, the term $v_{i,j}v^j$ can be disregarded, and then $a^i = \partial v^i/\partial t$.

Problems

1. Show that the equation characterizing an incompressible fluid can be written in the form

$$v^i_{,i} = \frac{1}{\sqrt{g}} \frac{\partial(\sqrt{g}\, v^i)}{\partial x^i} = 0, \qquad \text{where } g = |g_{ij}|.$$

2. Show that the Navier-Stokes equations can be written

$$\frac{\partial v^i}{\partial t} = \nu g^{jk}\left[\frac{\partial^2 v^i}{\partial x^j\,\partial x^k} + \begin{Bmatrix} i \\ lk \end{Bmatrix}\frac{\partial v^l}{\partial x^j} + \begin{Bmatrix} i \\ lj \end{Bmatrix}\frac{\partial v^l}{\partial x^k} - \begin{Bmatrix} l \\ jk \end{Bmatrix}\frac{\partial v^i}{\partial x^l}\right.$$

$$\left. + \left(\frac{\partial \begin{Bmatrix} i \\ lj \end{Bmatrix}}{\partial x^k} + \begin{Bmatrix} i \\ mk \end{Bmatrix}\begin{Bmatrix} m \\ lj \end{Bmatrix} - \begin{Bmatrix} i \\ lm \end{Bmatrix}\begin{Bmatrix} m \\ jk \end{Bmatrix}\right)v^l\right] - \frac{1}{\rho} g^{ij}\frac{\partial p}{\partial x^j}$$

$$- v^j\left(\frac{\partial v^i}{\partial x^j} + \begin{Bmatrix} i \\ lj \end{Bmatrix}v^l\right) + \frac{\nu}{3} g^{ij}\frac{\partial}{\partial x^j}\left(\frac{\partial v^k}{\partial x^k} + \begin{Bmatrix} k \\ lk \end{Bmatrix}v^l\right) + F^i,$$

where $\nu \equiv \mu/\rho$ is the kinematic viscosity.

3. Show that the equation of continuity can be written

$$\frac{\partial \rho}{\partial t} + \frac{\partial(\rho v^i)}{\partial x^i} + \rho v^i \frac{\partial \log \sqrt{g}}{\partial x^i} = 0.$$

Hint: Use the expression for $v^i_{,i}$ in Problem 1.

4. Show that the equation of continuity in cylindrical coordinates

$$[g_{11} = 1, g_{22} = (x^1)^2, g_{33} = 1]$$

is

$$\frac{\partial \rho}{\partial t} + \frac{\partial(\rho v^i)}{\partial x^i} + \rho\frac{v^1}{x^1} = 0,$$

and in spherical polar coordinates $[g_{11} = 1, g_{22} = (x^1)^2, g_{33} = (x^1)^2 \sin^2 x^2]$ is

$$\frac{\partial \rho}{\partial t} + \frac{\partial(\rho v^i)}{\partial x^i} + \rho\left(\frac{2v^1}{x^1} + v^2 \cot x^2\right) = 0.$$

5. The curl $\mathbf{v}$ of the velocity field $\mathbf{v}$ is equal to twice the angular velocity of rotation. The vector $\boldsymbol{\omega}$ such that curl $\mathbf{v} = 2\boldsymbol{\omega}$ is called the *vorticity vector*. Show that $\omega^i_{,i} = 0$. *Hint:* $\omega^i = -\frac{1}{2}\epsilon^{ijk}v_{j,k}$.

6. If the vorticity vector $\omega^i = 0$, the motion is called *irrotational*. Show that, if the motion is irrotational, the velocity vector $\mathbf{v}$ is the gradient of the *velocity potential* Φ.

7. Write out the approximate equations of motion of a viscous fluid when the motion is slow.

125. Remarks on Turbulent Flows and Dissipative Media

We conclude our brief survey of the elements of mechanics of continua with a few remarks on turbulent flows of fluids and on construction of models for media, in which the processes are irreversible.

Fluid flows in which the velocity components v^i experience complicated pulsating changes are called *turbulent*. In dealing with turbulent flows of liquids and gases it is natural to represent the velocity components in the form $v^i = \bar{v}^i + v'^i$, where $\bar{v}^i$ is the mean value of v^i over a suitable period of time and v'^i is the pulsating component of v^i. Similar resolutions into mean and pulsating components can be made for the pressure p and density ρ, so that $p = \bar{p} + p'$ and $\rho = \bar{\rho} + \rho'$. The development of the theory of turbulent flow crucially depends on the character of averaging processes used to compute $\bar{v}^i$, $\bar{p}$, and $\bar{\rho}$ and on the formulation of relations among these average quantities.

If one assumes, for example, that the pulsating components v', p', and ρ' are governed by the Navier-Stokes equations for an incompressible fluid, then one averaging process applied to Navier's equation leads to a set of equations obtained by Reynolds.[12] These equations involve not only the $\bar{v}^i$, but also the mean values of the pulsating components of velocity. Because of the presence of these latter components, the system of Reynolds equations is incomplete and new hypotheses, based on experimental evidence, must be introduced to complete the system.

It appears unlikely that a unified formulation of satisfactory models for turbulent flows of compressible viscous fluids or for viscoelastic and plastic solids can be constructed within the framework of classical mechanics and thermodynamics. The development of such models is likely to be based on statistical mechanics in which mechanical characteristics are viewed as probabilities and their values appear as mathematical expectations.

A discussion of models of plastic and viscoelastic materials, utilizing the principles of thermodynamics of irreversible processes, is contained in a monograph by L. I. Sedov, cited in footnote 12 and in A. Cemal Eringen's *Non-linear Theory of Continuous Media* (New York), 1962

[12] See, for example, H. Schlichting, *Boundary Layer Theory*, New York (1955), Chapter XVIII, and L. I. Sedov, *Introduction to Mechanics of a Continuous Medium*, Moscow (1962), pp. 213–217.

BIBLIOGRAPHY

P. Appell, *Traité de méchanique rationelle*, vol. 5 (Paris, 1926).
L. P. Eisenhart, *Riemannian Geometry* (Princeton, 1926).
T. Levi-Civita, *The Absolute Differential Calculus* (London, 1927).
A. S. Eddington, *The Mathematical Theory of Relativity* (Cambridge, 1930).
A. J. McConnell, *Applications of the Absolute Differential Calculus* (London, 1931).
O. Veblen, *Invariants of Quadratic Differential Forms* (Cambridge, 1933).
T. Y. Thomas, *Differential Invariants of Generalized Spaces* (Cambridge, 1934).
R. B. Lindsay and H. Margenau, *Foundations of Physics* (New York, 1936).
L. Brillouin, *Les tenseurs en mécanique et en élasticité* (Paris, 1938).
C. E. Weatherburn, *Riemannian Geometry and the Tensor Calculus* (Cambridge, 1938).
L. P. Eisenhart, *An Introduction to Differential Geometry* (Princeton, 1940).
P. G. Bergmann, *An Introduction to the Theory of Relativity* (New York, 1942).
A. D. Michal, *Matrix and Tensor Calculus* (New York, 1947).
J. L. Synge and A. Schild, *Tensor Calculus* (Toronto, 1949).
G. Y. Rainich, *Mathematics of Relativity* (New York, 1950).
D. J. Struik, *Lectures on Classical Differential Geometry* (Cambridge, Mass., 1950).
F. D. Murnaghan, *Finite Deformation of an Elastic Solid* (New York, 1951).
A. E. Green and W. Zerna, *Theoretical Elasticity* (Oxford, 1954).
I. S. Sokolnikoff, *Mathematical Theory of Elasticity* (New York, 1956).
J. L. Synge, *Relativity: The Special Theory* (Amsterdam, 1956).
J. L. Synge, *Relativity: The General Theory* (Amsterdam, 1960).
W. Prager, *Introduction to Mechanics of Continua* (Boston, 1961).
T. Y. Thomas, *Concepts from Tensor Analysis and Differential Geometry* (New York, 1961).
A. C. Eringen, *Nonlinear Theory of Continuous Media* (New York, 1962).
J. C. H. Gerretsen, *Lectures on Tensor Calculus and Differential Geometry* (Groningen, 1962).
L. I. Sedov, *Introduction to Mechanics of a Continuous Medium* (Moscow, 1962).

INDEX